II "International Theoretical and Practical Conference on Alternative and Smart Energy" (TPCASE 2020)

IOP Conference Series: Materials Science and Engineering
Volume 1035

Voronezh, Russia
16-18 September 2020

Editors:

I. I. Popov
M. I. Popova
T. A. Sadchikova

ISBN: 978-1-7138-4560-7
ISSN: 1757-8981

Printed from e-media with permission by:

Curran Associates, Inc.
57 Morehouse Lane
Red Hook, NY 12571

CURRAN ASSOCIATES INC. proceedings.com

Some format issues inherent in the e-media version may also appear in this print version.

Institute of Physics
Dirac House, Temple Back
Bristol BS1 6BE UK

Phone: 44 1 17 929 7481
Fax: 44 1 17 920 0979

techtracking@iop.org

Additional copies of this publication are available from:

Curran Associates, Inc.
57 Morehouse Lane
Red Hook, NY 12571 USA
Phone: 845-758-0400
Fax: 845-758-2633
Email: curran@proceedings.com
Web: www.proceedings.com

TABLE OF CONTENTS

IOP Conf. Series: Materials Science and Engineering 1035 (2021) 011001 doi:10.1088/1757-899X/1035/1/011001

Preface

From 16 to 18 September 2020, in the city of Voronezh, on the basis of the Voronezh State Technical University, The II International Theoretical and Practical Conference on Alternative and Smart Energy (TPCASE-2020) was held online aie.cchgeu.ru. The co-organizer of this scientific event was the Academy of Engineering Sciences named after A.M. Prokhorov.

Energy today is the most important driving force of world economic progress, and the well-being of billions of the world's inhabitants directly depends on its state. The steady increase in the number of people leads to an increase in energy consumption. Every day more and more reserves of natural resources (coal, gas, oil) necessary for the operation of traditional energy are being depleted, so the problem of switching from traditional sources of energy is becoming more and more urgent. In addition to the prospect of depletion of traditional sources, there is also an environmental problem, since the combustion of hydrocarbon fuels leads to harmful emissions into the atmosphere, worsening the human habitat, creating environmental problems. Therefore, it is important to find new energy sources that are safer from an environmental point of view.

The world has already seen significant progress in the use of alternative, renewable energy sources.

Problems in the fields of hydrogen energy, solar energy, thermoelectric energy, wind energy, nuclear energy, ecology, functional materials for alternative energy, energy-saving technologies, systems and materials, smart energy, as well as new educational programs on energy efficiency and alternative energy were discussed during the conference.

Scientists from Russia and the countries of near and far abroad (Voronezh, Novovoronezh, Moscow, St. Petersburg, Samara, Kazan, Volzhsk, Taganrog, Rostov-on-Don, Belgorod, Perm, Saratov, Khanty-Mansiysk, Magnitogorsk, Krasnodar, Nizhny Novgorod, Biysk, Irkutsk, Yekaterinburg, Orel, Komsomolsk-on-Amur and other cities of Russia, as well as participants from Italy, India, Vietnam, Egypt and Belarus) gathered to discuss new developments and achievements in the energy sector. More than 200 participants in total have participated in the event.

ZOOM platform was used for the conference. The platform allows one to conduct online conferences of up to 300 people in one session, recording all events, opening more sessions at the same time, discussing, raising your hand and chairing (presenter) and co-chairing (co-presenter).

For two days five parallel sections were held and more than 150 reports, including six plenary speeches were heard.

The conference was opened by the Temporary Acting Rector of Voronezh State Technical University, Dr. Tech. of Sciences, Professor Igor Drozdov at 10 a.m., September 16, 2020.

At the plenary sections the following reports have been presented:

1. Investigation of the intensification of heat transfer during boiling at macro-, micro- and nanoscale. Yu.A. Kuzma-Kichta (FGBOU VO "National Research University" MPEI ").
2. Digital energy: state, problems, and prospects (on the example of Voronezhenergo). A.P. Galitsky (Rosseti Center - Voronezhenergo).
3. Hydrogen technologies for energy storage. A.I. Schastlivtsev (Joint Institute for High Temperatures RAS).

4. Processes of heat and mass transfer during the phase and chemical transformations in micro-sized systems. Fundamental research and practical applications. V.V. Kuznetsov (Institute of Thermophysics SB RAS, Novosibirsk).
5. Prospects for the development of thermoelectric cooling systems and electric power generation. Yu.E. Kalinin (FSBEI HE "Voronezh State Technical University").
6. Plasma-beam methods for modifying materials. S.D. Fedorovich (FGBOU VO "National Research University" MPEI ").

The reports for the oral presentation were selected by the organizing committee.

Organizing and Program Committees were represented by the famous experts in the field:

Prof. Dr. I.G. Drozdov, vice-rector for scientific work of VSTU;
Prof. Dr. Yu. V. Gulyaev, Academician of the Russian Academy of Sciences, President of the Academy of Engineering Sciences named after A.M. Prokhorov
A.V. Sergeev, Associate Professor, Department of Solid State Physics, VSTU, Voronezh
V.V. Glotov, VSTU, Voronezh
E.V. Ievleva, VSTU, Voronezh
A.S. Samofalova, VSTU, Voronezh
Prof. Dr. D. A. Konovalov, Professor, Department of Theoretical and Industrial Heat Power Engineering, Voronezh State Technical University
Prof. Dr. Yu.E. Kalinin, Professor, Department of Solid State Physics, Voronezh State Technical University
Prof. Dr. A.S. Baghdasaryan, IRE named after V.A. Kotelnikov RAS, Moscow
Prof. Dr. V.I. Borzenko, JIHT RAS, Moscow
Prof. Dr. A.V. Burkovsky, VSTU, Voronezh
Prof. Dr. V.V. Bukhmirov, ISPU, Ivanovo
Prof. Dr. A.D. Danilov, VSTU, Voronezh
Prof. Dr. A.V. Dedov, corresponding member of RAS, MPEI, Moscow
Prof. Dr. V.M. Ievlev, Academician of RAS, Moscow State University, Moscow
Prof. Dr. Yu.A. Kuzma-Kichta, MPEI, Moscow
Prof. Dr. N.V. Mozgovoy, VSTU, Voronezh
Prof. Dr. E.N. Muravyov, AIN named after A.M. Prokhorov, Moscow
Prof. Dr. Sh.A. Piralishvili, RSTU named after P.A. Solovyova, Rybinsk
Prof. Dr. S.L. Podvalny, VSTU, Voronezh
Prof. Dr. A.T. Ponomarenko, ISPM named after N.S. Yenikolopova RAS, Moscow
Prof. Dr. V.I. Ryazhskikh, VSTU, Voronezh
Prof. Dr. N.F. Timerbaev, Kazan State Economic University, Kazan
Prof. Dr. S.V. Fedosov, NRU MGSU, Moscow
Prof. Dr. A.K. Fedotov, BSU, Minsk (Belarus)
Prof. Dr. R.U. Aliev, ASU, Andijan (Uzbekistan)
Prof. Dr. O. Manca, University of Campania, Naples (Italy)
Prof. Dr. Sai Sundara Krishnan, PSG College of Technology (India)
Prof. Dr. Mohamed Helmy Taman, Tanta University (Egypt)
Prof. Dr. Marco Pasetti, University of Brescia (Italy)
Prof. Dr. Nguyen Van Thin, HIDITECH (Vietnam)
Assoc. Dr. A.V. Dmitriev, Kazan State Economic University, Kazan
Dr. V.V. Kuznetsov, IT SB RAS, Novosibirsk
Dr. O.I. Shevaleevsky, IBCP RAS, Moscow

Editors:
I.I. Popov, Ph.D., VSTU, Voronezh

IOP Conf. Series: Materials Science and Engineering 1035 (2021) 011001 doi:10.1088/1757-899X/1035/1/011001

M.I. Popova, VSTU, Voronezh
T.A. Sadchikova, VSTU, Voronezh

Extended articles were sent for peer review to the experts in the field. Fifty articles were accepted for publication.

The organizing committee of TPCASE-2020 expresses deep gratitude to all participants and guests of the conference for their cooperation, their unusually hard work, and a high level of a high level of professional knowledge in their working areas.

Special thanks to IOP Conference Series: Materials Science and Engineering for publishing the best works.

IOP Conf. Series: Materials Science and Engineering 1035 (2021) 011002 doi:10.1088/1757-899X/1035/1/011002

Peer review declaration

Fifty-nine articles based on oral reports were selected by the conference organizing committee and sent for peer review to the reviewers from the corresponding areas of expertise. The peer review was carried out by leading experts in the field. After review, fifty articles were accepted.

- **Type of peer review: Single-blind / Double-blind / Triple-blind / Open / Other (please describe)**
- **Conference submission management system: e-mail**
- **Number of submissions received: 59**
- **Number of submissions sent for review: 59**
- **Number of submissions accepted: 50**
- **Acceptance Rate (Number of Submissions Accepted / Number of Submissions Received X 100): 84.75**
- **Average number of reviews per paper: 1**
- **Total number of reviewers involved: 12**
- **Any additional info on review process: -**
- **Contact person for queries: Ivan Popov, ipopov@vgasu.vrn.ru**

IOP Conf. Series: Materials Science and Engineering 1035 (2021) 012001 doi:10.1088/1757-899X/1035/1/012001

Automated system for cleaning solar panels based on a linear actuator

Yu N Zatsarinnaya[1,3] and D I Amirov[2]

[1]Department of Power stations named after V.K. Shibanov, Kazan State Power Engineering University, Krasnoselskaya str., 51, Kazan, 420066, Russia
[2]Department of Power stations named after V.K. Shibanov, Kazan State Power Engineering University, Krasnoselskaya str., 51, Kazan, 420066, Russia

[3]Corresponding author: waysubbota@gmail.com

Abstract. High volatility in the hydrocarbon market in tandem with an increase in anthropogenic impact on the environment has become one of the main reasons for the accelerated transformation of the global energy industry. Today we have witnessed a gradual increase in the share of renewable energy sources (RES) in the energy industry structure of leading economically developed countries. Solar energy is a promising area of renewable energy for most economies because of its availability and ease of use. Not only for commercial companies but also for the private sector. Pollution of a front surface of a module has become a pressing problem of generating electrical energy by using photovoltaic panels (PVP). For a solution to the problem of timely panels cleaning developers proposed several options of systems none of which were widely used. We started the development of the panel cleaning system which is based on the linear actuator operation principle. The system is planned to be controlled by using the Arduino UNO R3 microcontroller which will allow us to minimize the share of human participation in the cleaning of PVP.

1. Problem statement

Over the past few years, many countries have begun to change their attitude towards energy resources which is used. One of the essential prerequisites for this was the adoption of the Paris Climate Agreement in 2015 which combined the efforts of world powers to curb climate change on the planet. Many countries have adopted new strategies and developed roadmaps to attract prosumers on renewable energy sources. According to the International Energy Agency (IEA) [1], Europe expects an annual increase in electricity consumption by 1.4% until 2030. Roadmap of 2050 developed by them aimed at reducing greenhouse gases through the integration of distributed and renewable energy sources [2].

Russia follows a similar path as evidenced by the adoption of Federal Law 471, dated 27 Dec. 2019, «On Amendments to the Federal Law On Electric Power Industry regarding the development of microgeneration» according to which prosumers with a microgeneration facility such as a solar panel can sell surplus generated electricity. Solar energy is one of the most promising areas for households. However, the urgent task of solar energy is the issue associated with cleaning the front surfaces of panels. The problem leads to enormous economic losses. According to the analysis for Russia, the figure is about 15 million per year [3]. This is associated with a reduction of output power in the absence of cleaning.

2. Analysis of existing PVP cleaning methods

We have analyzed the advantages and disadvantages of various PVP cleaning methods from pollution. The analysis presented in table 1.

Table 1. Front surface cleaning methods of PVP.

No	Method name	Principle of operation	Advantages	Disadvantages
1	Manual cleaning	Cleaning PVP surface with brushes	Work with any kind of pollution	High cost; Chance of accidents among workers
2	Greenbotics robot [4]	Automated Panel Cleaning System; Robot moves along a panel and cleans the surfaces of PVP with microfiber	Protection against any pollution; Autonomy	High cost; Covers a large area; Complexity of execution; Requires electricity
3	Venturi method [5]	Principle of operation is based on the laws of aerodynamics. Differential pressure creates a directed air flow that blows loose snow.	Autonomy; Simplicity	Dimming; Suitable for windy places; Defence not from all type of pollution
4	SCROBBY [5]	Automated Panel Cleaning System; Robot is hung on a cable and when it receives a signal comes into action.	Autonomy	Dimming with hanging cables and the robot itself; Unreliable power supply
5	Cleaning system based on the principle of Ampere's law[6]	Alternating current is passed through the wires in different directions and as a vibrations result cleaning occurs.	Snow cleansing; Autonomy	Requires electricity; Dimming
6	Device based on a mechanical thrust film [7]	If PVP dims then part of the film is rewound	Autonomy; Cheapness; Simplicity	Risk of frequent repairs because of choice of low-quality film; Chance of dimming part of PVP surface

The conclusion that should be drawn from this table is that the main advantage that they strive for is the autonomy of a device.

3. Description of the operation principle of the automated PVP cleaning system

As part of this work, we have launched a project to develop a system for cleaning the front surfaces of solar modules. The work was divided into four main tasks: designing the measuring part, the logical part, the executive part and also device's power system (figure 1).

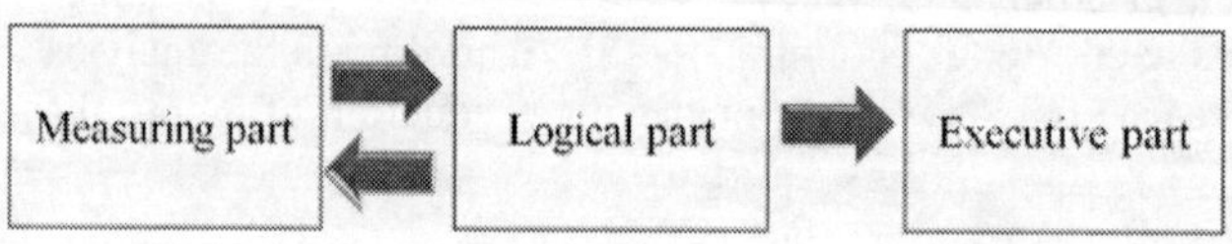

Figure 1. The structure of the panel cleaning system based on the Arduino Uno R3 microcontroller.

3.1. The executive part

The main component of the executive part is the Nema 17 stepper motor. Motors of Nema series have high torque and with the help of special drivers A4988, it is possible to rotate their shaft in both directions [8].

Cleaning is produced by using a brush which can be fixed to the carriage. There are six metric screw holes for mounting. The holes are located two opposite each other. This makes it possible to adjust the brush height in increments of 5 mm. The carriage can move along the plane of the solar panel along two guides one of which is a threaded rod and the second is an aluminum cylinder.

The operability of the entire system depends on the choice of guides. Characteristics of the motor used must be considered in the process of choosing a threaded rod. It is necessary that the thread pitch of the rod was as large as possible so the carriage quickly passes the specified distance. However as the larger the thread pitch the larger the diameter and mass of the rod made of the same material. This leads to an increased effort which the motor needs to overcome. This means it increases currents which in turn leads to large heating of the controller. We have not yet done any experiments on choosing the optimal threaded rod but so far we have preferred the M8 thread.

The executive part is based on the operation principle of a linear actuator (figure 2).

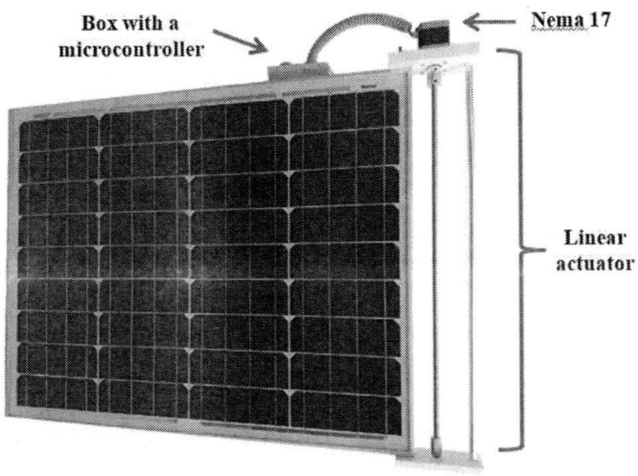

Figure 2. Linear actuator for PVP.

Control signals are sent to the motor through the A4988 driver. When the motor shaft rotates then a rod that is connected to the shaft by an adapter sleeve also starts to rotate. Because of an aluminum guide, the carriage does not turn but moves linearly. The brush mounted on the carriage removes dust, dirt and snow.

The carriage is supposed to move in one direction in order to save energy. After that, it is planned to check the surface of the PVP and if the pollution has not been eliminated then reverse the movement to the place of the initial start. If the pollution has been eliminated after the first passage of the carriage then in order to save energy it is left it alone until the next command.

3.2. The logical part

The logical part includes the software package which contains all operation settings and cycles for canceling the false operation of the executive part at dusk. Programs for the microcontroller are written in C ++ in the Arduino IDE development environment.

3.3. The measuring part

At this stage of development, the motor is switched on by using an IR (infrared) remote control and an IR receiver. The control involves three buttons. The first is responsible for the rotation of the shaft "to the right," the second - "to the left," the third - "stop." It is planned to use measuring sensors in the future.

Sensors for measuring the wavelength of solar radiation:

- Photoresistor;
- Light sensor;
- UV sensor;

Sensors for measuring the distance to an object:

- Ultrasonic distance sensor;
- Infrared barrier;
- Dust sensor.

Each element that can be used in the measuring part of the solar panel cleaning device has both advantages and disadvantages that limit the scope of application of certain sensors. Photoresistor - resistor whose resistance depends on the intensity of incident light. Its parameter is the nominal resistance. It may be equal to 200 kohms. In complete darkness, the resistance of such a resistor is 200 kohm, and in bright light it drops to almost zero. This effect can be used in a solar panel cleaning device. The Arduino Uno R3 microcontroller must be programmed that when the resistance of the photoresistor increases which may be caused by ingress of snow dirt or dust the executive part works and cleans the front surface of the panel and photoresistor. However, it should be noted that at night the resistance can reach the response value which will lead to a false switching on of the executive part. This fact limits the use of these components and it is a significant disadvantage. it's supposed to use a cycle in the microcontroller program that increases the response resistance depending on the time of day or use LED backlight in order to eliminate this effect.

The fraction of UV radiation in the solar spectrum is only 9% [9] but this does not exclude the possibility of using a UV radiation sensor as the measuring part. The principle of operation is the same as the photoresistor. But unlike photoresistors, most of these sensors have a larger measurement range. For example, the measuring range of the GYML8511 sensor from 400 to 280 nm and for photoresistors the range is from 450 to 600 nm. In turn, it is possible to combine two types of analog sensors to improve the measuring part.

The next example of a measuring device is an ultrasonic distance sensor. It could be HC - SR04. Important components of the sensor are two piezoelectric elements. One of them emits an ultrasonic wave when 15 microseconds pulse is applied, and the other piezoelectric element receives the same reflected wave from an obstacle. Then the delay time from transmission to the reception of the wave is measured and the distance is calculated. The frequency of the sound wave is within the ultrasound frequency which provides a concentrated direction of the wave because sound with a high frequency is less scattered in the environment.

The signal may reflect off drops when it rains or snows and resulting in a spurious echo. This phenomenon can be used as a logical unit for the device of the executive part. However, it will not be possible to build an automatic system that monitors the accumulation of dust on a solar panel using an ultrasonic distance sensor.

The next group of measuring devices is based on the use of infrared radiation. it includes an IR barrier and a dust sensor.

IR barrier is widely used in security systems. It is a sensor consisting of an infrared light source and an infrared receiver. When the beam path intersects the signal enters the logical element of the system. IR LED is used as a light source and an IR phototransistor as a receiver. This technology has a third element - the environment. The environment is a source of aerosol attenuation, resonance absorption and the presence of barriers in the form of precipitation.

A special case is a barrier in the form of snowfall. This environment is still poorly formalized and the action of IR channels is estimated on the basis of experimental data.

Using the IR barrier is perfect for protection from snow on the front of PVP.

Another infrared technology is a dust sensor. It uses optical sensing to detect dust. A photosensor and infrared LED as known as an IR LED, are optically located in the dust sensor module. A photosensor detects reflected infrared rays from LEDs that are reflected from dust particles in the air.

One of the most common dust sensors on the market is the GP2Y1010AU0F sensor. The GP2Y1010AU0F module can detect the smallest particles in the air, which makes it possible to detect even cigarette smoke. The high output pulse from the sensor is triggered when the dust is detected.

3.4. The power system

It is necessary to provide a reliable power supply for the drive to operate the motor in a linear actuator system. Electricity for the microcontroller can be obtained from a 9-volt crown battery (figure 3).

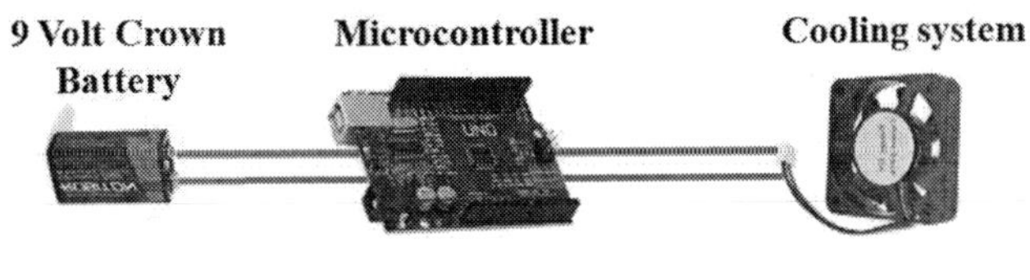

Figure 3. Microcontroller power system.

Electricity for shaft rotation can be obtained from a battery connected to the solar module through a charge controller (figure 4).

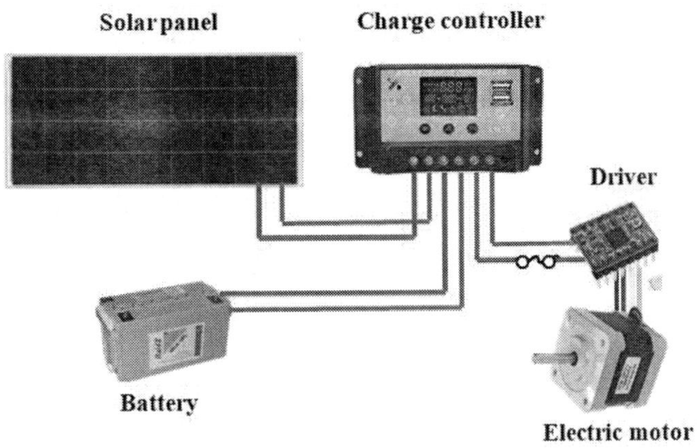

Figure 4. Motor power system.

A wide range of storage devices is presented on the market of batteries which differ not only in parameters but also in technology embedded in operation principle. This mean that the choice of batteries should be treated responsibly.

We have analyzed 221 projects using chemical energy storage devices that have been either implemented or are being implemented in the world in the last decade (the main source is The DOE Global Energy Storage Database, as well as other sources). Since 2015 more interest has been shown in lithium-ion batteries in renewable energy projects. In this regard, our project also plans to use a drive with a similar technology in the further development of the cleaner.

4. Conclusion

The analysis of scientific and technical literature and practical experiments revealed the following:

- Pollution of the solar panels surface causing economic damage to the energy industry is a task that can be solved in the near future;
- It's possible to assemble a cleaning system based on the operation principle of a linear actuator;
- It is possible to use the Arduino Uno R3 microcontroller as a system control;
- Batteries of various technologies can be used as a power source for an electric motor with a rated voltage of 12 volts. Lithium-ion batteries are in great demand given the experience of cases over the past 5 years.

  The battery can be powered by a solar battery through a charge controller.

References

[1] International Energy Agency 2018 *Renewables 2018 Analysys and Forecasts to 2023* (Paris: International Energy Agency)
[2] Sumper A *Micro and Local Power Markets* 2019 (Hoboken, NJ: John Wiley & Sons, Inc) p 272
[3] Zacarinnaja Ju N, Amirov D I, Zemskova L V and Rahmatullin R R 2019 Study of the solar panel efficiency with the influence of pollutants on it *Transactions of Academenergo* **1** 81-92
[4] *SunPower Cleans Up Solar With Acquisition of Greenbotics* available at - URL: https://www.greentechmedia.com/articles/read/sunpower-cleans-up-solar-with-acquisiton-of-greenbotics (accessed: 20.01.2020)
[5] Sanarov S V and Lebedev R I 2016 *Methods for Cleaning the Solar Battery. Introduction to Energy* p 148
[6] Ismagilov F R, Vavilov V E and Nurgalieva R A 2018 *Method for automated cleaning of solar panels* Patent RF no. 2018104009
[7] Vaskovskay T, Thakurta P G and Bialek J 2018 Contribution of transmission and voltage constraints to the formation of locational marginal prices *Int. J. Elec. Power* **101** 491-9
[8] 2013 *Exploring Arduino: Tools and Techniques for Engineering Wizardy* ISBN 978 – 1 – 118 – 54936 – 0 (John Wiley & Sons, Inc)
[9] Vissarionov V I, Derjugina G V, Kuznecova V A and Malinin N K 2008 *Solar Energy: A Study Guide for Universities* ed V I Vissarionova (Moscow: MPEI Publishing House) p 317

IOP Conf. Series: Materials Science and Engineering 1035 (2021) 012002 doi:10.1088/1757-899X/1035/1/012002

Textured solar cell modeling in TCAD

E YU Plotnikova[1], A V Arsentiev[1] and M E Harchenko[2]

[1] Voronezh State Technical University, 20-Letiya Octyabrya St., Voronezh, 394006, Russia
[2] Voronezh Semiconductor Plant - Assembly, Leninsky Avenue, 199a, Voronezh, 394033 Russia

Abstract. We investigate one of the promising methods for increasing the output characteristics of solar cells - surface texturing. The structures and characteristics of microcrystalline silicon solar cells are simulated. Solar Cells with different surface geometries are compared. The first type is the classic flat structure. The last and the most effective is Solar Cell with the surface in the form of microscopic pyramids separated by flat regions.

1. Introduction

There are two types of energy sources, non-renewable and renewable, which are widespread in a present-day world. Renewables include hydropower, wind, geothermal, biomass and solar energy [1]. Let's consider the features of solar energy as a renewable energy source.

The most widespread material from which solar cells have been made for many years is silicon. The first structures were made on monocrystalline silicon. Now it is polycrystalline/microcrystalline and amorphous silicon. In addition to silicon, colar cell structures are made on GaAs and its solid solutions; on CIS ($CuInSe_2$), CIGS (Cu (In, Ga)Se_2) structures; on transparent metal oxides (i.e. ITO, ZTO): on several types of organics, etc. [2].

A typical solar cell is inflexible and flat. But there are firm and flexible strictures, which are suitable for use on embossed surfaces, become more popular nowadays [3].

The surface of the cell is smooth in the classic version of the solar cell structure. It is covered with protective translucent/transparent films. Depending on the structure of the SC, the film can be dielectric or conductive. Classical dielectric films are SiO_2, SiO_2+TiO_2, etc. The typical conducting film is ITO [4].

To increase the output power, and, consequently, the efficiency of solar cells, various methods of changing its structure are used. One of these methods is to give the surface a texturing of various types [5].

Modern technologies make it possible to create solar cell surface relief with the best and the most power effective shapes. In some multi-crystalline solar cells, it was found that one of the best ways to achieve good material quality is the material deposition on a rough surface. Therefore, solar cells with a textured rear surface instead of a textured front surface have been investigated experimentally. In order to model the resulting effects, the influence of single surface texture was investigated

For the first time, we investigate the classical flat solar cell structure, with typical electrophysical parameters. Next step we modify the solar cell top surface and analyze its behavior in this section.

2. TCAD Modeling

In our investigation, we simulate a Solar Cell based on a silicon substrate and doped with boron (10^{14} cm^{-3}). A heavily doped region is formed on the backside for the anode contact (10^{18} cm^{-3}). The bottom contact is made from a transparent conductive metal oxide ITO (thickness 0.5 μm) by ion-beam sputtering.

A p-type region (10^{20} cm^{-3} on the top) with a thickness of about 0.5 μm is created in the upper part of the structure by ion implantation of phosphorus. A thin transparent conductive layer of ITO cathode (its thickness is 0.1 μm) is applied to it.

The dimensions of the simulated classical flat structure (in 2D) are 2 μm in width and 6 μm in height (figure. 1). The depth of the simulated structure is taken equal to 2 μm (the square of the surface is considered).

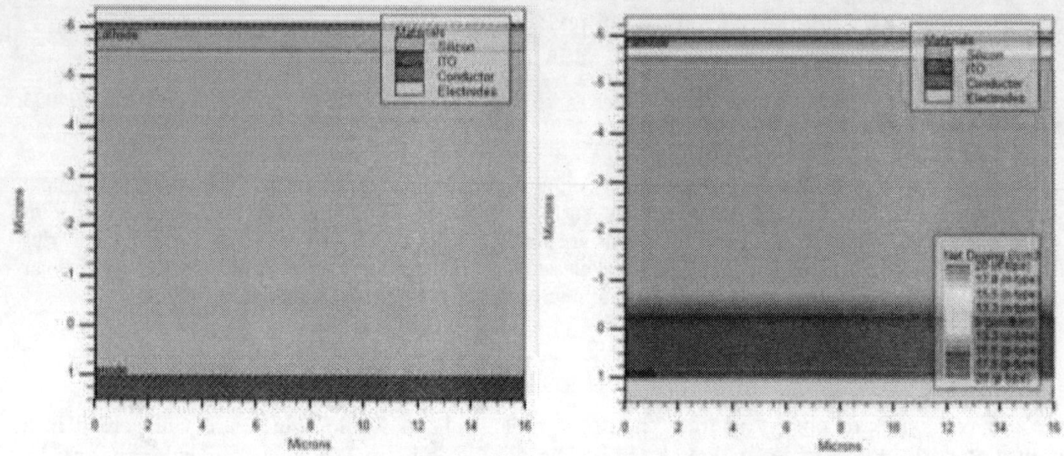

Figure 1. Flat Solar Cell Structure.

To form a relief on the surface of the structure, regions of a regular geometric shape — pyramids and parallelepipeds of various depths and heights of walls (table 1) — are etched before the p-type ion implantation process.

We analyzed several types of structures and discover that a structure with pyramidal peaks (call it "small peaks") of 0.5 μm peak height, 3 μm peak width and a valley between peaks width of 1 μm was recognized as a promising type of modification (figure 2).

For each structure specific material parameters were configured in TCAD. For silicon, the charge carrier mobility was specified as typical for the polycrystalline material of solar cells (mun = 20, mup = 1.5). The bandgap (eg300 = 1.9) and the concentration of intrinsic charge carriers (nc300 = 2.5e20, nv300 = 2.5e20) also were specified for each structure. To calculate the spectral characteristics of ITO, a table from the Sopra database was attached to the model. The resulting solar cell structure was maintained for further analysis.

In the next step the current-voltage characteristics (IV) were simulated for all constructed structures. For the modeling, standard Shockley-Reed-Hall models and Auger recombinations were used.

Under such conditions (without lighting), for each type of solar cell surface, a dark IV characteristic was calculated in the range from 0 to 2 V with a step of 0.1 V. It was taken as the basis for further calculation of the characteristics of solar cells under illumination.

It was assumed that the luminous flux corresponds to the spectrum of AM1.5. The radiation source is located above the middle of the structure. Wavelengths from 0.3 to 1.2 microns are analyzed. The data on the change in intensity was used for further analysis of the characteristics of solar cells.

After the light radiation modeling, the IV characteristic of the solar cell was calculated under its illumination (from 0 to 2 V in 0.1 V increments). The simulation results for epy flat SC and SC, which showed the best characteristics, are given below (figure 3).

IOP Conf. Series: Materials Science and Engineering 1035 (2021) 012002 doi:10.1088/1757-899X/1035/1/012002

Table 1. Solar cell surface structure.

	Solar Cell type					
	small peaks (best results)	large peaks	trapezoid shapes	jagged structure of various depths	flat	rectangular
	1	2	3	4	5	6
apex angle from normal	75°	55°	< 75°	< 45°	-	0°
width of bottom, um	3	3	< 3	< 2	-	2
width of top flat surface, um	-	-	< 2 um	-	-	2
Solar Cell Structure	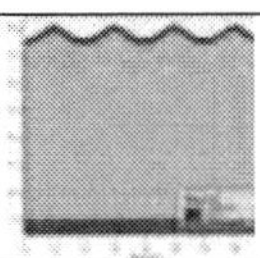	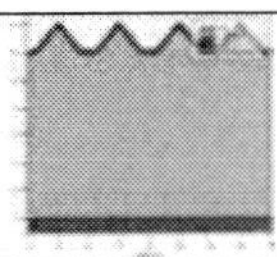	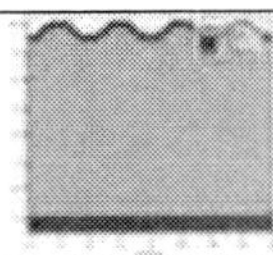	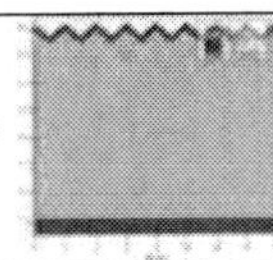	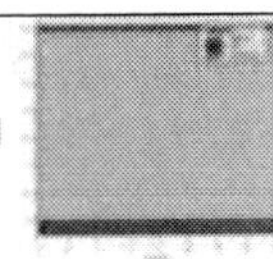	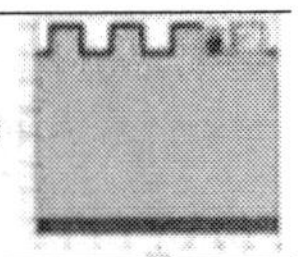

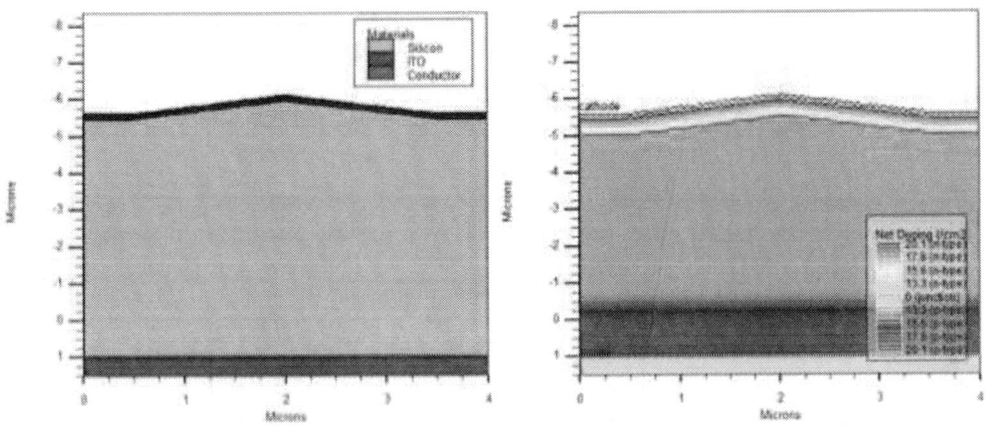

Figure 2. Solar Cell with small Peaks on top (materials and carrier concentration).

During the simulation, the basic parameters were calculated so we can analyze the textured solar cells' characteristics. Short circuit current and open circuit voltage were determined. The short circuit current was determined on the IV curve when illuminated at a point of zero on the X axis. The open circuit voltage was determined at zero on the Y axis.

The power of solar cells was calculated on the results of the IV characteristics under illumination and was a determining parameter for further analysis of the efficiency of solar cells.

3. Results and Discussion

When we changed the surface structure of the solar cells, texturizing them (figure 2), we increased its effective area but left the original cell`s dimensions. With the shape of the surface relief changing, the percentage of absorbed and reflected radiation changes too. Five types of surface topography were investigated (table 2). The results were compared with each other and with the original flat SC structure.

The experiment results were that during the formation of rectangular valleys, the maximum effective power emitted by SCs worsened compared to a flat structure (1.51 $W \cdot 10^{-9}$ for rectangular SC and 2.01 $W \cdot 10^{-9}$ for flat SC, respectively).

When a jagged structure of various depths was formed on the surface, the maximum effective power increased by 0.15 $W \cdot 10^{-9}$ compared to the flat SC. Other surface configurations showed similar power growth (structures 1-3 in Tab. 1-2). The best result of all modeled structures was shown by a solar cell with small peaks on the surface of the structure (structure 1 in table 1 and 2). The selection of the angle at the top of the pyramid ensured the greatest absorption of the incident light (figure 4).

The analyzed parameters to determine the most effective SC were its power (figure 5), short circuit current, open circuit voltage (table 2), and IV characteristics (figure 3). It can be seen that for all structures the open circuit voltage remained almost unchanged, while the short circuit current changed to a large extent. Consequently, the configuration of the IV and the power curves changed. Figures 3 and 5 compare the initial flat SC and the most optimized SC with small peaks. The discrepancy between the values corresponds to the calculated values of the short circuit current.

Table 2. Results of Modeling.

	Solar Cell type					
	small peaks (best results)	large peaks	trapezoid shapes	jagged structure of various depths	flat	rectangular
	1	2	3	4	5	6
$I_{sc} \cdot 10^{-9}$, A	3.028	2.97	2.98	2.84	2.82	2.33
V_{oc}, V	1.1	1.1	1.1	1.0	1.1	1.1
$P_{max} \cdot 10^{-9}$, W	2.16	2.15	2.12	2.02	2.01	1.51

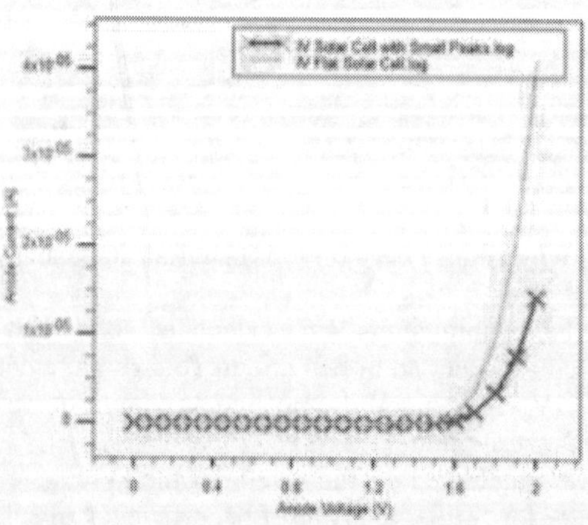

Figure 3. Comparison IV of solar cell (marked) with small peaks and flat solar cell (solid); anode current and cathode current Y-axis, respectively.

IOP Conf. Series: Materials Science and Engineering 1035 (2021) 012002 doi:10.1088/1757-899X/1035/1/012002

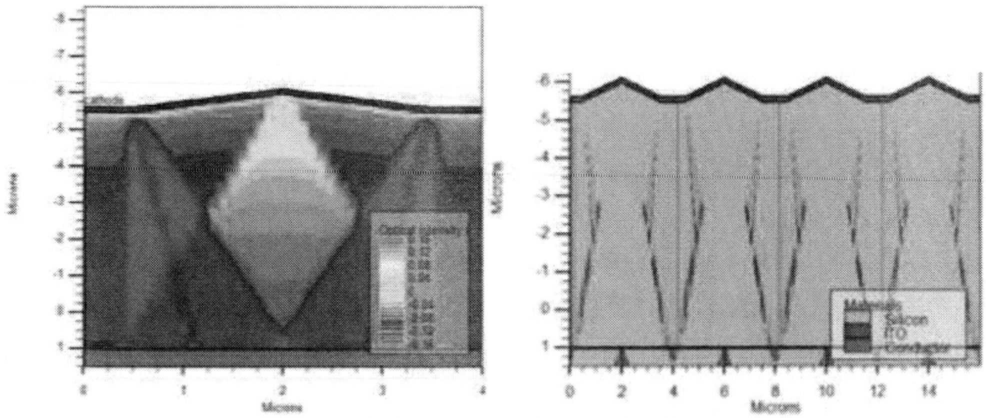

Figure 4. Optical intensity in Solar Cell.

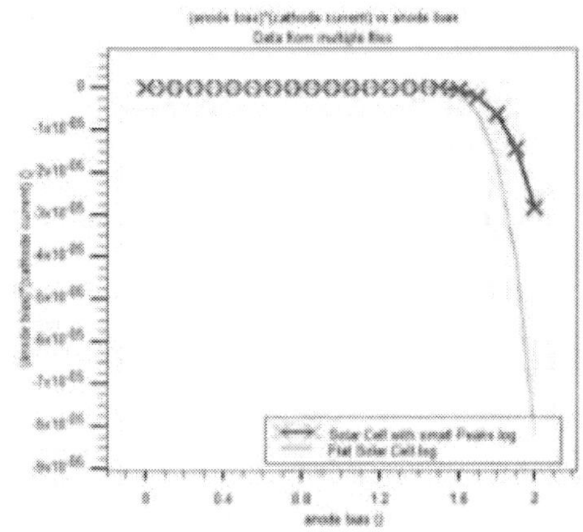

Figure 5. The efficiency of power generation.

4. Conclusion

The best performance results were obtained from solar cells with small and large peaks, while gaining a power gain of 7.5% and 7%, respectively, compared with a flat structure. Other surface options also give an increase in power due to an increase in the effective surface area per unit length of solar cells and optimal configuration of reflection and absorption, but last is not so strong.

References

[1] Hovel H J 1975 Solar cells *Semiconductors and Semimetals* **211**-6

[2] Green M A 2007 Thin-film solar cells: review of materials, technologies and commercial status *Journal of Materials Science: Materials in Electronics* **18**(1) 15-19

[3] Chopra K L, Paulson P D, and Dutta V 2004 Thin- film solar cells: an overview. *Progress in Photovoltaics: Research and Applications* **12**(2-3) 69-92

[4] Haase C and Stiebig H 2007 Thin-film Silicon Solar Cells with Efficient Periodic Light Trapping Texture *Applied Physics Letters* **91**(6) 0611161– 0611164.

[5] Shota H, Takashi M, Hideyuki T and Yoshihiro H 2006 Influence of Texture Feature Size on Spherical Silicon Solar Cells *Rare Metals* **25**(s1) 115–120

Development of an autonomous power supply for oil and gas fields based on a gas turbine unit driven by an active-reactive turbine

D A Bazykin[1,2] and A V Barakov[1]

[1]Voronezh State Technical University, Moscow Avenue, 14, Voronezh, 394026, Russia
[2]Financial industrial company «Kosmos-Neft-Gas», Voronezh, Russia

Abstract. The relevance of the task lies in the development of an autonomous source of electrical power for oil and gas fields, significantly remote from power lines and the necessary infrastructure. An autonomous source is made on the basis of a gas turbine unit driven by an active-reactive turbine, which fuel is associated petroleum, low-pressure, untreated natural gas without preliminary preparation. The developed installation does not require frequent maintenance, the consumption of lubricants, preliminary preparation of fuel gas, and has relatively small overall dimensions. Atmospheric air is used as a working fluid while working surfaces are not subject to pollution. The work provides information on mining equipment, which must be continuously supplied with electric energy, on the currently available analogs of the developed device and their shortcomings. A diagram of the proposed gas turbine installation is given, its structure and principle of operation are described. Information about the developed equipment is given. Preliminary calculations of the main parameters are made, the results are shown.

1. Introduction

Oil and gas fields, far removed from the necessary infrastructure, power plants, to a large extent require the use of autonomous power sources to ensure the uninterrupted operation of mining and technological equipment. Moreover, the most appropriate is the use of their own energy, incidentally emitted during production, as fuel for power sources. In this regard, there is a need to create local effective power plants that can solve this problem and contribute to the useful utilization of low-pressure, untreated natural gas, as well as associated petroleum gas (APG).

The composition of APG deposits has some differences and, depending on the conditions, may vary. The main components of associated petroleum gas are saturated hydrocarbons - methane homologs from CH_4 to C_6H_{14} [1]. In the vast majority of cases, associated petroleum gas is simply flared directly in the field. This is due to the remoteness from places where it is possible to process gas, the lack of transport infrastructure, as well as the lack of equipment with which it can be useful to dispose of.

The burning of associated petroleum, as well as low-pressure gases in flares, leads to a significant deterioration of the environmental situation in the area where the production takes place, and this also leads to large losses of valuable energy resources. A rational approach to this problem is required, consisting of the most useful use of the potential of raw materials.

The complexity of useful APG processing lies in the fact that it has an unstable component composition, and it also has a high content of hydrogen sulfide. These circumstances, in turn, require the purchase of expensive equipment for gas cleaning and preparation. There are currently many solutions to this problem.

There are facilities for APG utilization that do not require special gas treatment. The plants were created for the direct production, mainly of electric energy, which causes a significant reduction in the cost of energy supply for the developed fields, the replacement of external suppliers, electricity by own gas generation, and also allows to reduce the number of harmful emissions, which favorably affects the environment.

Recently, gas turbine units have been used primarily as autonomous sources of electrical power in the fields, with the help of which two problems can simultaneously be solved - the utilization of low-pressure, associated petroleum gas, and the power supply of field equipment [2, 3, 4].

2. Materials and methods

The main objects that consume electricity to a large extent in oil and gas fields are:
- rigs designed for good drilling;
- installations used in the extraction of oil and gas - pumps and compressors;
- technological equipment, the main task of which is the collection and pumping of oil and gas, which can include: group metering units (GMU), booster pump stations (BPS), booster compressor stations (BKS);
- submersible pumping units for oil wells;
- equipment used to maintain in-situ pressure;
- objects of electrochemical protection of the pipeline against corrosion;
- automatic control and monitoring systems in the field.

It is for the purpose of the continuous supply of equipment with electric power in significantly remote oil and gas fields that there is a need for autonomous power sources. The most technologically advanced, efficient and universal method of utilizing energy carriers emitted during production is the generation of electric energy based on the technological cycle of gas turbine units [5, 6, 7]. It was decided to take the gas turbine unit (GTU) as the basis of the designed source, the fuel for which would be the very energy carriers present in the fields.

Due to the remoteness of the deposits, the costs of building power plants are offset by a significant reduction in the costs of building networks and transformer substations, and the cost of generated electricity is 2-3 times lower than the established tariffs.

Several types of plants are presented on the oil and gas equipment market, the fuel of which can be associated gas, low-pressure gas and which can be used as autonomous power sources in the fields. As an example, here are some of them:

1. Closed steam turbine power plant (CSTPP) [8];
2. The Russian analog of CSTPP is an autonomous source of APE-4.48 manufactured by TEMZ OJSC, Tomsk;
3. GTU-003S manufactured by FSUE Moscow Machine-Building Production Enterprise “Salyut” [9];
4. Stirling engines [10];
5. Power generators based on microturbines [11].

The main disadvantages of the installations of paragraph 1 and paragraph 2 are:
- low efficiency (4%);
- large dimensions and weight;
- stringent requirements for transportation and installation at the field;
- the presence of a vacuum behind the turbine causes air to enter a closed loop, which leads to a loss of operability;
- toluene is used as a working fluid. Toluene is toxic, toxic, toluene vapors easily form explosive mixtures that ignite even from a spark of static electricity.

Installation paragraph 3 has higher efficiency, performed according to the classical scheme of a gas turbine with a regenerative cycle. Its disadvantage is expressed in the fact that there is a need for preliminary preparation of fuel gas for pressure and quality.

The disadvantage of the Stirling engine (paragraph 4) is that the working bodies are mainly helium or hydrogen, and therefore it is necessary to ensure the tightness of the structure with superfluidity of the working fluid. It also requires frequent maintenance of the units of this unit and constant consumption of lubricants, so the use of the engine in the conditions of the Far North fields is extremely difficult.

Power generators based on microturbines (paragraph 5) are made, like the installation of paragraph 3, according to the classical scheme of a gas turbine with a regenerative cycle. They use a lean premix combustion system to achieve low emission levels over the full power range. Working with dry premix requires working with a high air-fuel ratio in the primary combustion zone. A large amount of air is thoroughly mixed with fuel before combustion. This pre-mixing of air and fuel provides clean combustion at a relatively low temperature. Their disadvantage is also the need for preliminary preparation of fuel gas, as well as any other type of fuel used for pressure and quality since their technological scheme has an internal combustion chamber. Also, these power supplies are characterized by a high speed of microturbines, which reaches 100,000 rpm.

The gas turbine unit for processing associated petroleum gas into electricity was taken as the basis for the developed device [12]. The installation diagram is shown in figure 1. With this design of the installation, it is quite possible to convert associated petroleum, crude and low-pressure gas into electricity, since the working fluid is air, the gases are used only to heat it. However, the installation will not require frequent repairs and maintenance.

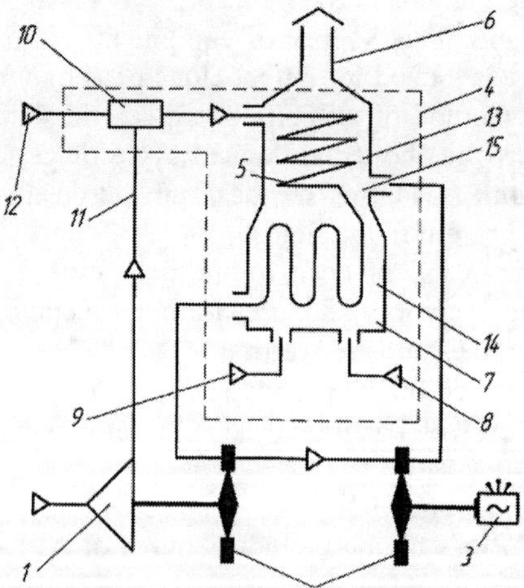

Figure 1. Diagram of a gas turbine installation for processing associated petroleum gas into electricity: 1 - compressor; 2 - turbine; 3 - electric generator; 4 - device for heating the air after the compressor; 5 - heat exchanger; 6 - an exhaust pipe; 7 - a combustion chamber made in conjunction with a pipe in the form of a ground flare unit for the associated petroleum gas combustion; 8 - pipeline for supplying gas to the combustion chamber; 9 - pipeline for supplying atmospheric air to the combustion chamber; 10 - injector; 11 - pipeline designed to connect the injector to the compressor; 12 - pipeline designed to connect the injector to an autonomous source of high pressure air; 13 - convection chamber; 14 - radiation chamber; 15 - the cavity between the chambers, in communication with the cavity exit of the working fluid from the turbine.

Evaluation calculations, taking into account the recommendations on the design of the air heating device given in the above documents, show the possibility of realizing the heating of the working air for the turbine to 750 °C. This temperature value occurs in various types of stationary gas turbines.

The disadvantage of the existing installation is the need to maintain high revolutions of the turbine, heating the working fluid to sufficiently high-temperature values, which entails an increase in the consumption of combusted gas.

Therefore, it is imperative that in the gas turbine installation under development all existing disadvantages of its analogs and prototype be eliminated. We have proposed the design of an autonomous power source based on a gas turbine unit driven by an active-reactive turbine. The proposed installation is characterized in that it does not have all of the above disadvantages, and APG and any low-pressure gas that does not require special preliminary preparation can be used as fuel.

Schematic diagram of the proposed device is presented in figure 2 [13].

The gas turbine installation contains an air compressor 1, a reactive module 2, an active module 3, an electric generator 4, a device 5 for heating the air after the compressor. Gas to the air heating device 5 is supplied through the pipeline 6, and air from the atmosphere is supplied via the pipeline 7. In addition, the air heating device 5 is equipped with an injector 8 connected by pipelines 9 and 10 to an autonomous high-pressure air supply source (figure 2 does not show it) and to the air compressor 1.

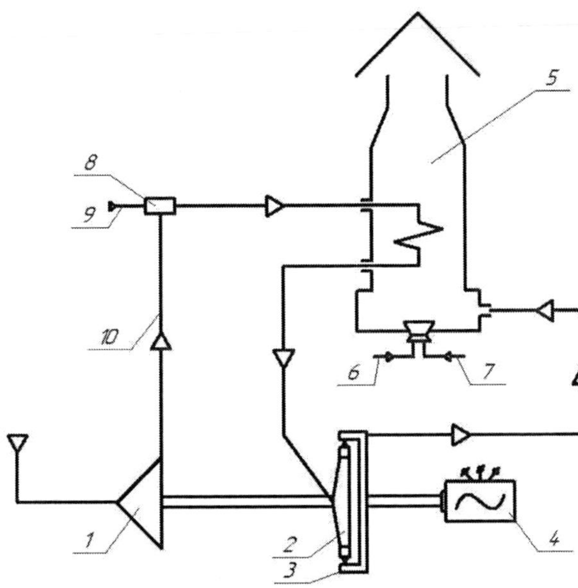

Figure 2. Schematic diagram of a gas turbine installation driven by an active-reactive turbine.

The operation principle of this installation. By supplying associated petroleum gas through pipeline 6 and through pipeline 7, air from the atmosphere to the air heating device 5, a combustion process is organized. The combustion products leaving in the atmosphere heat the heat transfer pipes (heat transfer channels) of the air heating device 5. Moreover, using the injector 8 at low pressure (less than 0,1 MPa), the pipes (channels) of the recuperator are purged. After heating the design of the device 5 to heat the air to a certain temperature using the injector 8, a flow of the working fluid (air) through the air compressor 1 is created, after which the air passes through the device 5. Then the heated air enters the reactive module 2. The presence of airflow through the reactive module 2 leads to its rotation, and accordingly, to the rotation of the shaft of the air compressor 1. At the outlet of the nozzles of the reactive module 2, the air is supplied to the channels made in the body of the active module 3, bringing it into rotation to the other side, and consequently, to the rotation of the shaft of the electric generator 4. An increase in the temperature of the air supplied to the reactive module 2 leads to an intensification of the rotation of the rotors, and at a certain temperature, the rotors go to the «self-propelled» mode (the mode when the power on the reactive module 2 becomes more powerful air compressor 1). After reaching this mode, the air supply to the injector 8 is stopped. By increasing the flow rate of fuel gas entering the air heating device 5, the gas turbine installation reaches its nominal mode.

In the proposed installation, the design of the reactive module can be made in the form of a «Segner wheel», the operation of which is based on the reactive action of the air stream. The design of the active module can be a disk with rotor blades located above the «Segner wheel» and forming channels into which air cyclically passes from the reactive module, thereby bringing the disk into rotation.

For this scheme of a gas turbine installation, two types of air heaters were developed in which the working fluid is heated by radiation and convective heat transfer methods [14, 15]. Both heaters are highly efficient, reliable and easy to manufacture. The working fluid - the air in them is heated using the products of fuel combustion. Their main advantages are:

- high productivity due to the developed heat exchange surface (the inner pipe is equipped with longitudinal ribs);
- ensuring the integrity of the structure during operation at high temperatures (it is possible to compensate for temperature elongations);
- no need to use an intermediate coolant.

In the design of the installation, an active-reactive turbine is used as a drive, the principle of operation of which is based on the reactive power of the outgoing air stream, while the turbine does not require maintaining high revolutions, and this design does not require heating the working fluid to high temperatures.

The design of the reactive part of the turbine is proposed to be carried out similarly to the design of the reactive centrifugal turbine shown in figure 3, with some modifications.

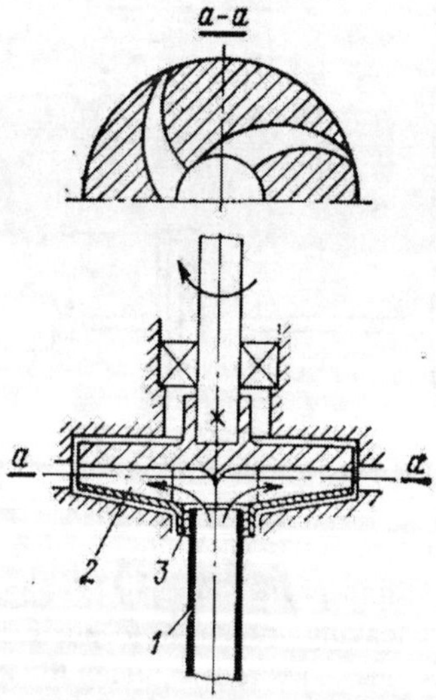

Figure 3. The design of a centrifugal reactive turbine: 1 - fixed tube; 2 - channels; 3 - labyrinth seal.

The active part of the turbine is proposed to be performed similarly to the turbine shown in figure 4.

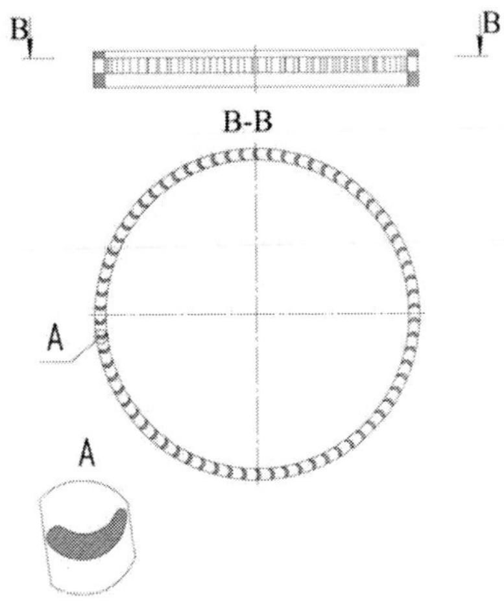

Figure 4. The design of the active part of the turbine.

In preliminary calculations of the main parameters of the installation, the following main calculated dependences were used.

Compressor power N_κ, kW is calculated by the formula

$$N_\kappa = P_1 \cdot Q_в \cdot \frac{\kappa}{\kappa - 1} \cdot \left[\left(\frac{P_2}{P_1} \right)^{\frac{\kappa-1}{\kappa}} - 1 \right] \cdot \frac{1}{\eta_\kappa} \cdot 10^{-3}, \text{ kW} \quad (1)$$

where P_1 – air pressure at the compressor inlet, Pa;

$Q_в$ – air volumetric flow, m^3/s;

κ – adiabatic coefficient;

P_2 – air pressure at the compressor outlet, Pa;

η_κ – compressor efficiency.

Heat flow rate Q, W of the heater

$$Q = \frac{m \cdot C_p \cdot (T_2 - T_1)}{\eta}, \text{ W} \quad (2)$$

where m – mass air flow, kg/s;

C_p – isobaric heat capacity, $\text{J/(kg} \cdot \text{K)}$;

T_2 – air temperature at the outlet of the heater, K;

T_1 – heater inlet air temperature, K;

η – heater efficiency.

Reactive force F, N determined by the formula

$$F = m \cdot V, \text{ N} \quad (3)$$

where V – air flow rate, m/s.

Torque M_t, $\text{N} \cdot \text{m}$ of turbine

$$M_t = F \cdot r, \ \text{N} \cdot \text{m} \quad (4)$$

where r – turbine impeller radius, m.

Turbine reactive power N_{RT}, W

$$N_{\mathrm{RT}} = \frac{M_t \cdot n}{\frac{2 \cdot \pi}{60}}, \text{ W} \qquad (5)$$

where n – turbine rotation speed, rpm.

Power of the active part of the turbine N_{AT}, W [16]

$$N_{\mathrm{AT}} = L_{at} \cdot \mathrm{G}_{at}, \text{ W} \qquad (6)$$

where L_{at} – disposable air expansion work in the turbine, J;

G_{at} – mass air flow through the active part of the turbine, kg/s.

3. Results

The proposed technical solution allows:

- ensure safety, increase the efficiency of the process and the environment;
- to heat the working fluid of the turbine in the heat exchanger to the required temperature, which ensures high uniformity of the temperature field of the working fluid in front of the nozzle of the turbine;
- organize the combustion process of associated petroleum gas with low pressure (less than 0,1 MPa) and without special preparation, which simplifies the technology, reduces the complexity and provides less harmful emissions.

The results of preliminary calculations of the main parameters of the developed gas turbine unit with a generated electric power of 10 kW are shown in table 1.

Table 1. Key parameters for preliminary calculations of the developed gas turbine unit with an electric power of 10 kW.

Parameter	Value
Compressor power, kW	37
Air consumption, kg/s	0,369
Air pressure in front of the compressor, Pa	101325
Air pressure after compressor, Pa	250000
Heater inlet air temperature, °C	50
Air temperature at the outlet of the heater, °C	400
Heater thermal power, W	258817
Heat transfer area, m^2	7,588
Fuel gas consumption, m^3/h	12
The temperature of the combustion products, K	1560
Active-reactive turbine power, kW	84
Received electric power, kW	10
Overall dimensions (Length x Width x Height), mm	1500x1500x1200

The dependence of the radius of the reactive turbine on the temperature of the heated air and the speed of rotation of the turbine is shown in figure 5.

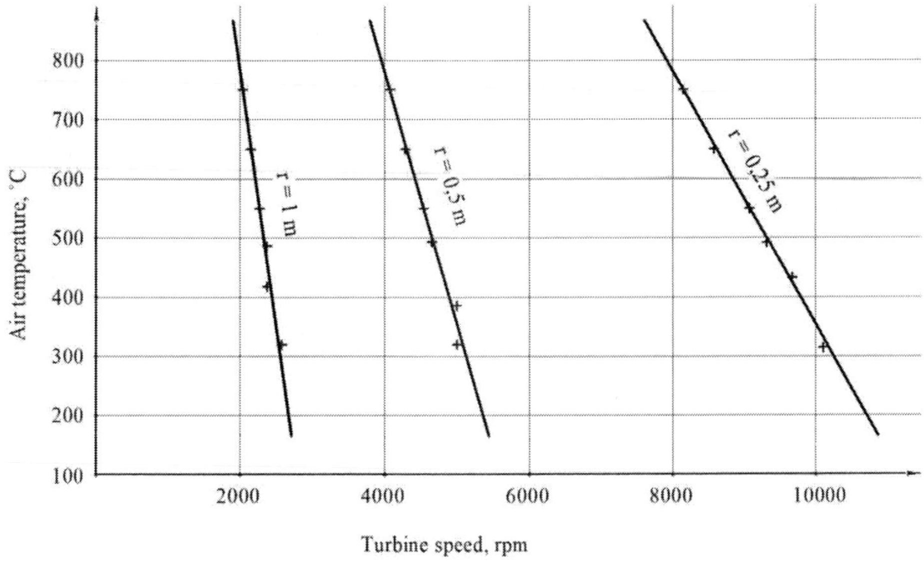

Figure 5. The dependence of the radius of the reactive turbine on the temperature of the heated air and turbine speed.

4. Conclusion

Thus, during the study, an autonomous source of electrical power for oil and gas fields was developed on the basis of a gas turbine unit driven by an active-jet turbine. Using the proposed installation, it is possible to process associated petroleum, as well as low-pressure, crude natural gas into electrical energy to ensure continuous operation of the equipment used in production. As the working fluid in the installation, atmospheric air is used, therefore the working surfaces are not subject to pollution. The installation does not require frequent maintenance, the consumption of lubricants, preliminary preparation of fuel gas, has a relatively small overall dimensions.

References

[1] Rybakov B A, Burov V D, Rybakov D B and Trushin K S 2008 Features of associated petroleum gas combustion in gas turbine plants *Turbines & Diesels* **2** pp 2–8

[2] Kazancheva A N 2017 *Prospects for the development of technical sciences* (Chelyabinsk, Russia) **1** pp 16–19

[3] Verevkin A P and Seleznev S B 2015 Disposal of associated oil gas based on electric power generation: problems and solutions *Scientific and technical journal Oil and gas business* **13** pp 56–62

[4] Arbekov A N 2014 Selection of the working medium for 6- to 12-kW closed organic-fuel-powered gas-turbine plants *High Temp* **52** pp 121–125

[5] Sukhov A I Lachugin I G Shevtsov A P Marushak G M Khokhlov V Y Aristov I V and Merkulova Y D 2013 Improved production of electricity by using gas turbine plant *Pumps. Turbines. Systems* **1** pp 22–26

[6] Gorshkov M V and Pashali D Y 2018 On the use of gas turbine power plants in oil fields to ensure the reliability of power supply and environmental friendliness *Innovation science* **8** pp 32–35

[7] Sadykov B H 2015 Autonomous power supply of oil fields *Science today. Key problems and development prospects* pp 91–92

[8] Bronicki L Y 2008 *ASME 2008 9th Biennial Conference on Engineering Systems Design and Analysis* (Haifa, Israel) pp 15–23

[9] Belyaev V E Kachaev S V and Tyulkov K V 2009 *The use of gas turbine units in the energy and industry. Abstracts of LVI scientific and technical conference* (Perm, Russia) pp 16–18
[10] Ahmadi M H Ahmadi M and Pourfayaz F 2017 Thermal models for analysis of performance of Stirling engine: A review *Renewable and Sustainable Energy Reviews* **1** pp 168–184
[11] Marco Antônio Rosa do Nascimento Lucilene de Oliveira Rodrigues Eraldo Cruz dos Santos Eli Eber Batista Gomes Fagner Luis Goulart Dias Elkin Iván Gutiérrez Velásques and Rubén Alexis Miranda Carrillo 2013 Micro Gas Turbine Engine: A Review *Progress in Gas Turbine Performance IntechOpen* pp 107–141
[12] Avdeev Y N Lachugin I G Shevtsov A P Sukhov A I and Chernoivanov D V 2019 Gas-turbine installation for processing of associated oil gas in electric power Patent No. 2686357 RF F 02 C 1/08 No. 2017102649 26.01.2017 Byul. **11** p 6
[13] Lachugin I G Shevtsov A P Sukhov A I Bazykin D A and Khokhlov V Y 2020 Gas-turbine installation for processing associated oil and various low-pressure gases in electricity Patent No. 2713785 RF F 02 C 1/04 No. 2019113542 29.04.2019 Byul. **4** p 4
[14] Bazykin D A Sukhov A I and Barakov A V 2019 Development of a heat exchanger device-regenerator for high-temperature gas heating *Power engineering: research, equipment, technology* **21** pp 38–49
[15] Lachugin I G Shevtsov A P Sukhov A I Bazykin D A and Khokhlov V Y 2019 Air heater Patent No. 2708175 RF F 24 H 3/08 No. 2018135351 05.10.2018 Byul. **34** p 9
[16] Matveev V N and Sulinov A V 2011 *Design calculation of single-stage and two-stage autonomous axial turbines of turbopump units of liquid rocket engines* (Samara, Russia: Samara economic university named after academic Korolyov S P (national research university))

IOP Conf. Series: Materials Science and Engineering 1035 (2021) 012004 doi:10.1088/1757-899X/1035/1/012004

Analysis of assessment of the technical condition of equipment of generating systems

M M Sultanov[1], N V Baydakova[1]and A V Afonin[1]

[1]Branch of the National Research University "MPEI", Lenin Avenue, 69, Volzhsky, 404110, Russia

Abstract. This work presents the results of the analysis of existing methods for assessing the technical condition of generating systems' equipment, the advantages of maintenance and repair systems by the actual condition are considered.

1. Introduction
Increasing the role of diagnostic methods and tools at enterprises of the Russian Thermal Power Complex (TPC) is associated with the need to ensure industrial safety in the current physical and moral wear and tear of equipment.

2. Methods
In operational terms, the main means of ensuring the reliability of the functioning equipment is a technical operation, however, to a certain extent, reliability also depends on production operation (on the modes of equipment use, degree of protection, etc.). Figure 1 shows the current operational maintenance structure [6]. This structure is fairly common and common to all countries. The difference is usually that different emphasis is placed on a type of repair.

Trends on an international scale are aimed at finding economic efficiency. The different approaches in different countries are due to structural and cultural differences. For example, in the USA, a large number of faults are allowed, and in Japan, the concept of "zero number of defects" is used, which leads to a high proportion of planned repairs in total maintenance costs.

The use of modern methods of monitoring and diagnostics in cogeneration plants of local value allows us to get away from the existing system of planned preventive repair (PPR) and build a system of equipment maintenance according to the actual technical condition of equipment (TCE), which will increase the resource and reliability of cogeneration plants, reduce downtime costs and repair.

The advantages of such a transition are obvious. During PPR, a large amount of repair work is carried out, including non-defective equipment, i.e. such equipment, the condition of which at the time of the PPR does not require urgent repair, but is subject to maintenance to ensure trouble-free operation during the overhaul period. Often, the residual resource of the equipment withdrawn for repair is significant. Statistics show that up to 50% of works carried out in accordance with the regulations are carried out without their actual need.

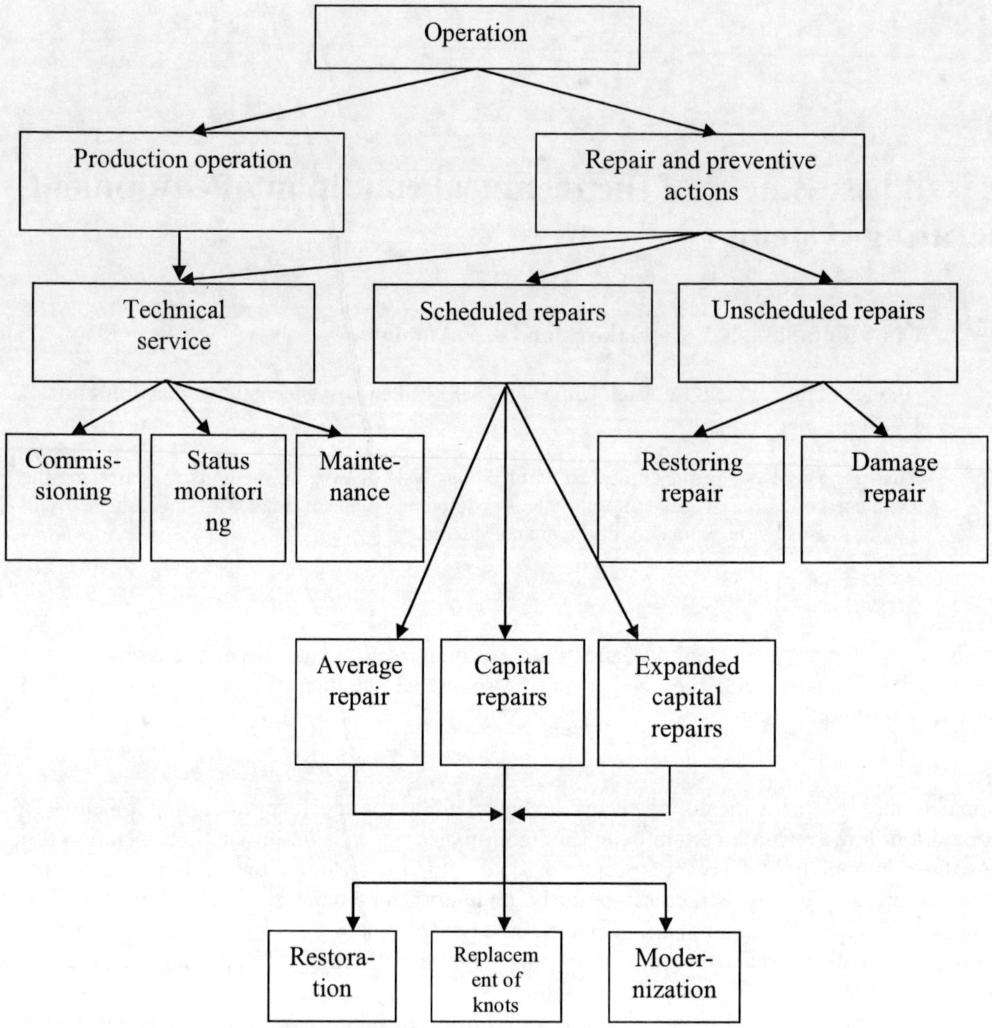

Figure 1. Structure of equipment operational maintenance.

3. Results and Discussion

Repairing on the actual condition of the equipment has the following advantages:

1. The efficiency of technical maintenance (TM) and repairs increases, at the same time equipment failures are practically excluded if there is constant information about the state of the units.
2. The forecasting and planning of possible volumes of technical maintenance (TM) and repairs of equipment are facilitated, costs by eliminating unnecessary types of work are reduced.
3. Effective planning and allocation of human and material resources.
4. Increase of between-repairs resource life.
5. Quality control of works.

The introduction of repairs' system based on the actual condition of the equipment requires a radical change in the principle of organization and maintenance of repairs. Additional technical development is needed, a large number of statistics on failures both random and non-random; creation of terminology, glossary; development of mathematical models; selection of diagnostic system; cost-effectiveness assessment. Special attention needs to be paid to the qualification of specialists, since constant or discrete monitoring of the equipment condition, application of various diagnostic methods, data analysis, interpretation of the resource of individual units are fundamentally different from the simple consolidation of defects into the bill of quantities.

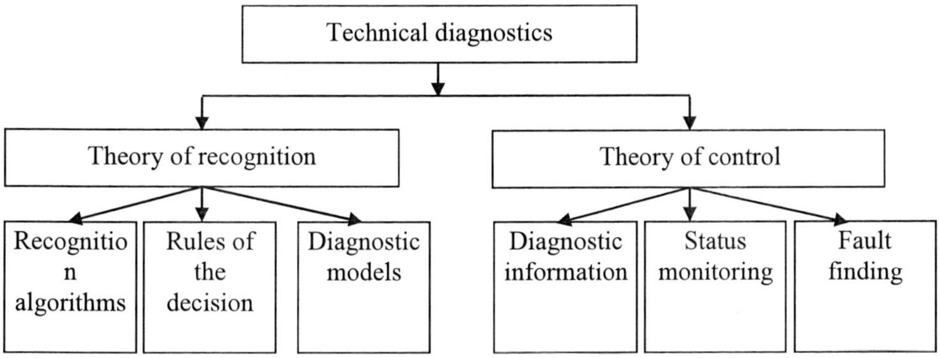

Figure 2. Technical diagnostics structure.

It is not rational to subject the entire set of units to detailed diagnostics, therefore, it is necessary to apply an individual approach to each group for which a state recognition system is determined using a sufficient minimum of information.

It is advisable to divide the equipment into several categories:

The I-st category. Large unit power units, unscheduled failures or accidents which lead to significant losses. These are reserve-free turbine units, pumps with a capacity of more than 1 MW.

The II-nd category. Units with an average capacity of 0.2 MW < N < 1 MW.

The III-rd category. Main units, in case of failure of which there is a danger of shutdown, but the main equipment will not suffer. For example, redundant tour units, pumps N < 200 kW.

The IV-th category. Auxiliary equipment requiring minor maintenance.

The V-th category. Non-critical auxiliary equipment: process piping, valves, pipelines.

Equipment of the I-st- and the II-nd categories is equipped with equipment for constant parameter monitoring. Category the III-rd - undergoes periodic monitoring. Categories the IV-th and the V-th generally do not require diagnostic.

Current and predictive diagnosis should be clearly defined when determining diagnostic methods. During the current diagnostics, the state of the object is determined at a certain point in time, which makes it is possible to determine the correctness of operation and the further possibility of performing diagnostic control before the next period, while the predictive method allows you to obtain initial data for interpolating possible changes in operation, as well as faults during the further operation of the equipment. Therefore, the predictive diagnostic method requires much more analytical work. To do this, several diagnostic methods are used simultaneously:

1. Functional diagnosing. This method allows you to detect a malfunction on the operating equipment and the possibility of online connection of the reserve, switching to another mode, without stopping the unit. The disadvantage is a one-time indicator in only one given mode of operation of the equipment. Fault detection in other modes may not be possible.

2. Test diagnosing. This method allows you to obtain complete information about the health of the object, about its technical condition. It should be borne in mind that test diagnostics are possible only on non-operating equipment or during repairing.

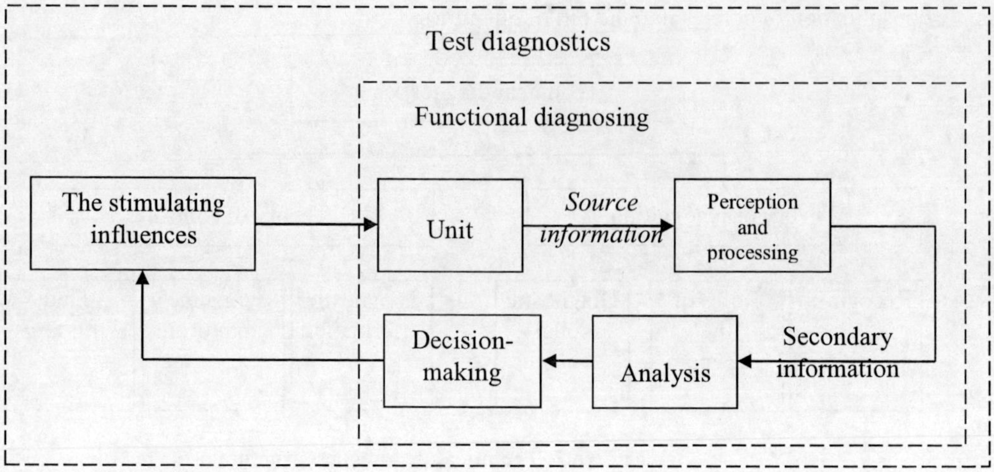

Figure 3. Diagram of basic operations of functional and test diagnostics.

The most objective data can be obtained by combining functional and test diagnostics. This complex method allows you to see the state of the equipment during operation and repair.

For predictive diagnostics, in addition to functional and test methods, retrospective data must be taken into account. The need to take into account them is due to the complexity of the mutual influences of various changes in the operation of the unit. For example, the natural wear of the shaft journal can affect the vibration characteristics both in the positive and negative direction, which, of course, affects the efficiency of the unit.

To date, equipment diagnostics within the design service life are limited to periodic monitoring of the technical condition of the hydraulic turbine, based on the requirements of the regulatory and technical documentation, thereby confirming the readiness of the equipment for further operation. The assessment of the technical condition at the same time is reduced to technical inspections, visual and measurement monitoring and other standard tests. When the equipment generates the design life, the main goal of diagnostics is to determine the possibility and feasibility of its continued operation.

Consider the main methods, approaches and opinions of specialists, as well as expert organizations that have experience in assessing the resource of hydroturbines and offer their solutions. The first group of hydroturbine resource estimation methods is based on a statistical analysis of operating experience data. Such methods include the following:

- estimation based on the calculation of failure rate [5];
- assessment of HA reliability on the basis of readiness coefficient (or operational readiness coefficient) and their modifications characterizing the ratio of HA operation duration at different operating modes and idle time [7, 8].

These methods have an undeniable advantage - this is the simplicity of their practical application and the ability to compare the resource of different GAs taking into account the selected parameter (failure rate, operational readiness factor) from the positions of integral reliability characteristics. They are applicable at the stage of the design life of the equipment, however, for unique powerful GAs outside this period, statistical models are not applicable. They can only be used as supporting evaluation techniques.

The second group of methods is based on a calculated estimate of the life of hydroturbines as a function of the accumulation of fatigue damages. The basis of this approach is given in the regulatory industry documents [9], developed to evaluate the resource at the design stage and ensure the necessary level of equipment reliability. In general, the calculation approaches involve the following main steps [10, 11, 12-17]:

- selection of resource-determining parameter (set of parameters), and criteria of limit state of the hydraulic turbine;
- construction of a mathematical model of investigated object, for example, GT impeller;
- calculation of the current value of the resource-determining parameter (set of parameters) according to the selected special method;
- forecast of resource-determining parameter change for the expected period of further operation or until the limit state is reached.

4. Conclusion

Similar approaches to estimating the resource of critical facilities are used in various sectors of energy, mechanical engineering, petrochemical industry and are spelled out in the corresponding industry NTD, which regulates the criteria for limit states, the volume, and procedure of calculations, and the values of reserve factors.

References

[1] Sultanov M M, Balandin O I, Skopova E M 2018 *Improving the efficiency of operation of power equipment of generating systems: training manual* (Volzhsky) p 85
[2] Sultanov M M 2016 *Reliability assessment, life extension and optimization of repair of TPP equipment and power systems* (Volzhsky) p 100
[3] Trukhanov V M, Sultanov M M, Skopova E M 2018 *Energy conservation and water treatment* **5 (115)** 10
[4] Uzakova L P, Kalandarov N O 2014 *Young scientist* **9** 3
[5] Georgievskaya E IN 2016 *Hydraulic machines, hydraulic pneumatic drives and hydraulic pneumatic automatics. Current status and prospects for development* 8
[6] Caps B AND, Lizard A AND 1999 *Productive operation, maintenance and repair of power equipment* 238
[7] Ivanchenko I P 2002 *Proceedings of CKTI* 290 8
[8] Ivanchenko I P, Prokopenko A N 2012 *Reliability and safety of energy* **3 (18)** 3
[9] Vasiliev Yu S, Petrenya Yu K, Georgievskaya E 2017 *Engineering. SPbUP scientific and technical statements. Science and Engineering* 23/2 20
[10] Ivanchenko I P, Potemkin A A 1986 *Energ y Engineering* **1** 40 p
[11] Kantor B I, Rzhevskaya I E, Strelnikova E A 2002 *Problems of mechanical engineering* **5/1** 6
[12] Ivanchenko I P, Topazh G I, Schur B A 2015 *Scientific and Technical Statements of St. Petersburg Polytechnic University* **2 (219)** 8
[13] Sudakov A B, Georgievskaya E In 2014 *Hydrotechnical construction* **2** 7
[14] Georgievskaya E B, Smelkov L L 2014 *Hydraulic engineering* **3** 4
[15] Georgievskaya E 2016 *Livability and structural materials science* 4
[16] Novkunsky A A 2015 *Bulletin of Science of Siberia* **15** 6
[17] Levina S M 2013 *V All-Russian meeting of hydropower engineering specialists. Strong Russia Powerful Updated Energy* 12

Acknowledgments

The work has been carried out with the financial support of the state task of the Russian Federation No. 0720-2020-0025 "Development of methods and analysis of methods to achieve a high level of safety and competitiveness of facilities of energy systems based on digital technologies."

IOP Conf. Series: Materials Science and Engineering 1035 (2021) 012005 doi:10.1088/1757-899X/1035/1/012005

Fast-acting shut-off and control ball valve application for control of gas flow rate to power plant boilers

I A Boldyrev [1], N V Kharitonov [1], A A Beokash [1]and P D Menshikov [1]

[1]Branch of the National Research University "MPEI", Lenin Avenue, 69, Volzhsky, 404110, Russia

Abstract. The work describes the design and algorithms for controlling the fast-acting shut-off and control ball valve calculation of the reliability of the natural gas flow control system for the power plant boiler. Its application makes it possible to improve the quality of fuel combustion control, as well as to reduce the quantity of shut-off and control valves of gas channel, which in turn also leads to improvement of reliability of the system as a whole.

1. Introduction

The proposed fast-acting shut-off and control ball valve (FSCBV) is designed for use in gas storage systems of thermal power plants for control of natural gas consumption to burners of steam or hot water boiler.

Application of FSCBV allows to significantly reduce quantity and nomenclature of shut-off and control valves of gas line, as well as to improve the quality of fuel combustion control in boilers.

The design of the article includes elements: ball crane, fast-action insert, electromagnetic clutch, electric drive [1]. One of the differences of the device from the known one is the use of the position indicator of the valve locking element, which makes it possible to assess the state of the device in any mode of operation.

Gas consumption networks of thermal power plants (TPP) in the section of gas pipelines within boilers include several threads, each of which consists of elements: gate valve at the inlet, control valve of pressure of the 1 stage, silencer, control valve of pressure of the 2 stage, silencer and gate valve at the outlet. Safety discharge valves and output gate valve with an electric drive are arranged on the outlet gas pipeline.

On the straight section of the gas pipeline within the boiler, the following is installed:

- Orifice plate for measurement of gas flow rate per boiler;
- Shut-off valve - general-purpose safety shut-off valve connected with process interlocking and boiler protection;
- Gas supply valve (GSV) - for control of gas pressure on the boiler.

A manual gate valve and plug valve with an electric drive are installed on branch to each burner of the boiler. Besides, the boiler gas pipeline has flanges for plug installation, compressed air tie-in for gas pipeline blowdown and blowout plugs.

The structure of the described diagram is shown in figure 1.

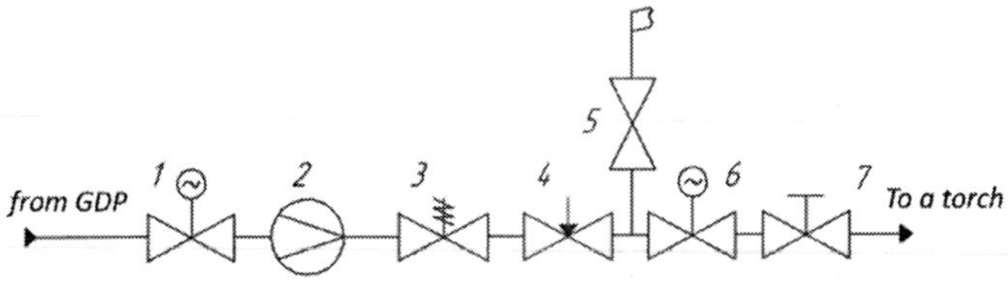

1 - Motor-operated gate valve; 2 – contraction; 3 – general-purpose safety shut-off valve; 4 - gas supply valve; 5 - valve with safety plug; 6 - plug valve with the electric drive; 7 - manual gate valve.

Figure 1. Typical gas supply diagram.

Rules [1] allow alternative versions of implementation of gas supply schemes with the use of electrified gate valves and electrically driven fast-acting shut-off valves (FSV) before the burner provided that the common-cell safety valve is maintained.

The circular of the Department of Science and Technology "UES of Russia" C 03-97 (T) by item 1.13 provided for the possibility to develop industrial production of electrically driven fast-acting shut-off and control valves (FSCV), combining the functions of GSSV and CV and their installation for reconstructed operating gas supply systems of boilers.

2. Methods

The scheme is proposed with the use of FSCBV (fast-acting shut-off and control valve), which ensures compliance with the requirements and minimizes the costs of adjusting the protection and interlocking schemes to the current rules.

When using FSCBV, the following requirements are met [1]:
- The tightness of "A" class lock;
- Shut-off function response speed up to 1 sec;
- Compliance with the requirement "power supply shutdown shall cause valve closing (opening) without additional power supply from other external sources";
- GSSV material is made of steel based on operating gas pressure up to GDP.

At the same time, the costs for ensuring explosion safety of boilers during reconstruction works on gas pipelines within the boiler, protection schemes and interlocks are minimized, as:
- Works on the design of pipelines within the boiler are excluded by a specialized organization;
- Costs for purchase of fast-acting shut-off valves (AMAX type) and technical means of digital state control of shut-off and control valves are excluded;
- Costs for purchase of certified pipelines, electrodes, bends, etc., are excluded;
- Costs for project expertise, acceptance commissions of Rostekhnadzor are excluded;
- Costs for installation and adjustment works are significantly reduced.

FSCBV design is shown on figure 2: ball valve consisting of housing and ball locking element installed in it, electric drive with the drive shaft, fast-action insert connecting spindle of ball element, and the drive shaft of electric drive, position alarm mechanism. Fast-action insert design provides rigid a connection of spindle and drives shaft in gas flow rate control mode, as well as disconnection of the drive shaft and spindle with the subsequent forced fast rotation of ball locking element due to torsion spring at the removal of power supply voltage from the fast-action insert.

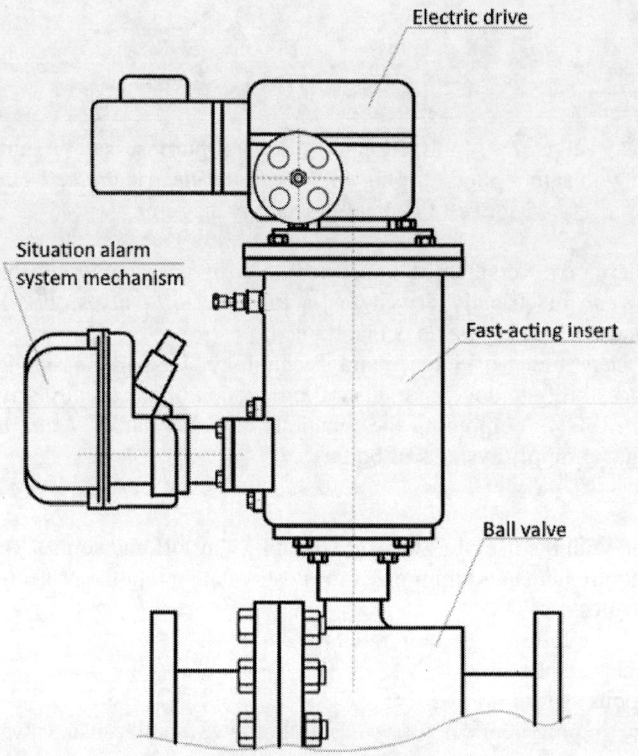

Figure 2. FSCBV design.

FSCBV is controlled by electric signals remotely and from the local control board. The main modes are:
- Change of ball shutoff element position during boiler normal operation
- Opening of electromagnetic clutch inside the fast-action insert for fast closing of the FSCBV at actuation of the boiler unit protection or at the check of the protection circuit
- Initialization of the position of the locking ball element at the first actuation of the FSCBV or after protection actuation.

The authors have developed and tested algorithms for control of FSCBV in all modes of boiler operation.

An example of a mnemonic diagram for the automated workplace monitor of the heat board is shown in figure 2. The mnemonic diagram shows the information on the position of shut-off and control valves, as well as values of process parameters (pressure, temperature, gas flow rate).

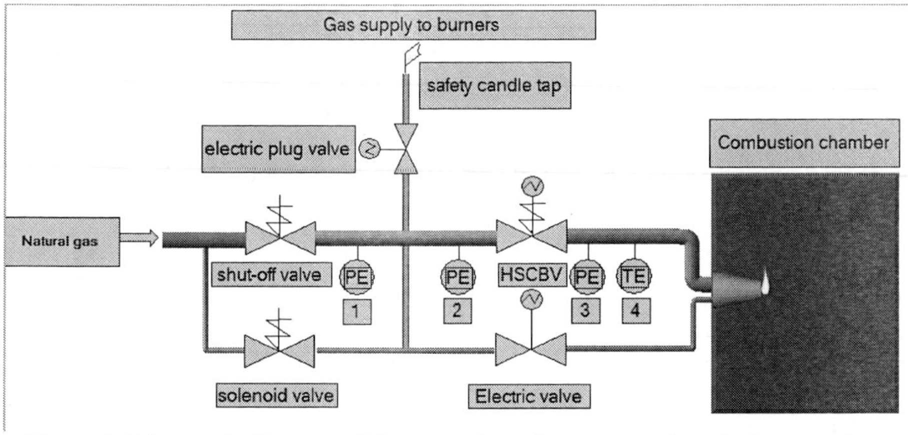

Figure 3. Mnemonic diagram of the gas unit on the automated workplace monitor.

In accordance with the system reliability requirements [2], the availability factor of the Software Technical Complex (STC) shall be at least 0.99 with a maintenance period of 1 month. The diagram of reliability calculation using FSCBV is shown in figure 4.

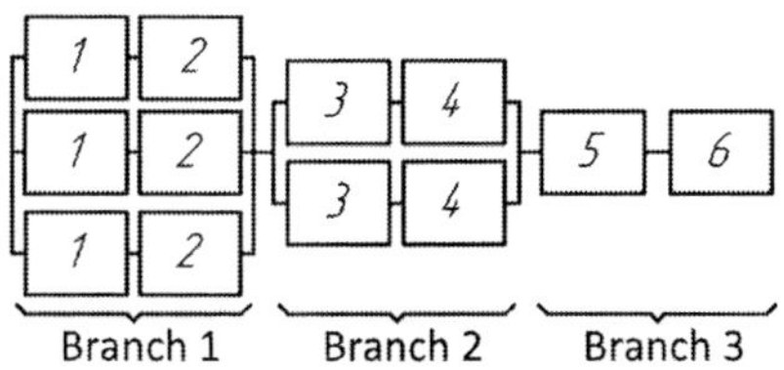

1 - power supply unit; 2 - pressure transmitter; 3 – PLC; 4 - I/O module; 5 - fast-action insert; 6 - ball valve.

Figure 4. System reliability diagram using FSCBV.

As a result of the calculations, the probability of failure-free operation of the system is determined to be 0.991.

In order to determine the required materials for making the fast-action insert elements, equivalent stresses acting in the article elements are determined. Figure 5 shows the finite element split grid for strength calculation. The maximum length of the element side is 5 mm, the number of elements 6724. The number of knots is equal to 2109.

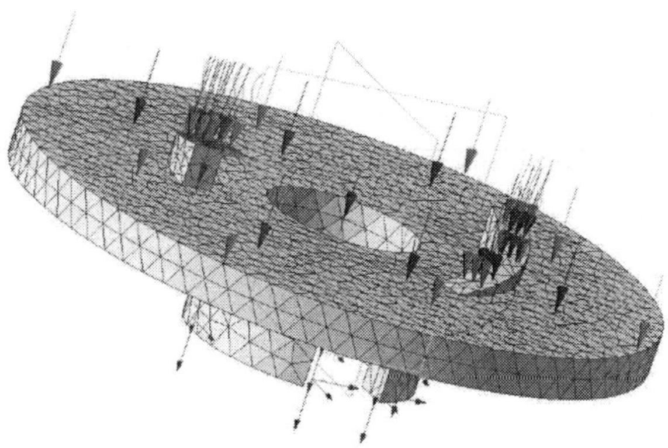

Figure 5. Finite element mesh for strength calculation of pressure plate.

As a result of static calculation of strength, the distribution of equivalent stresses by Mises is obtained, at the same time minimum and maximum values are 2.24 MPa and 38.26 MPa respectively.

Application of special design of electromagnetic coupling in the insert speed of operation makes it possible to manually control the position of shut-off element in place by rotation of electric drive hand wheel without power supply need. The specified function allows us to safely perform adjustment of limit switches and position alarm mechanism, as well as manual control of the product after actuation of the protection circuit and absence of power supply.

When the voltage is released from the control circuit of the FSCBV, the coupling is disengaged, which in turn opens the shafts of the ball valve and the electric drive, and the valve returns to the closed position by means of the spring.

It is shown that in the flow rate range corresponding to the nominal operation mode of the boiler, the flow characteristic is linear. At low angles of FSCBV opening, the flow characteristic has a smoothly increasing character, which allows using of FSCBV for gas flow control in the boring mode without using additional bypass gas pipelines.

The FSCBV tests were carried out on the bench, the structure of which is shown in figure 6. The test bench verifies the correct control of the position of the shut-off element, the correct installation of the limit switches, as well as the life of the control and shut-off function.

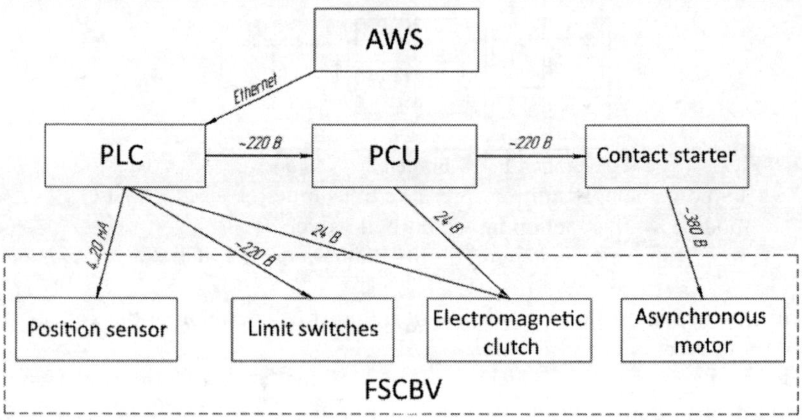

AWS - automated workstation; PLC - programmable logic controller; PCU - prototype control unit
Figure 6. Structural diagram of the test bench.

The prototype control unit (PCU) allows us to manually act on the FSCBV: to open or close the valve, to start or stop the power supply to the electromagnetic coupling.

The PIM position alarm mechanism installed on the housing generates a current signal of the ball shutoff element position in the range of 4-20 mA, which is supplied to the PLC and a switch indicator for visual monitoring of the prototype position.

The electromagnetic coupling as part of the fast-action insert has a power supply voltage of 24 V DC and is controlled with PLC and PCU.

Figure 7 shows block diagrams of FSCBV control in the manual (a) and automatic mode (b).

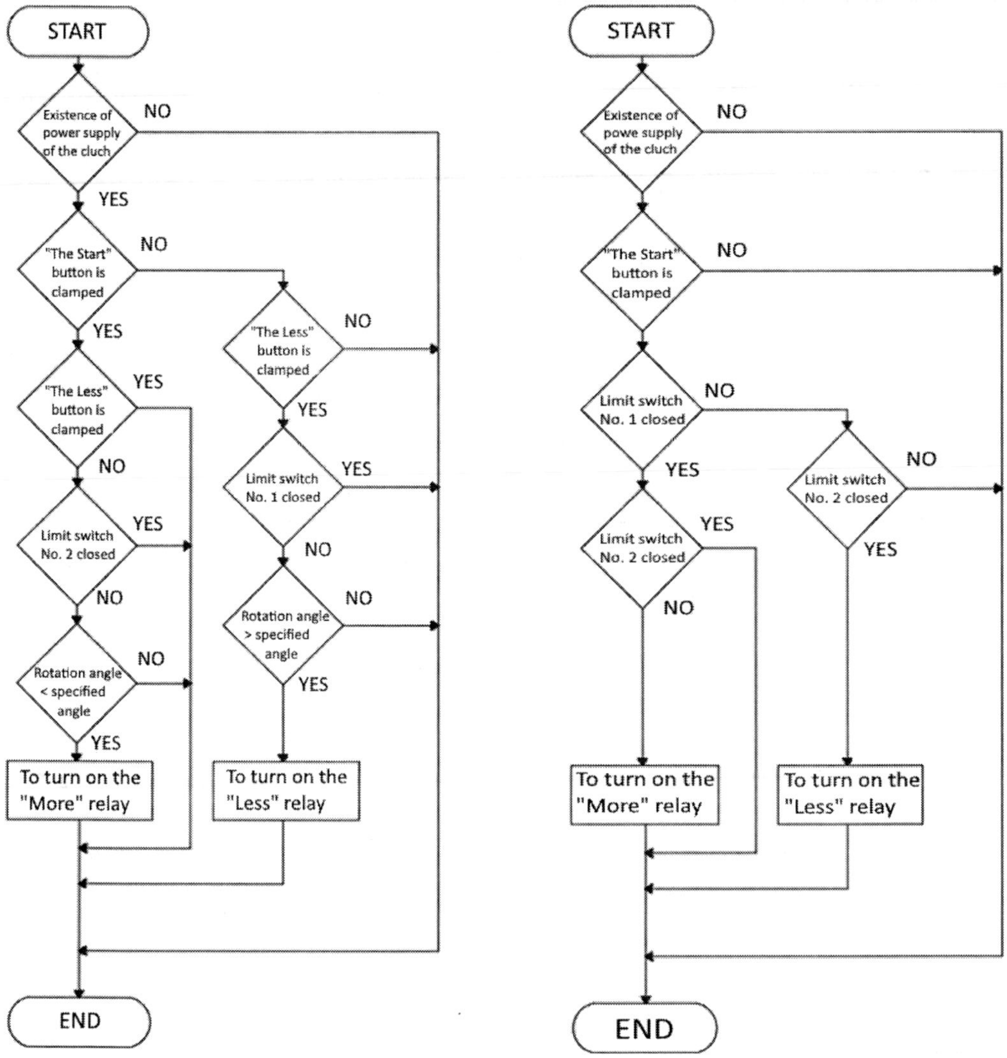

(a) (b)

Figure 7. Block diagrams of FSCBV control algorithm.

3. Results and Discussion

As a result of the tests, it was shown that the reliability of the product and the actual characteristics of the resourcefully meet the requirements of the regulatory documents and the indicators specified in the Design Statement of Work.

4. Conclusion

The use of FSCBV allows to significantly reduce costs for modernization of gas supply schemes of power units, improve operating characteristics of boilers, as well as simplify maintenance of gas units.

References

[1] Boldyrev I A, Haritonov N V, Vovenko A N, Afonin A V, Menshikov P D and Beokash A A *Fast-acting shut-off and control ball valve* Patent RU 194 386 Published: 09.12.2019 Bul. No 34

[2] Guidance document "Technical requirements to the subsystem of process protections made on the basis of microprocessor equipment" 2001 RD 153-34.1-35.137-00 Moscow RAO UES of Russia 65 p
[3] Guidance Document "Technical Conditions for Process Protections and Interlocks When Using Fuel Oil and Natural Gas in Boiler Plants in Accordance with Explosion Safety Requirements" 2001 RD 153-34.1-35.108-2001 Moscow RAO UES of Russia 48 p
[4] Eismont V P 2019 *Pipeline Safety Valves* Moscow Infra-Engineering
[5] Gorobchenko S L 2019 Evolution of valves *Pipeline valves and equipment* **3** (102)
[6] Tarasiev Y I 2019 On Classification of Control Valves *Pipeline valves and equipment* **3** (102)

Acknowledgments
The research is funded by Russian Federation public contract №0720-2020-0025 "Technique development and method analysis for ensuring power system object security and competitiveness based on the digital technologies."

Investigation of gas-sensitive properties of cobalt oxide films

V V Bespoludin[1,3], V V Polyakov[1], V V Petrov[1], A V Nesterenko[1] and Z E Vakulov[2]

[1]Southern Federal University, Taganrog, Russia
[2]Federal research Centre the southern scientific Centre of the Russian academy of sciences, Rostov-on-Don, Russia

[3]Corresponding author: bespoludin@sfedu.ru

Abstract. This paper presents a study of the gas-sensitive properties of cobalt oxide films formed by rapid thermal annealing on a sitall substrate. Cobalt films were formed on a pre-cleaned sitall substrate by vacuum thermal evaporation. The thickness of the deposited films was 470 nm. Rapid thermal annealing (RTA) was used to form cobalt oxide films. RTA of cobalt films was carried out at temperatures of 500°C, 600°C and 700°C. The resistivity of the films after RTA at 500°C, 600°C and 700°C was 3.6×10^{-2} Ohm·cm, 1.2×10^{3} Ohm·cm and 5.8×10^{3} Ohm·cm, respectively. The surface morphology of the obtained films was investigated by the AFM method. The gas sensitivity of cobalt oxide films was studied for gases like CO, NO_2, CH_4, C_3H_6O (vapors of acetone), C_3H_8O (vapors of isopropyl) and $NH_3 \cdot H_2O$ (vapors of ammonium hydroxide 25%). The gas sensitivity was measured at 300°C. The results of gas sensitivity measurements showed significantly lower sensitivity to gases such as CO, NO_2 and CH_4. It was found that cobalt oxide films have good sensitivity to gases such as C_3H_6O, $NH_3 \cdot H_2O$ and C_3H_8O. Sensitive elements based on cobalt oxide films are interesting as gas sensor elements for the detection of C_3H_6O, $NH_3 \cdot H_2O$ and C_3H_8O.

1. Introduction

Currently, metal oxides are one of the most promising materials for use as a sensitive layer in gas sensors due to a number of advantages such as high gas sensitivity, selectivity, low cost, short response and recovery time, etc. Metal oxide gas sensors have all these advantages that make them promising for use in various fields of gas analysis [1-8]. Sensor elements based on metal oxides are capable of detecting a large number of toxic and explosive gases. The principle of operation of metal oxide gas sensors is based on a change in electrical resistance as a result of the adsorption of gas molecules on the surface of the sensitive layer [9-11].

One of the promising metal oxide gas sensors is a gas sensor based on cobalt oxide. This is due to the fact that cobalt oxide has a good response to a large number of industrial gases, selectivity and long-term use [1-2].

There are two main types of cobalt oxide, Co_3O_4 and CoO. Both cobalt oxides are p-type semiconductors [2]. There are various methods for the formation of cobalt oxide on a dielectric substrate [12-19]. One of the promising methods is rapid thermal annealing (RTA) by halogen lamps in air. The main advantages of rapid thermal annealing are high-speed processing of semiconductor wafers and various microstructures, relatively low energy consumption, low cost, cost of equipment and its operation, high heating rate, wide heating range, reproducibility and uniformity of processing, high productivity and small sizes of equipment for rapid thermal annealing.

The use of RTA allows us to form a cobalt oxide film from a cobalt metal film. The deposition of metal films on a substrate by vacuum thermal evaporation with subsequent annealing these films by RTA allows us to achieve good results in the formation of cobalt oxide films.

2. Experiment details

Cobalt films were formed on a pre-cleaned sitall substrate by vacuum thermal evaporation (UVN-72).

The thickness of cobalt films was 470 nm. Electrical properties were measured using the Hall effect measurement system HMS-3000 (Ecopia Corp., Korea). The resistance of the formed cobalt films was 2.43 Ohm, and their resistivity was 5.48×10^{-5} Ohm·cm. The resistivity of the formed cobalt oxide films was investigated after rapid thermal annealing. The measurement results are presented in table 1. The fabrication process of the gas sensor based on a cobalt oxide film is shown in figure 1. In the first step, the substrate is cleaned before deposition of the metal (figure 1a). After that, cobalt films were deposited on a cleaned sitall substrate by vacuum thermal evaporation (figure 1b). Further, in order to form oxide, cobalt films were subjected to RTA by halogen lamps in the air (figure 1c). RTA was carried out at temperatures of 500°C, 600°C and 700°C. The next step was the formation of cobalt metal contacts (figure 1d).

Table 1. The dependence of resistance and resistivity on rapid thermal annealing temperature.

Cobalt oxide film	Resistance, Ohm	Resistivity, Ohm·cm
RTA at 500°C	1.6×10^{3}	3.6×10^{-2}
RTA at 600°C	5.6×10^{7}	1.2×10^{3}
RTA at 700°C	2.6×10^{8}	5.8×10^{3}

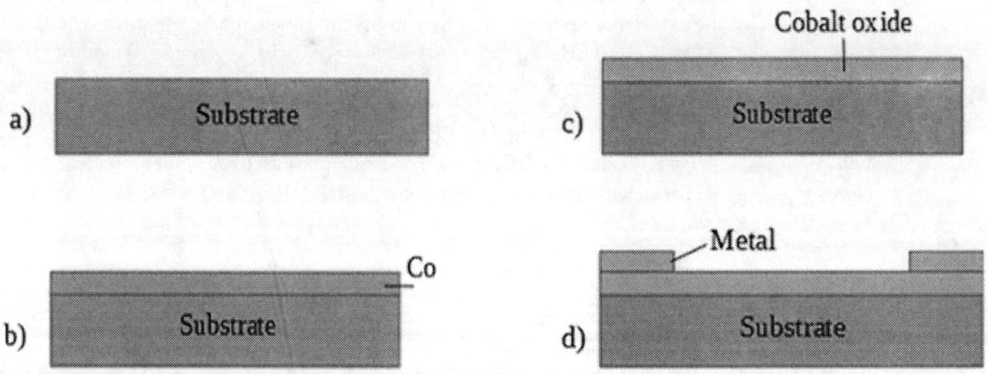

Figure 1. The fabrication process of the gas sensor based on the cobalt oxide film: (a) cleaning the substrate before deposition of the metal; (b) deposition of the metal; (c) the formation of cobalt oxide by RTA; (d) deposition of the metal contacts.

The gas sensitivity of cobalt oxide films formed by RTA was studied for gases like CO, NO_2, CH_4, C_3H_6O (vapors of acetone), C_3H_8O (vapors of isopropyl) and $NH_3\cdot H_2O$ (vapors of ammonium hydroxide 25%). The gas sensitivity of thin cobalt oxide films was measured at a temperature of 300°C (for all the measurements). The gas sensitivity coefficient S (%) was determined by the following equation:

$$S = \left|\frac{(R_{gas} - R_{air})}{R_{air}}\right| * 100$$

where R_{gas} – film resistance in the presence of gas, R_{air} – film resistance in the air.

3. Results and discussion

The morphology of the surface of cobalt films without annealing and with RTA annealing was investigated by the AFM method using the Ntegra Vita probe nanolaboratory, figure 2. Figure 2 (a) shows the surface morphology of the film of metallic cobalt without annealing. The surface morphology of the films formed by rapid thermal annealing at 500°C, 600°C and 700°C is shown in figures 2 (b), (c) and (d), respectively. From the results, it can be seen that the use of RTA makes it possible to substantially change the surface of the film, to significantly increase its roughness and grain size in comparison with a film of metallic cobalt. It was found that the average surface roughness for a film without annealing is 3.5 nm, and for films after RTA at 500 °C, 600 °C, and 700 °C the average surface roughness is 28.9 nm, 25.9 nm, and 18.7 nm, respectively.

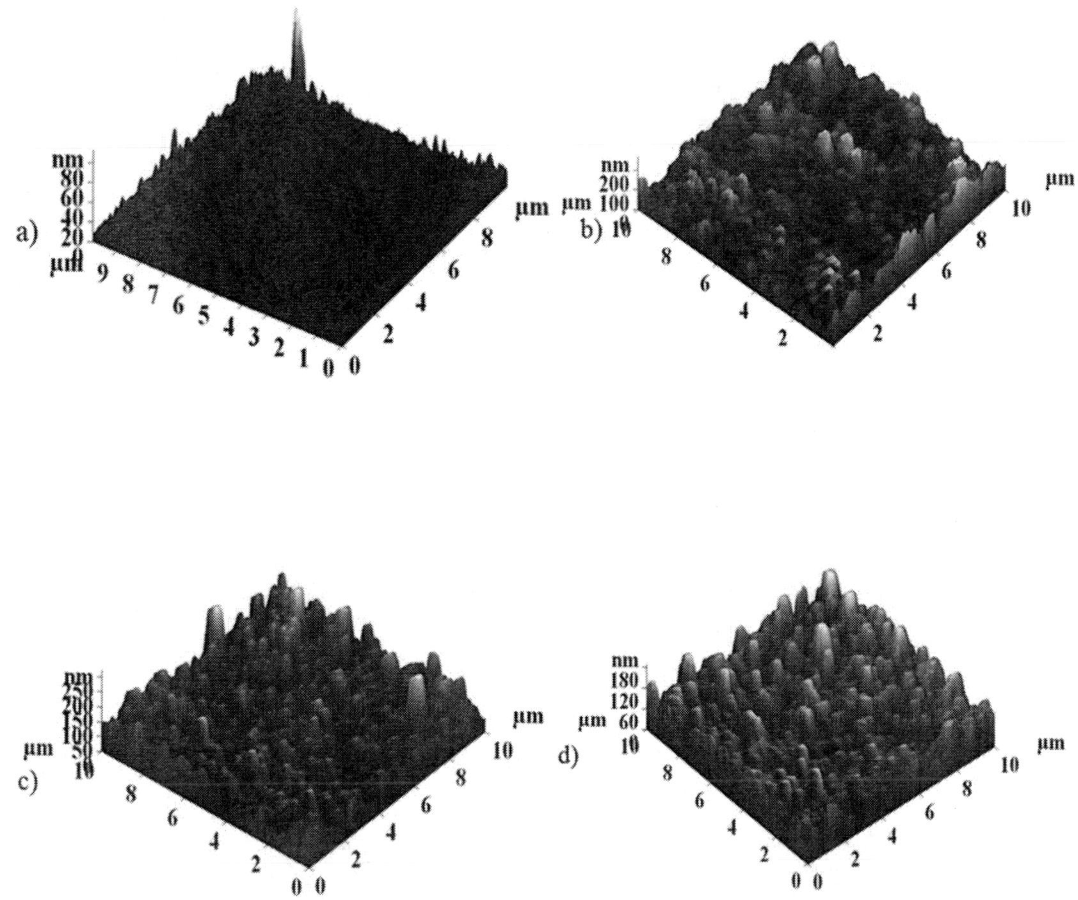

Figure 2. The morphology of the surface of the Co film without annealing (a) and after RTA at temperatures of 500°C (b), 600°C (c) and 700°C (d).

Figure 3 shows the response of cobalt oxide films formed by RTA at 500°C, 600°C and 700°C. From the results, it can be seen that the cobalt oxide films are most sensitive to C_3H_6O vapors compared to other gases. In addition, cobalt oxide films show good sensitivity to $NH_3 \cdot H_2O$

vapors (vapors of ammonium hydroxide 25%) and to C_3H_8O vapors. The results obtained are in good agreement with published data [1].

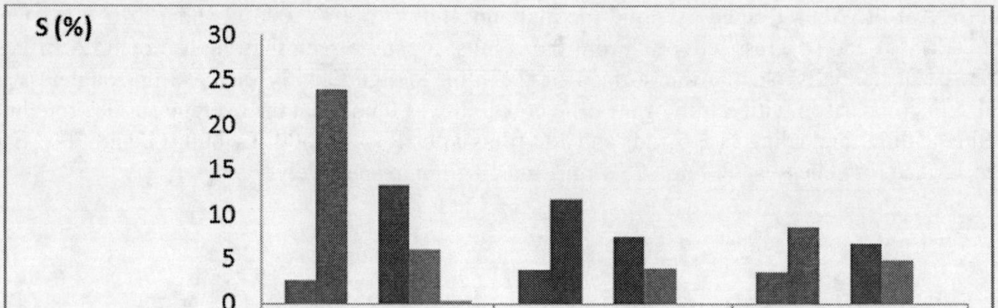

	Annealed at 500 °C	Annealed at 600 °C	Annealed at 700 °C
CO (9 ppm)	2.6	3.7	3.47
C_3H_6O (9 ppm)	24	11.7	8.6
CH_4 (5 ppm)	0.16	0.03	0.04
$NH_3 \cdot H_2O$ (10 ppm)	13.21	7.52	6.7
C_3H_8O (7 ppm)	6	3.95	4.84
NO_2 (9.5 ppm)	0.4	0.1	0.03

Figure 3. The dependence of the response of cobalt oxide films on the annealing temperature and various gases.

4. Conclusion

In summary, it was found that cobalt oxide films, which were formed by rapid thermal annealing, show significantly lower sensitivity to gases such as CO, NO_2 and CH_4. The cobalt oxide films formed by RTA at 500°C show the highest gas response, and only response to CO was slightly lower than the response of films formed by RTA at 600°C and 700°C.

Based on these results, it can be concluded that cobalt oxide films formed by RTA have good sensitivity. Cobalt oxide films are interesting as gas sensor elements for the detection of C_3H_6O, $NH_3 \cdot H_2O$ and C_3H_8O.

Acknowledgments

Formation of cobalt oxide samples was carried out with the partially financial support of the Ministry of science and higher education of the Russian Federation in the framework of the state task in the field of scientific activity No. 0852-2020-0015.

References

[1] Patil V, Pawar S, Chougule M, Raut B, Ramesh M and Sen S 2011 *Sens. & Trans. J* **128** 100–114

[2] Balouria V, Kumar A, Samanta S, Bhattacharya S, Singh A, Debnath A K, Mahajan A, Bedi R K, Aswal D K and Gupta S K 2012 *AIP Conf. Proc.* **1451** 289–291

[3] Fan C, Sun F, Wang X, Huang Z, Keshvardoostchokami M, Kumar P and Liu B 2019 *Nanomaterials* **9** 1277

[4] Gao L, Yin C, Luo Y, and Duan G 2019 *Nanomaterials* **9** 493

[5] Hotovy I, Rehacek V, Siciliano P, Capone S, Spiess L 2002 *Thin Solid Films* **418** 9–15

[6] Huang F, Yang W, He F, and Liu S 2016 *Sensors and Actuators B: Chemical* **235** 86–93

[7] Teimoori F, Khojier K and Dehnavi N Z 2017 *J Theor. Appl. Phys.* **11** 157–163

[8] Ma X X, Pan Z Y, Guo L, Li J H, Wu Z Y, and Yang S H 2012 *Chin. Sci. Bull.* **57** 4019–4023

[9] Zhang M M and Jiang G S 2007 *Chin. J. chem. Phys.* **20** 315–318
[10] Akamatsu T, Itoh T, Masuda Y, Shin W, Matsubara I and Kida M 2020 *Journal of Asian ceramic societies* **8** 138–148
[11] Kärkkänen I, Floren A, Mändar H, Avarmaa T, Jaaniso R 2009 *Procedia Chemistry* **1** 654–657
[12] Ramesh S, Lee Y J, Msolli S, Kim J G, Kim H S and Kim J H 2017 *RSC Advances* **7** 50912–50919
[13] Han N, Pan G, Zheng J, Wang R, Wang Y 2019 *Materials Research* **22**
[14] Mandal S, Rakibuddin M and Ananthakrishnan R 2018 *ACS Omega* **3** 648–661
[15] Akamatsu T, Itoh T, Izu N and Shin W 2013 *Sensors* **13** 12467–12481
[16] Alev O, Kılıç A, Çakırlar C, Büyükköse S and Öztürk Z 2018 *Sensors* **18** 956
[17] Lin T, Lv X, Hu Z, Xu A and Feng C 2019 *Sensors* **19** 233
[18] Makhlouf M Th, Abu-Zied BM and Mansoure T H 2013 *Journal of Nanoparticles*
[19] Li B, Zhou Q, Peng Sand Liao Y 2020 *Frontiers in Chemistry* **8** 321

The application of solar collectors in district heating systems: a case study in Krasnodar Krai

E A Buzoverov[1] and A Z Zhuk

Joint Institute for High Temperatures of the Russian Academy of Sciences, 13/2, Izhorskaya str., 125412, Moscow, Russia

[1]Corresponding author: teoconsult@inbox.ru

Abstract. Though renewable energy sources are widely used around the world, the cost of "green" energy in most cases is still uncompetitive with the "traditional" one. This is most pronounced in countries and regions with a developed resource base, including Russia. However, renewable energy sources in some cases can be successfully implemented in district heating systems using natural gas. This paper reports on the effective use of simple design solar collectors in a gasified central heat supply system in the city of Gelendzhik, Krasnodar Krai, with the cost of heat generated does not exceed 0.01 USD/kWh. The conditions for reproduction of this project are defined following the results of its operation.
Key words: solar collector, district heating, cost efficiency

1. Introduction

Renewable energy production is experiencing a sharp rise in the world over the past decade. The growth rate of heat and electricity generation via renewable energy sources is up to 14% per year [1], their contribution to global output varies from 10% to 60% in different countries [2]. The development of renewable energy systems is often supported at the state level around the world. There are projects for the complete transition of economically developed countries from fossil fuels to renewable energy sources by 2050 [3]. The power engineering in the near future will undoubtedly focus on utilizing renewable energy sources and the development of associated environmentally friendly technologies.

Solar energy is one of the most promising branches of renewable energy. This is due to the versatility and ubiquitous availability of the solar resource in contrast to hydrothermal and wind resources. In addition, solar energy causes minimal environmental damage compared to other technologies.

The history of solar energy utilization in Russia dates back more than a century, units of domestic manufacture continue their successful operation, the feasibility of their use is obvious due to attractive quality-price indicator [4-6]. The intensity of solar emission over certain regions of Russia is comparable, and in some places exceeds this indicator for European countries, including Germany and Austria, where facilities for collecting and converting solar irradiation have long been widely implemented [7]. The availability of free space for the equipment of almost any capacity as well as recent developments in the field of highly efficient materials also contributes to the further growth of Russian solar energy sector [8].

Over the past decade, solar energy technologies have moved from the development phase fraught with high costs and low technical and economic indicators to the level of commercial implementation.

IOP Conf. Series: Materials Science and Engineering 1035 (2021) 012007 doi:10.1088/1757-899X/1035/1/012007

Solar energy sector comprises two general concepts: solar collector units, where the solar irradiation is transferred to heat carrier, and photovoltaic panels, where the direct generation of electric energy is carried out. There are also projects combining these two concepts in devices where photovoltaic energy is implemented for generating heat [9, 10]. Herewith, the price of photovoltaic panels for the indicated period has decreased by an order of magnitude, the total surface of devices installed is growing incrementally, the cost of electricity they produce is approaching the market price of that for retail consumers in centralized power systems.

Initially, photovoltaic panels, as a high-tech product, were significantly more expensive than solar collectors. But at the moment, the photovoltaic panels' production process improvement, intensive support from government agencies, and the transition to mass production have provided a breakthrough in this direction. Results of calculations in published reports show that at present the cost of energy produced by photovoltaic panels and then converted into the heat of water is comparable to the cost of energy directly absorbed by solar collectors for the same application.

This circumstance caused a reduction in the global market for solar collectors. The rate of growth of the global solar collector market decreased from 18% in 2011 to 6% in 2015 and further to 5% in 2016 [11]. Forecasts based on these statistics predict a further drop in their market demand.

However, like any technology, solar collectors have certain specific applications in which they can show efficiency and competitiveness. This study reflects one specific area where solar collectors are successfully used in traditional heat supply systems based on gas boiler plants, even in conditions of relatively low cost of natural gas for Russian consumers.

2. The Gazprom Teploenergo experience in Krasnodar Krai

Krasnodar Krai holds one of the leading positions in introducing renewable energy sources in Russia. These include solar panels, wind generators and geothermal power stations. The construction of solar systems for the domestic hot water supply of residential buildings, resort facilities can save up to 1.5 million tonnes of standard fuel annually, which is approximately equal to 10% of the region's total consumption of heat and energy resources per year.

As a proactive experiment held in 2017 and 2018, the Gelendzhik branch of OOO Gazprom Teploenergo Krasnodar equipped several boiler plants in the city of Gelendzhik with a solar collector system. Unlike most previously implemented projects, where the cost efficiency of operational solutions is not high enough and requires significant subsidies and other measures of government support, this project, based on cheap solar collectors, showed sufficient energy and economic efficiency, since it required minimum investment and operating costs.

The city of Gelendzhik is a popular resort center in Krasnodar Krai: the population is 76 thousand people, while the flow of tourists reaches 2-3 million people per year. The climate is mild, close to the Mediterranean, but in winter the temperature drops to –10°C, and strong winds are usual, so there is a seasonal need for full-fledged heating systems. The city is fully gasified, there are 3 block boiler plants with a capacity of more than 30 MW, several dozens of municipal and private boiler plants of lesser capacity, as well as individual heat sources. The city's construction is dense, there is a significant number of apartment buildings, generally located in the district heating zone. Resorts, spas and the private sector are mainly equipped with individual heat sources.

Two district heating substations, that were equipped with solar collectors, are located in a range of boiler plant #3. They carry loads of district heating systems and domestic hot water supply of blocks located in the Severny building estate. The design of the system downstream from district heating substation is four-pipe (2 pipes for heating, 2 pipes for hot water), the domestic hot water load connection is closed-type, it is fed through heat exchangers, the value of hot water load is 110 kW. The boiler plant #3 utilizes trunk natural gas as a fuel; there is a reserve fuel oil facility, but it is out of operation.

The flow diagram of the solar collector is shown in figure 1.

IOP Conf. Series: Materials Science and Engineering 1035 (2021) 012007 doi:10.1088/1757-899X/1035/1/012007

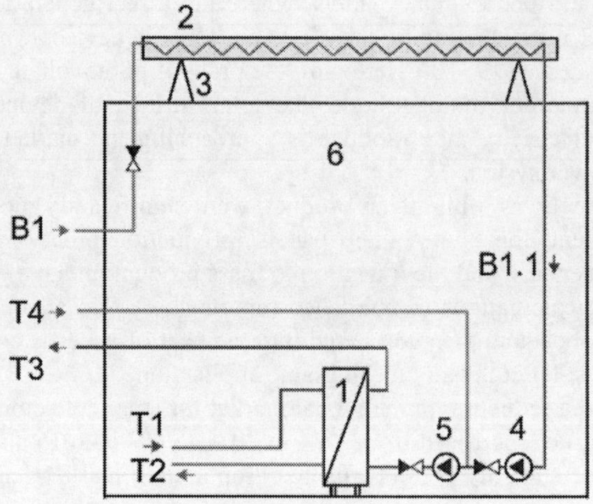

Figure 1. Flow diagram of the solar collector.
T1 — heat supply delivery pipeline; T2 — heat supply return pipeline; T3 — hot water delivery pipeline; T4 — hot water system circulation pipeline; B1 — cold water pipeline; B1.1 — cold water pipeline after the first stage of heating from a solar collector; 1 — hot water system heat exchanger in the district heating substation; 2 — solar collector; 3 — solar collector mount; 4 — cold water system booster pump; 5 — hot water system circulation pump; 6 — district heating substation building.

Solar collectors of a simple design were installed on the roof of the central heat substation; they are fed with cold water where its preliminary heating is carried out, then it enters a hot water system heat exchanger, in which additional heating is performed up to the standard temperature of + 65°C.

The design of solar collectors under this case is based on black polyethylene pipes with an outer diameter of 20 mm, their length within one district heating substation is 1 km (10 loops of 100 m each, connected to a single manifold in accordance with overhead piping flow layout, which eliminates the need for hydraulic adjustment of individual loops). The collector pipes are installed on the roof of the district heating substation openly, so the energy efficiency of the system can be increased by adding shields for preventing wind blowing over solar-heated pipes. In the cold season — from November to March — the system is taken out of operation and drained, so the use of antifreeze is not required. All these features facilitated mitigation of the project capital costs: the total cost of materials and equipment used was about 1,300 USD, the construction and installation work was done by the company's own staff without the involvement of contractors.

It is assumed that, provided the work is performed by a third-party organization, given the equipping collectors with protective steel shields, the cost of a system of this size will not exceed 5,500 USD.

The proposed design of solar collectors has the following advantages compared to existing analogs:

1. The water fed to collector initially has a relatively low temperature, which enables you to utilize not only direct solar radiation but also the heat of the surrounding air. Therefore, the solar collector also acts as a heat exchanger with a temperature drop of up to 30°C.

2. In the collector system implemented, the water undergoes only the preheating procedure, and the final heating is carried out in the heat exchange unit of the district heating substation. This scheme provides for complete utilization of the heat received by the collector during its period of operation with predicted performance and does not require the use of storage tanks as well as control valves or any automation.

The above advantages allowed to reduce the capital cost of the project by almost an order of magnitude toward the comparables, and at the same time achieve high energy efficiency. That's why, as will be shown below, even at a low cost of heat generated using natural gas, solar collectors can work effectively in district heating systems.

3. Calculation of the solar collectors' efficiency

The flow rate of cold water through within one district heating substation is 100 m^3/day, the average temperature of cold water during the non-heating period at the inlet of the station is +20°C. Measurements showed that solar collector increases water temperature by 7°C on the average, the period of collectors operation is 214 days per year. Based on these features of the system, heat supply from the collector to the network is estimated at 174 thousand kWh/year.

With the boiler plant (the general heat source of a district heating substation) fuel efficiency of 0.83, the natural gas economy reaches 26 thousand n.m^3/year, which in monetary value would be 1,960 USD/year, given the natural gas price of 0.07 USD/m^3.

The forecasted payback period of collectors at their cost of 5,500 USD is 2.8 years. This value is considered attractive in the field of projects for rehabilitation of district heating systems.

Key project performance parameters are listed in table 1.

Table 1. Technical and economic indicators of the solar collector application project in the Krasnodar Krai.

Indicator	Unit	Value
Solar collector surface area	m^2	60
Investments	USD	5,500
Annual benefits	USD/year	1,960
Pay back period	years	2.8
Substitute cost of heat	USD/kWh	0.01
Specific daily heat output	kWh/m^2	13.5
Specific annual heat output	kWh/(m^2·year)	2,900
Specific capital cost for the construction of collectors	USD/m^2	107
Levelized cost of heat transferred from collectors	USD/kWh	0.005
Proportion of domestic hot water load substituted by solar collector	%	7

The Gelendzhik branch of OOO Gazprom Teploenergo Krasnodar operates 24 boiler plants and 9 district heating substations. The total roof area of facilities with a domestic hot water load (all central heat substations and some boiler plants) is estimated at 2.4 thous. m^2. The enterprise's total heat generation potential via solar collectors can be estimated at 7 thous. MWh annually, and the annual fuel economy — 827 thousand n.m^3/year of natural gas, which is about 1% of the enterprise's consumption.

The economic efficiency of the solar collector under the study can be even 3-4 times higher at non-gasified facilities, for example, at boiler plants using fuel oil, diesel fuel, or electricity.

In the above example, cold water is fed to the district heating substation through the Troitsky group water supply system from a reservoir located more than 100 km away, which ensures its relatively high temperature due to natural heating during transportation. Greater efficiency of solar collectors will be achieved at lower temperatures of cold water fed to them, e.g. if it is extracted from the wells located not far from the place of solar collector operation. In this case, the temperature drop in the water heating process, and, consequently, heat generation, maybe even more significant.

It should also be noted that the warm climate of Gelendzhik is a favorable factor that contributed to the efficiency of the implemented solar collector scheme. This provides an acceptable temperature drop between the environment and the water heating system. And considering that the period of negative temperatures in this region is short, it is feasible to operate solar collectors for more than 7 months a year.

4. Conclusions

- In general, the usage of solar collectors for generating heat becomes impractical, fading for photovoltaic panels.
- The application of solar collectors for domestic hot water pre-heating in district heating systems can be effective even when the fuel price is relatively low.
- The Gelendzhik project turned out cost-effective yet technically simple. It does not require any sophisticated gear or automation equipment. The cost of generated heat is 0.005 USD/kWh while the levelized cost of heat transferred from collectors is 107 USD/m^2, which provides for less than 3 years payback period.
- The system under research can contribute up to 7% of the heat required for domestic hot water supply purposes; the potential fuel economy for the heat supply organization in total is estimated at 827 thousand n.m^3 annually.
- The implementation of solar collectors of the proposed design will be most effective under the following conditions: the high cost of fuel for the district heating system, low temperature of cold water fed to the collector, warm climate at the collector location.

References

[1] Ermolenko G V, Tolmacheva I S, Ryapin I Yu, Fetisova Yu A, Matsura A A and Reutova A B 2016 *Spravochnik po vozobnovlyaemoj ehnergetike Evropejskogo soyuza. Analiticheskie obzory instituta ehnergetiki* [*European Union Renewable Energy Handbook. Institute of Energy Analytical Reviews*] Moscow: Institute of Energy: Higher School of Economics 96 p (in Russian)

[2] Enerdata, 2019 *Statisticheskij Ezhegodnik mirovoj ehnergetiki* [*Statistical Yearbook of World Energy*] Online: https://yearbook.enerdata.ru [Available online: 25.12.2019] (in Russian)

[3] Fischetti M 19.11.2015 "139 Countries Could Get All of their Power from Renewable Sources" *Scientific American* Online: https://www.scientificamerican.com/article/139-countries-could-get-all-of-their-power-from-renewable-sources1 [Available online: 25.12.2019]

[4] Frid S E and Lisitskaya N V 2018 Sovremennye solnechnye kollektory: tipichnye parametry i tendentsii ikh izmeneniya [Modern solar collectors: typical parameters and trends] *Geliotekhnika* [*Journal of Solar Engineering*] **2** pp 27–38 (in Russian)

[5] Butuzov V A 2019 Stoletnij opyt raboty rossijskikh nauchnykh shkol solnechnogo teplosnabzheniya [A hundred years of experience in Russian scientific schools of solar heat supply] *Ehnergiya: ehkonomika, tekhnika, ehkologiya* [*Energy: Economics, Technology, Ecology*] **2** pp 16–29 (in Russian)

[6] Butuzov V A 2018 Sovetskoe i rossijskoe solnechnoe teplosnabzhenie — nauchnye i inzhenernye shkoly [Soviet and Russian solar heat supply — scientific and engineering schools] *S.O.K.* [*Plumbing. Heating. Conditioning*] **8** (200) pp 89–99 (in Russian)

[7] Frid S E, Mordynskij A V and Arsatov A V 2012 Solnechnye vodonagrevateli akkumulyatsionnogo tipa [Storage-type solar water heaters] *Teploehnergetika* [*Journal of Heat Power Engineering*] **11** pp 69–76 (in Russian)

[8] Frid S E and Lisitskaya N V 2018 State-of-the-Art Solar Collectors: Typical Parameters and Trends *Applied Solar Energy* **54** 4 pp 279–86 DOI: 10.3103/S0003701X18040059

[9] Frid S E and Lisitskaya N V 2018 Fotoehlektricheskie generatory dlya goryachego vodosnabzheniya [Photovoltaic generators for hot water supply] *Intellektual'naya ehlektrotekhnika* [*Intelligent Electrical Engineering*] **4** pp 52–62 (in Russian)

[10] Frid S E and Lisitskaya N V 2018 Konkurentosposobnost' solnechnykh batarej pri generatsii tepla [Competitiveness of photovoltaic panels in heat generation] *Proc. of REENCON–XXI congress "Vozobnovlyaemaya ehnergetika XXI vek" [Renewable Energy of 21st Century]* pp 219–22 (in Russian)

[11] Frid S E and Tarasenko A B 2018 Ispol'zovanie fotobatarej dlya goryachego vodosnabzheniya — opyt i perspektivy [Utilizing photovoltaic batteries for hot water supply — experience and prospects] *Al'ternativnaya ehnergetika i ehkologiya* [*Journal of Alternative Energy and Ecology (ISJAEE)*] **16–18:23–38** (in Russian)

Improving the characteristics of thermoelectric generator batteries based on bismuth telluride by optimizing the parameters of hot pressing n-$Bi_2Te_{2.4}Se_{0.6}$

A A Grebennikov[1,3], A I Bocharov[1], V A Makagonov[2], O V Kalyadin[2], V A Yur'yev[2] and K G Korolev[2]

[1]Basic research and educational center «Physics and technology of thermoelectric phenomena», Voronezh State Technical University, Moscow Avenue, 14, Voronezh, 394026, Voronezh, Russia

[2]Department of Solid State Physics, Voronezh State Technical University, Moscow Avenue, 14, Voronezh, 394026, Russia

[3]Corresponding author: anton18885@yandex.ru

Abstract. The effect of hot pressing modes (pressing pressure and holding time under pressure) on the thermoelectric properties of n-type bismuth telluride $Bi_2Te_{2,4}Se_{0,6}$ doped with Hg_2Cl_2 was investigated. Samples were obtained by powder metallurgy technology - the synthesis of a chemical compound followed by hot pressing. A change in the hot pressing modes does not significantly influence the value of the thermo-EMF and conductivity of the samples. A change in the hot pressing mode significantly influences on the value of thermal conductivity. Both the increase of pressing pressure and the increase of the holding time under pressure leads to a decrease in the thermal conductivity of the material. Thus, the thermoelectric figure of merit of bismuth telluride can be increased by increasing the pressing pressure, holding time under pressure, or both parameters simultaneously. The increase of the thermoelectric figure of merit was 15% in the investigated samples. As a result of the tests of thermoelectric generator batteries, it was found that the output power of the battery made from a material with a high figure of merit was 27 W at temperatures 70 °C on the cold side, 300 °C on the hot side. The output power of the battery which was made from the material with a lower figure of merit was 25 W at a similar temperature regime.

1. Introduction

The use of generating equipment that is safe, environmentally friendly, use alternative energy sources is an actual task for the energy sector. Thermoelectric power generators have a number of advantages over traditional electric generators. The simplicity of design and the absence of moving parts provides high reliability - the most important criterion for the autonomous energy supply of remote and inaccessible objects such as technological communications and telemetry facilities, field systems for collecting and preparing gas for transport and cathodic protection systems for gas and oil pipelines [1-3].

The characteristics of thermoelectric generators substantially depend on the parameters of thermoelectric materials from which thermoelectric batteries and modules are made. Currently, bismuth telluride and solid solutions based on it are widely used for the manufacture of branches of thermoelectric generator batteries. This is due to the high values of the thermoelectric figure of merit in the temperature range of ~ 200 - 350 °C and high values of thermoelectromotive force [4-6]. Bismuth telluride and solid solutions based on it currently have no alternative [7] for the production of thermoelectric generators despite the active search of cheaper and environmentally friendly thermoelectric materials. Therefore, the search for technologies to improve the thermoelectric properties of bismuth telluride is an extremely urgent and practically significant task.

The possibilities of increasing the thermoelectric figure of merit by obtaining low-dimensional structures [8,9] are being actively studied in recent years. The use of low-dimensional structures such as thin films [10, 11], superlattices [10, 12], whiskers [13], nanoscale structures [14, 15], quantum wells [16, 17], quantum wires [18, 19, 20], etc. leads to a decrease in the thermal conductivity of the lattice or to an increase in the density of electronic states [21]. However, there are significant technological difficulties in obtaining such materials. What is more, their mechanical properties are practically unstudied. In this regard, low-dimensional structures are interesting only for scientific research. It is prematurely to consider the possibility of their introduction to industrial production.

The idea of increasing the thermoelectric figure of merit by using graphite-based nanomaterials proved unjustified. Thus, it was shown in [22, 23] that the introduction of carbon nanotubes, as well as amorphous carbon into the bismuth telluride matrix, leads to a significant decrease in the electrical conductivity, thermoEMF, and thermoelectric figure of merit. The introduction of fullerene C60 into bismuth telluride leads to an increase in thermoEMF and a decrease in the thermal conductivity, the electrical conductivity and thermoelectric figure of merit [24].

Powder metallurgy methods are most popular for obtaining thermoelectric materials based on bismuth telluride and its solid solutions. The most promising among them are melt spinning (MS) methods for producing powders of synthesized materials and spark plasma sintering (SPS) for consolidating the obtained powders. The figure of merit (ZT) of bismuth telluride samples obtained using these methods exceeds 1.0 [25-28]. At the same time, their introduction into production is associated with difficulties, the most significant of which are changes in the existing technology process and the re-equipment of rather expensive technological equipment. The problem of structural stability at elevated temperatures of materials obtained by this technology is also unresolved.

Hot pressing of pre-synthesized powder of thermoelectric material is simpler in comparison with MS and SPS methods. This implementation of powder metallurgy technology is used in production as a rule. However, the results obtained by different authors depend on the synthesis conditions of the initial powders of the thermoelectric material and its sintering and can significantly differ (see [29-33] for example). Therefore, the determination of the optimal parameters of the technological cycle that can allow us to obtain thermoelectric materials with the highest figure of merit by powder metallurgy technology is an actual problem.

The effect of hot pressing on the thermoelectric properties of bismuth telluride is investigated. It is shown that a change in pressing pressure or holding time under pressure affects the electrical conductivity and thermal conductivity of the material, which leads to a change in the thermoelectric figure of merit.

2. Methods

Samples of $Bi_2Te_{2,4}Se_{0,6}$ doped with calomel were used as objects of study. The samples were obtained using a two-stage technology of powder metallurgy, including the synthesis of a chemical compound followed by hot pressing. The modes of hot pressing are shown in table 1, where T is the pressing temperature, p is the pressing pressure, t is the holding time under pressure.

Table 1. Hot pressing Formatting modes of $Bi_2Te_{2,4}Se_{0,6}$.

Sample No.	Hot Press Mode		
	T, °C	p, t/sm^2	t, min
1	400	5	5
2	400	5	20
3	400	5,5	10

The thermal conductivity of the samples was investigated using Netzsch LFA 467, the conductivity and the thermo-EMF coefficient was investigated using Netzsch SBA 458 in the temperature range of 30 - 300 °C.

3. Results and discussion

The temperature dependences of the thermo-EMF of $Bi_2Te_{2.4}Se_{0.6}$ samples are shown in figure 1. The resulting dependencies are similar. Curves are typical for degenerate semiconductors [34].

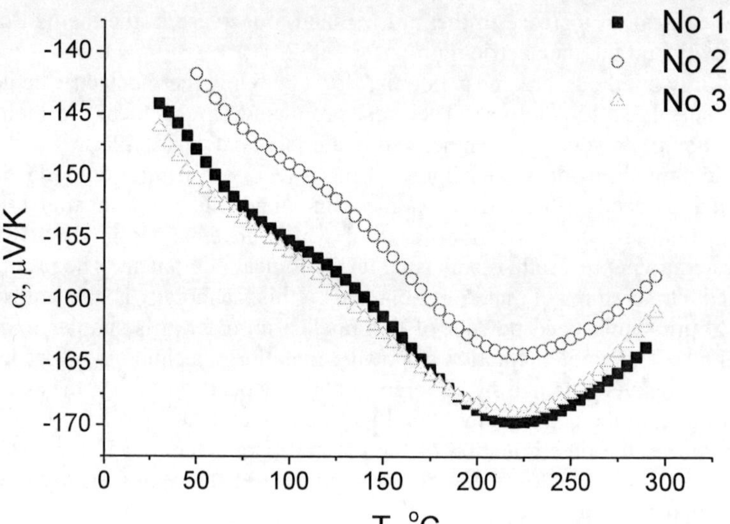

Figure 1. Temperature dependences of thermo-EMF of $Bi_2Te_{2,4}Se_{0,6}$.

Negative values of thermo-EMF mean that the main charge carriers are electrons, therefore, the samples are n-type thermoelectric material. The increase of absolute values thermo-EMF with increasing temperature is associated with an increase in the energy of the main charge carriers [34]. At temperatures above 200 °C, intrinsic conductivity appears, which leads to a decrease in the absolute value of thermo-EMF.

The values of thermo-EMF (α) of samples No. 1 and 3 are close over the entire temperature range. The α values of sample No. 2 are on average 3.5% less (in absolute value) than values of No. 1 and 3, which may be due to a higher concentration of free charge carriers.

The temperature dependences of the conductivity of samples No. 1.3 practically coincide (figure 2). The conductivity of sample No. 2 is greater than that one of No. 1 and 3, which is consistent with the results of measuring thermo-EMF. As is known [34], large values of conductivity at lower values of thermo-EMF are associated with a higher concentration of charge carriers in doped semiconductors.

The conductivity of all samples decreases with increasing temperature. Such a dependence is observed in heavily doped semiconductors [35]. In this case, the semiconductor degenerates at cryogenic temperatures and the material begins to exhibit properties close to the ones of metal. The concentration of impurities is so high in a degenerated semiconductor that the intrinsic properties are practically not manifested, but mainly the properties of the impurity.

An analysis of the temperature dependence of conductivity allows us to determine the temperature ranges in which impurity or intrinsic conduction mechanisms predominate, as well as determine the mechanisms of charge carrier scattering. In doped semiconductors in a degenerated state, a change in the conductivity with a change in temperature occurs in the same way as in metal, inversely proportional to the temperature, so the dependence σ(1/T) is linear.

The dependence σ(1/T) for sample No. 1 is shown in figure 3. Up to the temperature of 145 ° C (indicated by a dotted line in figure 3), the dependence σ(1/T) is linear. Impurity conductivity is realized in this interval. A deviation of the dependence on the line is observed at temperatures above 145 °C, which corresponds to the appearance of intrinsic conductivity in the sample. The temperature of occurrence of intrinsic conductivity for samples No. 2 and 3 was determined similarly. Its values were 155 and 100 °C respectively.

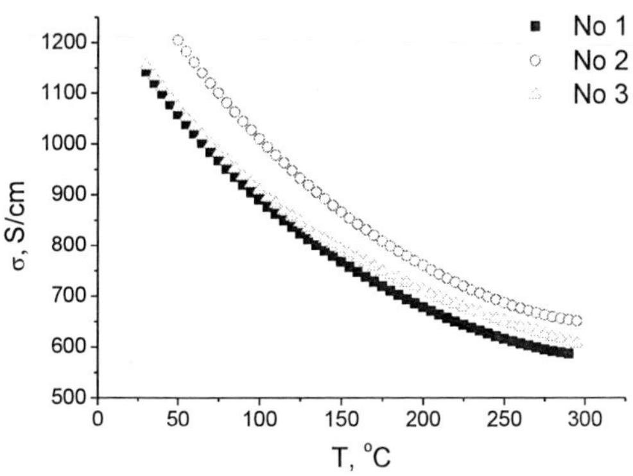

Figure 2. Temperature dependences of conductivity of $Bi_2Te_{2,4}Se_{0,6}$.

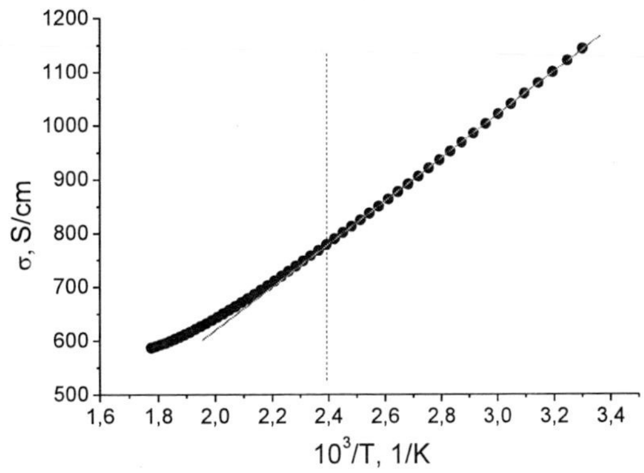

Figure 3. Dependence of conductivity from inversed temperature of the sample No. 1. The temperature of appearance of intrinsic conductivity is shown by the dotted line.

The temperature dependences of the thermal conductivity of samples No. 1 and 2 almost coincide (figure 4). Thermal conductivity values of sample No. 3 in the entire temperature range are lower than for No. 1 and 2.

As it is known, thermal conductivity (λ) consists of the electronic (λ_e) and phonon (λ_p) components [35,36]. The value of the electronic component of thermal conductivity λ_e can be determined according to the Wiedemann-Franz law, which for degenerated semiconductors has the form:

$$\lambda_e = 2,44 \cdot 10^{-8} T\sigma \quad (1)$$

Then the value of the phonon component of thermal conductivity is:

$$\lambda_p = \lambda - \lambda_e \quad (2)$$

The results of calculating the λ_e and λ_p are shown in figures 5 and 6.

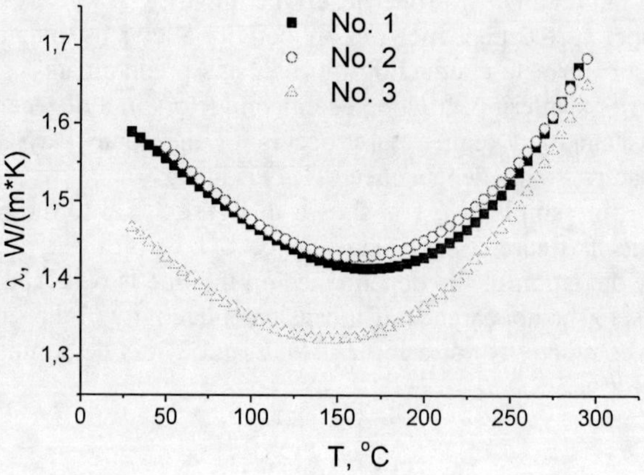

Figure 4. Temperature dependences of thermal conductivity of $Bi_2Te_{2,4}Se_{0,6}$.

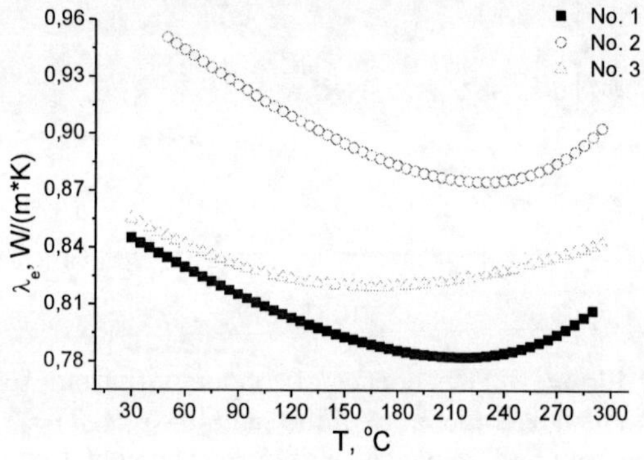

Figure 5. Temperature dependences of the electronic component of thermal conductivity of $Bi_2Te_{2,4}Se_{0,6}$.

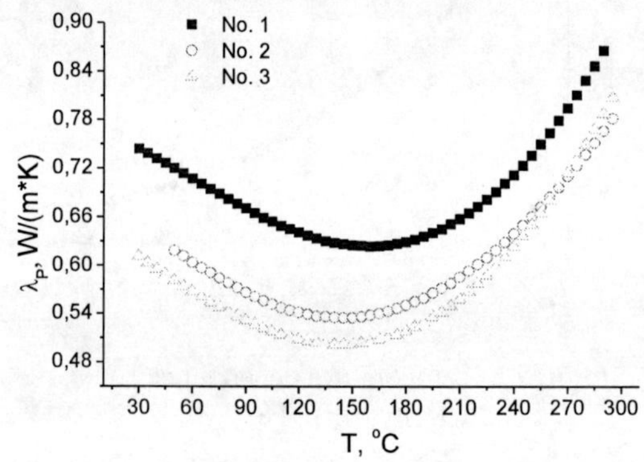

Figure 6. Temperature dependences of the phonon component of the thermal conductivity of $Bi_2Te_{2,4}Se_{0,6}$.

As it is shown in figures 5 and 6, a change in the hot pressing mode leads to an increase (in comparison with sample No. 1) of the electronic and a decrease in the phonon component. The increase of the electronic component is due to an increase in the mobility and concentration of charge carriers. The $\lambda_p(T)$ dependences of samples No. 2 and 3 differ insignificantly, whereas in comparison with sample No. 1 their average integral values λ_p are less by 12 and 16% respectively. The pressing technology of samples No. 1 and 2 differs only in the exposure time under pressure, for samples No. 1 and 3 differs by the exposure time and pressing pressure (table 1). It follows that the reduction of the phonon component is achieved, first of all, by increasing the exposure time of the material under pressure, the influence of the pressing pressure is not high.

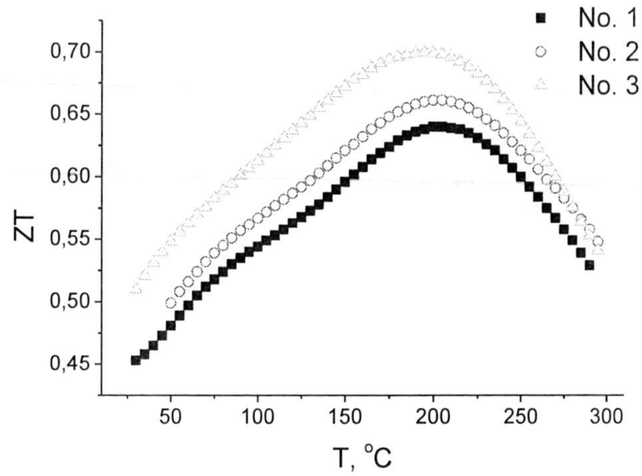

Figure 7. Temperature dependences of thermoelectric figure of merit of $Bi_2Te_{2,4}Se_{0,6}$.

The difference in the electrical and thermal properties of the studied samples caused a difference in the thermoelectric figure of merit (figure 7). A simultaneous increase in pressing pressure and holding time under pressure (samples No. 1 and 3) has the greatest effect on the value of ZT. As a result, ZT increases by an average of 10.0%.

Materials for the branches of thermoelectric generator batteries were obtained by the proposed technology to verify the influence of the properties of materials on the operational characteristics of the products. Two thermoelectric generator batteries based on a material with high ZT (sample No. 3) and standard material (sample No. 1) were made for testing. As a result of tests of thermoelectric generator batteries, it was found that the output power of the battery made from a material with high ZT was 27 W at temperatures 70 °C on the cold side, 300 °C on the hot side. The output power of a battery which was made from a material with a lower ZT was 25 W at a similar temperature regime.

4. Conclusion

An increase in the holding time under pressure during hot pressing of n-type bismuth telluride leads to a decrease in thermo-EMF and an increase in electrical conductivity. An increase in holding time with a simultaneous increase in pressing pressure does not lead to a significant change in thermo-EMF and electrical conductivity

The increase in exposure time under pressure during hot pressing does not influence on the thermal conductivity of n-type bismuth telluride. An increase in holding time with a simultaneous increase in pressing pressure leads to a decrease in thermal conductivity.

The selection of optimal pressing modes allows us to increase the value of ZT without changing the composition of materials or the main stages of the technological cycle of obtaining materials. As a result, an increase in the output power of thermoelectric generator batteries is achieved without any significant change in the technological cycle of their production.

References

[1] RIF – Thermoelectric generators: www.rifcorp.ru/products/termoelektricheskie-generatory
[2] Kriotherm – Thermoelectric generator GTEG: kryothermtec.com/thermoelectric-generators-operating-on-gas-fuel.html
[3] Thermointech – Thermoelectric generator for the oil and gas industry: thermointech.ru/en/products/generator-termoelektricheskiy-gte
[4] Goltsman B M, Kudinov B A and Smirnov I A 1972 *Thermoelectric Semiconductor Materials Based on Bi_2Te_3* (Moscow: Nauka)
[5] Eibl O, Nielsch K, Peranio N and Volklein F 2015 *Thermoelectric Bi_2Te_3 Nanomaterials*

(Wiley–VCH)
[6] Maciá-Barber E 2015 *Thermoelectric Materials: Advances and Applications* (CRC Press)
[7] Rowe D M 1995 *Thermoelectrics* (CRC Press)
[8] Riffat S and Ma X 2003 *Applied Thermal Engineering* **23** 913
[9] Heremans J P 2005 *Acta Physica Polonica A* **108** 609
[10] Ezzahri Y, Zeng G, Fukutani K, Bian Z and Shakouri A.A. 2008 *J. Microelectronics* **39** 981
[11] Venkatasubramanian R, Siivola E, Colpitts T and Quinn B O 2001 *Nature* **431** 597
[12] Venkatasubramanian R, Colpitts T, Watko E, Lamvik M and El-Masry N 1997 *Journal of Crystal Growth* **170** 721
[13] Funahashi R and Matsubara I 2001 *Appl. Phys. Lett* **79** 362
[14] Bulat L P, Pshenaĭ-Severin D A 2010 *Physics of the Solid State* **52** 485
[15] Lin H 2005 *Phys. Rev. B.* **72** 174113
[16] Harman T, Taylor P, Walsh M and LaForge B 2002 *Science* **297** 2229
[17] Tavkhelidze A 2009 *Nanotechnology* **20** 405401
[18] Boukai A 2008 *Nature Letters* **451** 168
[19] Hochbaum A 2008 *Nature Letters* **451** 163
[20] Keyani J and Stacy A M 2006 *Appl. Phys. Lett.* **89** 233106
[21] Shevelkov A V 2008 *Russ Chem Rev* **77** 1
[22] Trawinski B 2018 *Materials Research Bulletin* **99** 10
[23] Bark H, Kim J S, Kim H, Yim J H and Lee H 2013 *Current Applied Physics* **13** S111
[24] Kulbachinskii V A, Kytin V G, Blank V D, Buga S G and Popov M Yu 2011 *Semiconductors* **45** 1194
[25] Ivanova L D 2017 *Semiconductors* **51** 1002
[26] Drabkin I A 2012 *Electronic Materials* 3 18
[27] Bhame S D, Pravarthana D, Prellier W and Noudem J G 2013 *Appl. Phys. Lett* **102** 211901
[28] Xie W 2013 *J Mater Sci* **48** 2745
[29] Hu L P 2012 *Acta Materialia* **60** 4431
[30] Zhai R 2017 *ACS Appl. Mater. Interfaces* **9** 28577
[31] Kim D H, Kim Ch, Heo S H and Kim H 2011 *Acta Materialia* **59** 405
[32] Han M K, Jin Y, Lee D H and Kim S J 2017 *Materials* **10** 1235
[33] Ge Z H, Ji Y H, Chong X, Feng J and He J 2018 *Scripta Materialia* **143** 90
[34] Stilbans L S 1967 *Semiconductor Physics* (Moscow: Sovetskoye radio)
[35] Ioffe A F 1960 *Physics of semiconductors* (New York: Academic Press)
[36] Tritt T M 2011 *Annu.Rev.Mater.Res.* **41** 433

Smart energy systems and smart metering systems in them

Yu N Zatsarinnaya[1,3] and M O Grigoreva[2]

[1]Department of Power stations named after V.K. Shibanov, Kazan State Power Engineering University, Krasnoselskaya str., 51, Kazan, 420066, Russia
[2]Department of Power stations named after V.K. Shibanov, Kazan State Power Engineering University, Krasnoselskaya str., 51, Kazan, 420066, Russia

[3]Corresponding author: waysubbota@gmail.com

Abstract. The authors of the article evaluated the level of digitalization and the achieved effects during using digital technologies by means of the digitalization index at present and its predicted values. Tools were given to support and stimulate prosumers as active participants in smart energy s3ystems. The integration into the power supply system of 2 prosumers is considered – a logging plant and a shopping center. Their consumption of electricity from the network is analyzed in the case when the generation of electricity from its own sources is insufficient from 2018-2019. The calculations were carried out in several price categories. The calculations showed that in the conditions of accurate planning, in the absence of emergency situations, prosumers who are legal entities can save. To switch to these calculations, it is necessary to install metering devices that allow you to measure hourly consumption. Thus, the smart energy system allows the consumer to choose a price category that suits him based on the energy consumption data of previous months.

1. Problem statement

Energy companies consider smart grids as the basis for their sustainable development, which is based on innovation. According to the Regulation of PJSC ROSSETI "On a Unified Technical Policy in the Electric Grid Complex", approved by the Board of Directors of the company (Protocol No. 378 of 11.08.2019), one of the purposes of the technical policy is the transition to risk-based management based on the introduction of digital technologies and big data analysis.

A distinctive feature of equipment, materials technology and integrated systems is the availability of smart electronic devices. This also applies to electric energy metering systems. The purpose of the technical policy in this area is the formation of unified approaches to the creation of automated electricity metering systems.

A smart energy system is understood to mean an electric power network consisting of producers, consumers, prosumers, capable of smartly integrating the actions of all users, to ensure reliable and safe electricity supply. The smart energy system is controllable, automated, compatible with existing systems [1]. To create such systems in the Russian energy sector, its transition to digital technologies is necessary. In smart energy systems, a new role appears for already familiar participants – the role of the prosumer, transforming the familiar role of the consumer to create new services for the electricity market.

The link that binds the automation levels of the technological model of digital smart grids is the CIM model. The Common Information Model is a standard that enables the interchange of information for managing elements regardless of their supplier or manufacturer. This is a single information model.

2. Current level of electric grids Digitalization in Russia and its perspectives

According to calculations carried out by the largest Russian transmission and distribution system operator PJSC ROSSETI the digitalization index for Russian power grid has amounted to 14.5%, which indicates the scale of the forthcoming digitalization as about 85.5%. In addition, during the digital transformation, the planned values of key performance indicators should be reviewed. Fig. 2 shows the current and calculated forecast values for digitalization index I_D corresponding to the digitalization of technological information systems, corporate systems and cybersecurity systems, and digitalization index I_{DT} as a component of I_D corresponding to digitalization of technological information systems and directly related to smart energy system infrastructure [2,3].

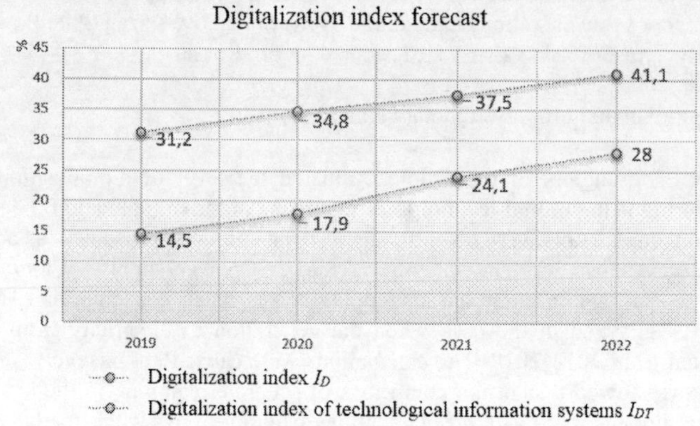

Figure 1. Current and forecast values of the digitalization index for power grid in Russia.

The concept of the development of digital technologies is relevant not only for Russia, but also for the EU countries. The state structures of these countries consider Smart Grid as the ideology of national programs for the development of the electric power industry, borrowed funds with a low interest rate are attracted to implement these technologies, as a rule, and there are no talks about payback periods in this case.

In Russia, the digital transformation is based on three strategic planning documents: Decree of the President of the Russian Federation dated 05.05.2018 No. 204, the Digital Economy of the Russian Federation Program, and the Strategy for Scientific and Technological Development of the Russian Federation.

At the value level in society, as well as at the level of state policy, in comparison with the EU and the USA Russia lacks the main message of the digital transformation which is decarbonization. Taking into account the total amount of power based on renewable energy sources, Russia in comparison with other countries is an outsider in this direction. Therefore, it is not necessary to link together all types of energy sources in the country. It's more rational to install solar power plants, wind farms, hydroelectric or geothermal power plants only in areas with a large number of energy sources to develop distributed generation [4]. The introduction of Smart Grid technology will solve the problem of inefficient use of power lines and the capacity of transformers [5, 6]. In particular, according to the statistics the actual load of the grid transformers today does not exceed 30% of their nominal capacity.

Since 2019, each subsidiary of ROSSETI should constitute at least 20% of the entire investment program annually as tools to support and stimulate the digitalization of the electric grid complex of Russia. Thus, the budget, which is aimed at these goals, is formed from tariffs for the transmission of electric energy and differs depending on the region. Indirectly, these investments can be increased by using equipment manufactured on the territory of the said region to build power grid infrastructure. Since the seller of the products, in this case, returns part of the funds in the form of a tax to the budget of the region, from which in the future, in particular, the investment program of the company purchasing the products is formed [7].

Expensive power is the main challenge for the Russian power industry, which is determined by the constant costs for the operation of the power system. The total cost does not depend on the volume of consumption and includes not only the payments for generating capacity but also the payment for the maintenance of electrical networks. The growth of forecast values indicates the scale of the upcoming digitalization.

3. Integration of prosumers in the power supply system

However, the strategic integration of prosumers in the power supply system is currently a problem for Russia. Firstly, the mass installation of smart metering systems is necessary, for which Federal Law No. 522 of 27.12.2018, "On Amending Certain Legislative Acts of the Russian Federation in Connection with the Development of Electricity (Power) Metering Systems in the Russian Federation," was passed. Secondly, it is necessary to carefully consider an effective tariff setting for the prosumer. This will make it possible to get additional profit and adjust its consumption, and will also contribute to the emergence of a competitive retail energy market (REM) [8-10]. The authors of the article suggest calculating the electricity tariff options for a prosumer on the REM by analogy with the calculation of legal entities in the wholesale market of electric energy and power (WMEP) [11, 12]. Table 1 presents the characteristics of price categories for the consumers of the WMEP.

Table 1. The characteristics of price categories for the consumers.

Price category	Metering	Consumption planning	Transmission tariff
1	Month	Don't need to plan	Single-rate tariff
2	Zones of the day	Don't need to plan	Single-rate tariff
3	Hour	Don't need to plan	Two-rate tariff
4	Hour	Don't need to plan	Two-rate tariff
5	Hour	Need to plan	Single-rate tariff
6	Hour	Need to plan	Two-rate tariff

In calculating prices, it is not profitable for a prosumer to use price categories for which independent planning of consumption by the enterprise is provided (fifth and sixth), because it is not possible to plan hourly consumption for a day ahead with a 100% probability, resulting in deviations from the declared volume, which must be paid to the energy supplier [13, 14].

Table 2 presents the electricity and power consumption of the shopping center. The main costs of electrical energy are due to the costs of lighting, air conditioning and ventilation. In the period from November 2018 to October 2019, the power regime was 196-297 thousand kWh and 315-497 kW of active power per month.

Table 2. The electricity and power consumption of the shopping center.

Year	Month	Electricity, kWh	Power, kW
2019	January	270527	452
2019	February	232461	414
2019	March	256972	410
2019	April	237720	371
2019	May	248856	453
2019	June	268374	497
2019	July	196352	360
2019	August	197063	315
2019	September	245148	398

2019	October	264690	431
2018	November	247591	439
2018	December	279114	480

Table 3 presents the electricity and power consumption of the logging plant. The main costs are due to lighting and technical equipment. In 2018, the power regime amounted to 448-667 thousand kWh and 857-1106 kW of active power per month.

Table 3. The electricity and power consumption of the logging plant.

Month	Electricity, kWh	Power, kW
January	592 415	1 071
February	472 250	1 106
March	506 548	1 086
April	542 744	1 089
May	490 239	1 006
June	515 535	1 048
July	506 686	1 006
August	533 470	938
September	527 574	857
October	584 756	909
November	447 758	913
December	666 589	1 004

Both legal entities are consumers with a relatively even load. Based on the analysis of data for several years, it is possible to predict the schedule of consumption of electric energy and power [15, 16]. This will allow the consumer to switch to calculating the costs in 5 or 6 price categories.

In terms of accurate planning, in the absence of emergency situations, this will help legal entities save. To switch to these calculations, it is necessary to install metering devices that allow you to measure hourly consumption [17, 18]. Thus, the smart energy system allows the consumer to choose a price category that suits him based on the energy consumption data of previous months.

4. Conclusion

The Smart Energy system is intended to become a driver of the transformation of the traditional energy complex as well as its implementation involves the creation of single information space for the Russian energy sectors. In calculating prices, it is not profitable for a prosumer to use price categories for which independent planning of consumption by the enterprise is provided (fifth and sixth), because it is not possible to plan hourly consumption for a day ahead with a 100% probability, resulting in deviations from the declared volume, which must be paid to the energy supplier.

However, the commissioning of smart energy systems can help in this regard. Based on monthly data, a consumer with a relatively even load is able to predict their consumption. Thus, the smart energy system allows the consumer to choose a price category that suits him based on previous data.

References

[1] Sumper A 2019 *Micro and Local Power Markets* (Hoboken N J: John Wiley & Sons, Inc.) p 272

[2] ROSSETI 2018 *Concept "Digital transformation 2030"*, Moscow: PJSC Russian Networks

[3] The innovative development program of PJSC ROSSETI for the 2016–2020 years period with perspective until 2025 Moscow: PJSC "Russian Networks" 2019

[4] Eroshenko S A, Samoylenko V O and Pazderin A V 2016 Renewable energy sources for perspective industrial clusters development *2nd International Conference on Industrial Engineering, Applications and Manufacturing,* Chelyabinsk, May 19-20, 2016 vol 1 (NY: Curran Associates) p 821

[5] Rogalev N D et el. 2013 *Fundamentals of the Economics of the Fuel and Eergy Complex* Moscow: Publishing House MPEI

[6] Nikolaev M V, Ovsyannikov A Yu and Marchinsky I Yu 2016 Experience of using "smart grids" in Russia *Smartgrid Technology Young Scientist* **28** 2 pp 63-68

[7] Staroverova N A, Shustrova M L, Satdarov M R 2019 Development of a cyber-physical system for the specialized on-track machine operators training *Cyber-Physical Systems: Industry 4.0*

[8] Dronova Yu V and Krasnova A O 2017 A new model of the retail electricity market: implications for the regional economy *Bulletin of PNRPU Electrical engineering, information technology, control systems* **2** pp 63-78

[9] Vaskovskay T, Thakurta P G and Bialek J 2018 Contribution of transmission and voltage constraints to the formation of locational marginal prices *International Journal of Electrical Power & Energy Systems* **101** pp 491-499

[10] Oganyan R G, Narakidze N D, Shaykhutdinov D V, Kirievskiy E V and Kostinskiy S S 2018 Digital substation conceptual model for the complex full-scale-model diagnostics *IOP Conference Series: Materials Science and Engineering* **441** pp 1-6

[11] Zatsarinnaya Yu, Logacheva A and Grigoreva M 2020 Electricity in retail markets as a commodity in smart energy systems *IOP Conference Series: Materials Science and Engineering* **1** p 79

[12] Balzamov, Akhmetova I, Balzamova E, Oykina G, and Coman G 2019 An analysis of the viability of implementing steam screw machines at the facilities of energy generation enterprises to reduce the energy costs for their own need *E3S Web of Conferences (SES-2019)* **124** pp 1-4

[13] Bulatov Y N, Kryukov A V and Suslov K V 2017 Multi-agent technologies for control of distributed generation plants in the isolated power systems *Far East Journal of Electronics and Communications* **17(5)** 1197

[14] Balobanov R N, Zaripov D K and Akhmadeev A A 2019 The device for monitoring the LED display high-voltage insulators state *IOP Conference Series: Materials Science and Engineering* **552** pp 1-9

[15] Mikhalkov A V 2011 Long-term forecasting of demand for capacity as an instrument of high-quality investment planning for the development of distribution electric networks of the subject of the Russian Federation St. Petersburg: *JSC IDGC Holding*

[16] Chernova A 2018 Designing database for decision support system for development of electrical grid *International Ural Conference on Green Energy (UralCon, IEEE Conferences)* Chelyabinsk pp 181 – 186

[17] Vaskovskay T, Thakurta P G and Bialek J 2018 Contribution of transmission and voltage constraints to the formation of locational marginal prices *International Journal of Electrical Power & Energy Systems* **101** pp 491-499

[18] Dmitriev A V, Dmitrieva O S and Zinurov V E 2018 Influence of elements thickness of separation devices on the finely dispersed particles collection efficiency *International Conference on Modern Trends in Manufacturing Technologies and Equipment (ICMTMTE 2018)* Sevastopol **224** p 0207

IOP Conf. Series: Materials Science and Engineering 1035 (2021) 012010 doi:10.1088/1757-899X/1035/1/012010

Multi-unit modular wind farm for areas of low wind potential

S S Dorjiev, E G Bazarova, M I Rosenblum and K S Morenko

Federal State Budgetary Scientific Institution "Federal Scientific Agroengineering Center VIM", 1st Institutsky proezd, 5, Moscow, 109428, Russia

Abstract. Many dispersed sites are remote from the power grid in agriculture. The centralized power supply of these objects with electric energy from the power grid is unprofitable. During the year, tests were conducted in full-scale conditions of wind power plants with different parameters under conditions of power shortage. The article proposes the development of a multi-unit wind power plant with an additive block, where the energy from multiple winds receiving devices is summed. The article shows that the wind turbine operates in a stable mode in an extended range of wind speeds of 4500 hours per year and generates rated power even in conditions of power shortage due to energy storage and operation of the wind turbines with different parameters. It is two times longer than the power of the wind turbine with separate generators during the year. Studies have shown that at low wind speeds of 4-5 m/s, a 1 kW wind power plant generates up to 840 W at wind speeds of 4-6 m/s, the efficiency of a multi-unit wind power plant is 4 times higher than the overall efficiency of individual wind power plants operating on their own generators.

1. Introduction

Existing wind power stations with a capacity of up to 20 kW operate efficiently at average wind speeds of 6.5-7 m/s. Most Russian territory has the average periodic wind speeds of less than 4-5 m/s. High-speed few-blade wind turbines, designed for speed ratio of Z=6-9, in areas with an average annual wind speed of 4-7 m/s operate in the design mode of about 152 to 720 hours or from 2 to 8 % per year [1]. Wind turbines are designed for a narrow range of wind speeds. To develop a power supply module, it is necessary to expand the range of wind speeds used, since about 80% of the time the wind speed is less than 5 m/s [2].

The purpose of this work is to study the operation of wind turbines with different parameters and to compare the power generated by a multi-unit installation working in additive mode and the sum of the power generated by individual wind turbines with the same parameters in a wide range of wind speed.

2. Materials and methods

The concept of multi-rotor wind turbines is increasingly found in world research and development in the field of wind energy. For example, researchers from the Indian Institute of technology, Aarhus (Denmark), Durham (UK) and Stanford (USA) universities model the aerodynamics of multi-rotor wind turbines using high-resolution digital modeling. They have found that a wind turbine with four rotors on a single foundation creates less turbulence and has higher energy production [3]. Since 2016, the Danish company Vestas has conducted 2.5-year tests of a 900 kW multi-rotor unit equipped with four V29-225kW wind generators with a rotor diameter of 29 m at a test site located near Roskilde (East of Denmark) [4]. An increase in capacity of 1.5% was noted during the tests. Also in 2014, the French company New Wind presented the concept of the wind turbine "Arbre à Vent" in the form of a tree equipped with vertical mini-turbines that can produce up to 3 kW even at low wind speeds [5].

The idea of a multi-wind installation is not new. The multi-wind frame wind farm of A. G. Ufimtsev is well known. Having 10 wind wheels mounted on a single frame, each wind wheel is connected to a separate generator by a gearbox. There are also known variants of other inventors, where the wind wheels are connected by long shafts to a single generator [6]. The patent study shows that the presented concepts of multi-wind turbines are aimed at increasing the efficiency of large megawatt wind power plants and increasing the ability to resist wind gusts [7].

Currently, a multi-unit wind installation that operates in an additive mode under a power shortage of one of the wind turbines is being developed. It is considered possible to connect wind turbines with different parameters with a generator using an additive module.

The essence of the multi-unit installation is that the mechanical energy from some wind receiving devices with different parameters is hydraulically summed in an additive block. The summed energy is transferred to the electric generator. The power of the planned installation is 1 kW. Such a system works in an extended range of wind speeds because the wind receiving devices with different parameters at different wind speeds and the accumulation of energy will compensate the lack of power at one of the wind receiving device. For example, if the wind speed is 5 m/s, a low-speed unit will produce 70% of the rated power, a high-speed one will produce 10% and a three-blade one will produce about 20%. In total, one will get 1 kW and the losses on the power generator will be minimal. In addition, the excess energy in the form of hydraulic energy of the working fluid is sent to the accumulator. Figures 1 and 2 show conventional schemes of wind turbines with separate generators and a multi-unit installation.

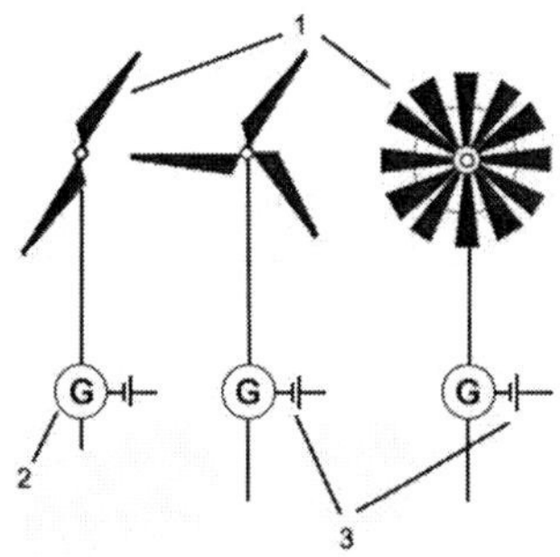

Figure 1. Design of wind turbines with separate generators: 1 – wind receiving devices, 2 – generators, 3 – accumulator.

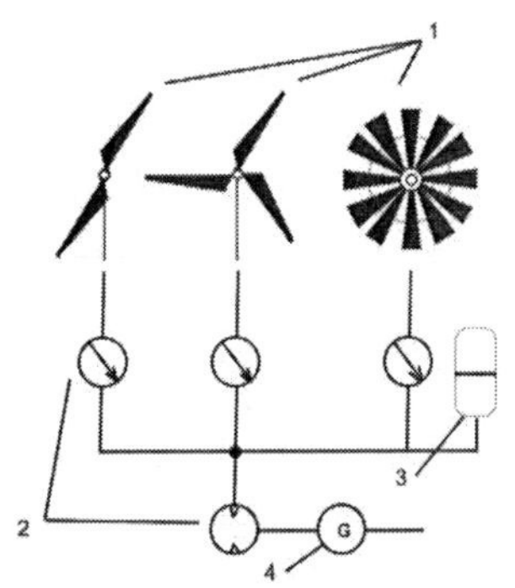

Figure 2. Scheme of a multi-unit installation with a single generator: 1 - wind receiving device, 2 – hydraulic block, 3 -accumulator, 4 – generator.

Figure 3 shows a multi-unit modular installation under development, which is being installed in the wind farm of the FSBSI FSAC VIM in Istra, Moscow district, Russian Federation.

Figure 3. Experimental sample of multi-unit wind farm.

At the wind farm of the Federal Scientific Agroengineering Center "VIM" in Istra, Moscow region, during the year, research was conducted on the operation of wind turbines of different parameters. Observations were made on the operation of 3-, 6–, 9-blade horizontal-axis and 6-blade vertical-axis wind turbines with diameters of 2, 3, 3.9, 4.15 m, with a nominal generator power of 1 kW. There were recorded simultaneously the power generated by the wind turbine on the generator shaft and the rotation speed of the wind turbine depending on the wind speed.

The obtained data are grouped by wind speed from 4 to 14 m/s, and the average annual values of the generated capacity are calculated by groups.

Using the obtained data, calculations of the power of the installation were performed. The output power of the wind receiving device P (W) is calculated using equation 1:

$$P = \xi\rho\nu^2S^2 \quad (1)$$

ξ – wind power utilization factor;
ρ – air density 1,225 , kg/m³;
S – wind turbine area, m²;
V – wind velocity, m/s.

$$S = \pi D^2 \quad (2)$$

D – wind turbine diameter, m.
Output power of the generator P_G (W):

$$P_G = P\eta \quad (3)$$

P – Output power of the wind turbine, W;
η – generator efficiency, %.
Output power from the generator of a multi-unit wind turbine P_M (W):

$$P_M = \sum P\,\eta_h\eta_G \quad (4)$$

P – Output power of the wind turbine, W;
η_G – generator efficiency, %;
η_h – hydraulic block efficiency, %.
The rotation speed of the wind turbine n [rpm] is calculated using the equation:

$$n = 60\nu Z 2\pi R \quad (5)$$

Z – speed ration;
R – wind turbine radius, m;
V – wind speed, m/s.

IOP Conf. Series: Materials Science and Engineering 1035 (2021) 012010 doi:10.1088/1757-899X/1035/1/012010

Along with this, calculations were made for the number of hours a year the wind installation and individual units will operate at full capacity, i.e. generate 1 kW of electricity. With an average wind speed of 4 m/s in the Istrinsky district of the Moscow region, where tests were conducted, according to the wind repeatability table of M. M. Pomortsev [8]. The wind speed is less than 5 m/s for about 5062 h a year, the wind speed is 5 m/s for about 1445 h, the wind speed is 6 m/s for about 1070 h, the wind speed is 7 m/s for about 640 h, the wind speed is 8 m/s for about 315 h, the wind speed is 9 m/s for about 152 h, the wind speed is 10 m/s for about 52 h and the wind speed is 11m/s for about 26 h. The article presents the calculations of excess energy storage. In the case of multi-unit installation, accumulation is hydraulic, the mechanical energy from the wind receiving devices converts to the hydraulic energy by the pumps and directs it to the accumulator, where the summation of energy occurs, and the surplus energy is stored. Accumulation and storage of excess energy of wind power plants with separate generators are electric using batteries.

3. Results and discussion

Table 1 shows the data obtained over the one-year data collection on the generated power of 3-, 6 -, 9-blade horizontally-oriented and 6-blade vertically-oriented wind turbines. The diameters of the wind wheels are 2, 3, 3.9 and 4.15 m. The wind energy utilization coefficient of 3-, 6 -, 9-blade horizontally-oriented and 6-blade vertically-oriented wind turbines are equal to 0.4, 0.35, 0.3 and 0.28.

Table 1. Powers of wind units with separate generators.

v (m/s)	Unit power (W)				Generator output power (to the load and the battery) (W)								Sum of powers at generator outputs (W)	To the battery (W)
	3-blade horizontal	6-blade horizontal	9-blade horizontal	6-blade vertical	3-blade horizontal	to the battery	6-blade horizontal	to the battery	9-blade horizontal	to the battery	6-blade horizontal	to the battery		
4	48	94	140	148	14	0	28	0	42	0	44	0	129	0
5	94	183	273	289	28	0	55	0	82	0	87	0	252	0
6	162	316	472	500	49	0	95	0	142	0	150	0	435	0
7	257	502	750	794	154	0	301	0	450	0	476	0	1000	0
8	384	750	1120	1185	230	0	450	0	672	0	711	0	1000	0
9	547	1068	1594	1688	492	0	961	0	1000	0	1000	519	1000	519
10	750	1465	2187		675	0	1000	318	1000	968			1000	1286
11	998	1950	2910		898	0	1000	75	1000	619			1000	1374

				5		
12	1296	100 0	16 6		1000	166
13	1648	100 0	48 3		1000	483
14	2058	100 0	85 2		1000	852

In the case of individual units, losses on the generator are about 70% at wind speeds up to 7 m/s and are about 40% at wind speeds of 7-8 m/s due to the low power generated on the wind generator shaft, the rpm is not enough for the generator to operate at full power. At a wind speed of 9 m/s, losses in the generator are 10%. At the same time, in the case of individual wind farms, excess generated capacity is not used.

Table 2 shows the of the multi-unit wind power plant at different wind speeds.

Table 2. Powers of multi-unit wind power plant.

v m/s	Wind receiving device power, W	To the battery, W	Power taking into account losses in hydraulic, W	Generator output power, W
4	570	0	553	506
5	890	0	875	840
6	1450	290	1378	1000
7	2303	1143	2188	1000
8	3439	2279	3267	1000
9	4896	3736	4651	1000
10	4401	3241	4181	1000
11	5858	4698	5565	1000
12	1296	136	1231	1000
13	1648	488	1565	1000
14	2058	898	1955	1000

The losses of the multi-unit plant are 15%, 5% at the hydroelectric unit and 10% at the generator since the energy summed in the additive block is transferred to the generator. Up to 6 m/s, the plant generator also does not work at full power, so the efficiency is about 70-80%. The output of the generator does not reach the rated value at wind speeds of 4-5 m/s, but more than 2 times higher than in wind turbines with individual generators, as can be seen from figure 4: at wind speeds up to 7 m/s, the power of the multi-unit wind turbines is approximately 2-3 times higher than the sum of the powers of the individual wind turbines.

The multi-unit plant works at full capacity at an average wind speed of 4 m/s for about 4500 hours per year, including the use of accumulated energy. This is 51% of the total time of the year. At a wind speed of 4-5 m/s, the multi-unit plant produces up to 840 watts. Under the same conditions, separated wind farms in total provide 1 kW of electricity for only 1185 h (13%), which is 3-4 times less.

The calculation of the operating time of wind power plants at 4, 5 and 6 m/s shows that in wind installations with separate generators (table 1) the amount of electricity generated is 130 W, at 5 m/s - 250 W and at 6 m/s-435 W. The multi-unit installation produces 500 W of electricity at 4 m/s, 840 W at 5 m/s, and 1 kW at 6 m/s when up to 300 W is accumulated (table 2). To compensate for the lack of energy at 4-5 m/s, it is possible to use the excess energy in the form of working fluid energy from the accumulator.

In wind power plants with separate generators at 4-5 m/s, only multi-blade wind receiving devices work fully, but they produce small power due to the low wind power utilization, which is insufficient for the generator operation. Wind farms in total accumulate up to 200 kWh per year. Accumulation is possible from 9 m/s.

According to the wind repeatability of M. M. Pomortsev, it is known that the average speed of 4 m/s is observed 1610 h per year, 5 m/s is observed 1445 h and 6m/s is observed 1070h. To calculate the operating time (T, h) of wind turbines with separate generators and multi-unit installation, we use the equation 6:

$$T = t_1 + t_2 + t_3 \tag{6}$$

t_1, t_2, t_3 - the number of hours per year with a certain wind speed of 4, 5 and 6 m/s according to M. M. Pomortsev.

For example, for multi-unit installation T=1610+1445+1070=4125h. To compensate for the lack of energy at 4-5 m/s, it is possible to use the excess energy in the form of working fluid energy from the accumulator.

In wind power plants with separate generators at 4-5 m/s, only multi-blade wind receiving devices work fully, but they produce small power due to the low wind power utilization, which is insufficient for the generator operation: $T=t_3=1070H$.

The annual accumulated energy (Q, kWh) is calculated using equation 7:

$$Q = \sum P_i t_i \tag{7}$$

P_i-power at a certain wind speed from tables 1 and 2, W;

t_i - number of hours per year with a certain wind speed according to M. M. Pomortsev, h.

Multi-unit installation accumulates 2619 kW h of energy per year: Q=290w*1070H+1143w*640H+2279w*315h+3736w*152h+3241w*52H+4698w*26h=2619kw*h.

Wind farms with separate generator in total accumulate up to 200 kWh per year. Accumulation is possible from wind speed of 9 m/s.

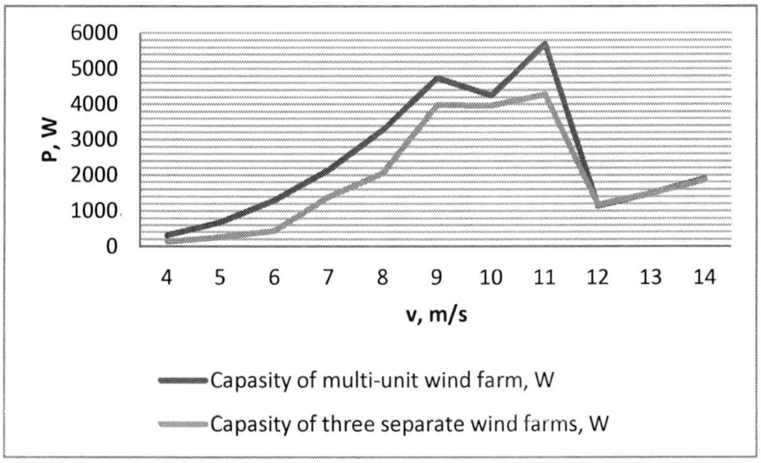

Figure 4. Comparison of capacities of a multi-unit and three separate wind farms with losses taken into account.

4. Conclusion

At low wind speeds, the torque and rotation frequency of low-blade wind receiving devices are low, therefore the frequency of the generator and the efficiency are low. However, summing energy of the individual units with different parameters that work efficiently at different wind speeds by using the additive unit, one can power the generator with energy from a number of wind receiving devices using hydraulic motors in an additive block, increasing the speed and efficiency of the generator, keeping the rotation speed at nominal.

Calculations have shown that the wind power plant at an average wind speed of 4 m/s operates at full capacity of 4500 hours per year or 51 % of the total time of the year. At 4-5 m/s, the multi-unit installation produces up to 830 watts. Under the same conditions, wind farms with separate generators in total provide 1 kW of electricity for only 1185 h (13%), which is 4 times less than the wind installation and there is no possibility to use the energy reserve.

At the same time, 2619 kWh of energy is accumulated in multi-unit installation, and it is possible to accumulate it at a wind speed of 6 m/s. If there is a shortage of wind power, the accumulated energy compensates for the lack by maintaining a constant power output of the generator.

References

[1] Kharitonov V P 2010 *Osnovy Vetroenergetiki [Fundamentals of Wind Energy]* (Moscow: VIESKH)

[2] Dorjiev S S, Bazarova E G, Pimenov S V, Rozenblum M I 2018 *J Phys Conf* vol 1111. 012053

[3] Hovgaard T G, Jørgensen S B and Jørgensen J B 2013 *Wind Energ.* vol 00:1–12

[4] Vestas Multi Rotor concept demonstator ar Risø Available at: https://www.vestas.com/~/media/files/multirotor%20fact%20sheet/ (accessed 15 February 2020)

[5] The Wind Tree Available at: https://newworldwind.com/en/wind-tree/ (accessed 25 January 2020)

[6] SHefter YA I Rozhdestvenskiy I V 1957 *Izobretatelyu o Vetrodvigatelyakh i Vetroustanovkakh [To Inventor About Wind Turbines and Wind Units]* (Moscow: Ministry of Agriculture publ.) chapter 2 pp 35–36

[7] Patents RF no. 2187019 2002, no. 2239092 2004, no.2581304 2016, no.2366829 2009, GB no. 2443886 2016, EP no.1407139 2009

[8] Pomortsev M M 1894 *O Zakone Raspredeleniya Skorostey Vetra [On the Law of Distribution of Wind Speeds Units]* (St Petersburg: Marine Government Printing House) Chapter 2 pp 35–36

Application of machine learning methods for optimizing the technical and economic performance of generating systems

E K Arakelyan[1], I A Boldyrev[2], K V Evseev[1,2] and Y A Gorban[1,2]

[1]National Research University Moscow Power Engineering Institute, Moscow, 111250, Krasnokazarmenaya str., 14, Russia
[2]Volzhsky branch of National Research University Moscow Power Engineering Institute, Volzhsky, 404110, Lenin Avenue, 69, Russia

Abstract. The paper is on the technical and economic performance optimization technique for thermal and power generating system using machine learning methods. The possibility of using regression analysis for parameter influence evaluation when calculating technical and economic performance in order to reach better generating unit efficiency is described. The approach to evaluate the parameter influence of a large distributed control system is presented.

1. Introduction

When producing power energy, it's necessary to assess and take into account the generating system performance. In this paper, the particular case of a thermal power plant generating unit, which includes such components as steam boiler, turbine, condenser, pumps, etc., is concerned [1]. During power equipment operation, assurance of its efficiency, maintenance schedules and commissioning and running tests are crucial. This information can be obtained by calculating technical and economic efficiency indices (TEI). Most commonly the data to calculate TEI is transferred to the distributed control system (DCS) via information and measuring channels (IMC) from field devices and sensors, that usually represent a lot of parameters and values. Though its number can be great, it does not always lead to more precise results, but it often tends to complicate the assessments [2].

The number of parameters and signals in modern generating units can be roughly at least 1000, however not all of them can affect the result the same way, which means that it is possible to select the most important ones, which affect the result of TEI assessments and power equipment monitoring and control most, among all. It means that the more precise and actual these values, the more precise the DCS monitoring and control, thus the more efficient the generating system [3].

The goal of the research is a development of the technique for generating system influential process parameter selection, thus it is possible to take into account, which ones are the most important for an information and control system efficiency. The problem can be solved with the machine learning methods to select the parameters among the set of input values. The historical and real-time data can be used as training data for a system state prognosis model. The feature of machine learning is that it is capable to solve classification and regression analysis tasks on data that are not exactly linearly separable [3]. Using such a technique as machine learning is also useful for calculating the operational technical and economic performance of a thermal power plant generating unit for which 10-15 minutes prognosis is necessary for ensuring high efficiency [2].

2. Machine learning methods

Machine learning (ML) is a tool for solving a wide range of applied problems (pattern recognition, equipment diagnostic, state analysis and prognosis, etc.) based on mathematical statistics, optimization methods, probability theory, graphs theory and other mathematical techniques.

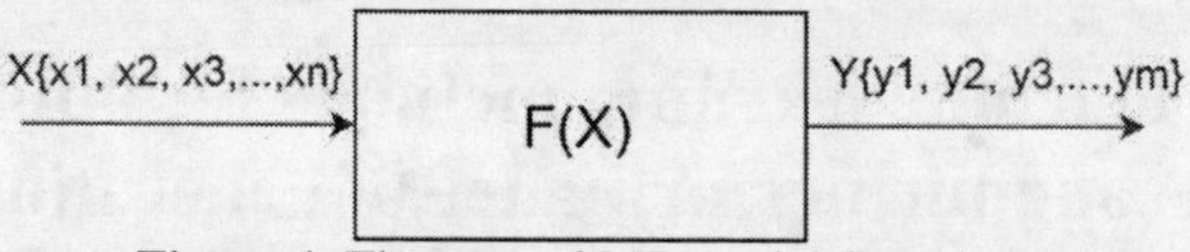

Figure 1. The general ML model flowchart.

The feature of ML is an indirect approach of getting the result of calculations, it requires building a certain model with a specific architecture and algorithms. After that step it should be trained, i.e. its properties (parameters) should be adjusted for solving the specific problem. Training is the main process in ML and it means finding the optimal parameters to maximize the output value, the result to correct answer based on dataset values [3]. In other words, the model should generalize the obtained data to calculate the result for new data that is not in the dataset. Using of any model can be considered a function (figure 1), which takes certain values as input and returns output values. Depending on the desired output it can be a continuous value (regression) or discrete (classification, clusterization) (figure 2) [3].

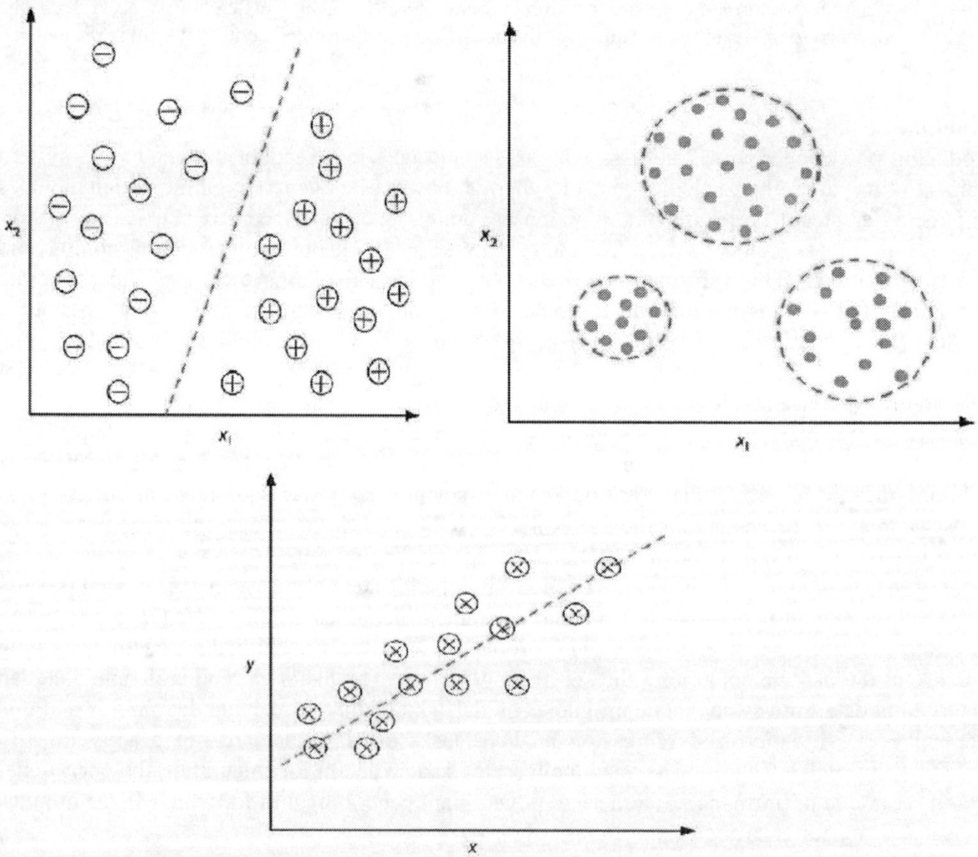

Figure 2. ML models. a - classification, b - clasterization, c - regression.

There are a few ML training approaches - supervised learning and unsupervised learning (figure 3). The former one requires both input (X) and output (Y) value (ground truth) sets when the last one generalizes the data by only vector X. In general, all ML approaches require dataset for training, another dataset can be used for the testing model, its accuracy.

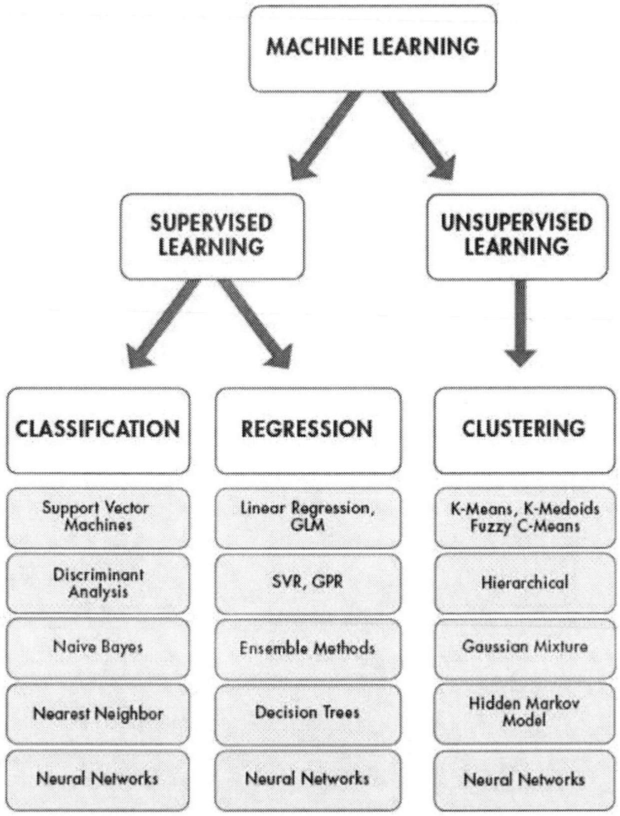

Figure 3. ML models and approaches.

Therefore, ML allows finding dependencies between values based on experimental data, however, it should be mentioned that getting an absolutely precise model is nearly impossible in most cases for there are always outliers in data values and during training, the training algorithm tries to find the parameters that minimize the error (the difference between ground truth Y and predicted output $\hat{Y}$) [3]. In other words, it requires finding the loss function global minimum to highest possible accuracy, which could be a challenge when the number of input features is great. Nevertheless, ML is a powerful tool in data analysis.

3. Use of ML methods for optimizing technical and economic performance

During the operation of thermal power plant generating units, it is necessary to ensure their efficiency, how well it transforms primary fuel to energy. To measure this value there are algorithms that take as input process parameters. It should be mentioned that distributed control systems operate a huge amount of values, however not necessarily all of them affect the generating unit performance equally, so there are parameters that are more important to monitor and control in order to reach higher efficiency. Thus it could be useful to select them and place greater focus on them what could be important for performance prediction within a certain range of time.

If the parameter values and results are given it is possible to analyze the data and find dependencies between them, how the former ones affect the result with regression analysis. One of the main purposes of regression analysis is determining the influence quantity of input values (predictors) on output ones and result from prediction under certain conditions [4].

The mentioned technique can be used for parameter selection from the great number of process variables, the output value is represented by technical and economic performance estimation of heat and power equipment. Nevertheless, in the case when there is a lot of input parameters it could be inconvenient to carry out regression analysis on all the parameters at the same time, thus it is reasonable to split the one complex task into a set of small ones with fewer input features and solve it separately. Such an approach could be represented as an oriented weighted graph, where each start node is an input parameter, end node is the result, performance estimation, other ones are intermediate outcomes (figure 4). The graph weights are determined with regression analysis and represent parameter influence. The sum of all incoming weights of a node equals 1, so first it should be normalized before the next step. Thus it shows the relative influence between parameters, the highest weight has the biggest influence on the node value.

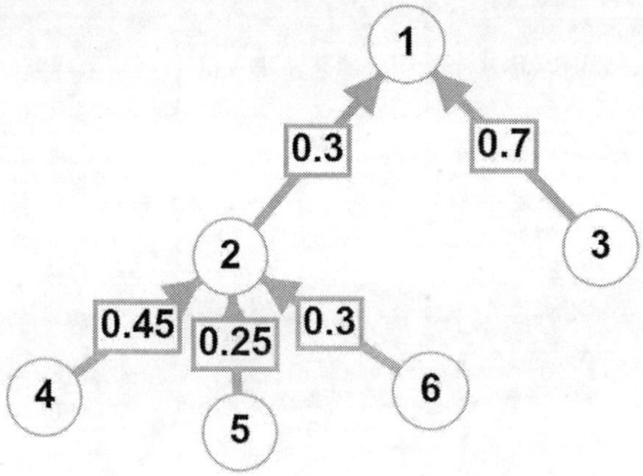

Figure 4. Parameter influence graph.

The input parameter influence on the result can be found as a product of weights during the graph traversal [5]. The result of the procedure is the vector of each input parameter influence on output value (figure 5). In cases when one parameter is used in calculations more than once, the highest influence is recorded. The results allow concluding which parameters monitoring and control are more important for it leads to higher generating unit efficiency. Thus it could be necessary to improve the DCS measuring channels in order to provide it more data considering these parameters. This problem is a part of the intelligent distributed control system design, which means more efficient monitoring and control by data flow optimization and uniting multiple DCS levels within one framework, that allows considering an object performance overall [6, 7].

4. Conclusions

Application of machine learning methods for technical and economic optimization of generating unit allows analyzing and evaluate distributed control system process parameters in order to select the most influential ones, thus it is possible to increase generating unit efficiency by improving the information content of equipment operation;

The proposed approach of splitting one complex calculation task into many simple for parameter influence estimation is reasonable in the case when it is necessary to operate with a great number of input parameters.

References

[1] Pletnev G P 2005 *Automation of Technological Processes and Production in Heat Power Engineering* (Moscow: MPEI)
[2] Arakeljan E K and Galustjan M K 2012 Methodological approaches to the calculation of annual technical and economic indicators of power equipment of an autonomous complex for the production of electricity and heat *New in Russian energy sector* **5** 16-28
[3] Raschka S and Mirjalili V 2019 *Python Machine Learning: Machine Learning and Deep Learning with Python, Scikit-learn and TensorFlow 2,* 3rd Edition (Birmingham: Packt)
[4] Draper N R and Smith H 1998 *Applied Regression Analysis, Third Edition* (New York: John Wiley & Sons, Inc.)
[5] Diestel R 2017 *Graph Theory* (New York: Springer)
[6] Arakeljan E K, Vasil'ev E D and Khurshudjan S R 2014 Aumomated control system issue and a possible way of its decision *Vestnik MPEI* **1** 15-20
[7] Evseev K V and Arakelian E K 2020 Information support of the intellectual ACS TP *Radio engineering, electrical engineering and power engineering* (Moscow: Moscow Power Engineering Institute) p 919

Acknowledgments

The research is funded by RNF grant № 19-19-00601.

Study of work efficiency of the energy system at the introduction of small hydropower plants

O I Zhelyaskova[1], A A Smirnov[2] and M M Sultanov[1]

[1] Branch of the National Research University "MPEI", Lenin Avenue, 69, Volzhsky, 404110, Russia
[2] The National Research University "MPEI", Krasnokazarmennaya street, 14, Moscow, 111250, Russia

Abstract. The paper describes a comprehensive approach to assessing the effectiveness of introducing renewable energy sources by the example of switching on small hydropower plants in a certain section of the energy system. The study combines an algorithm for selecting the composition of the included generating equipment in the case of connecting new consumers, and a developed program for calculating and constructing generation and load schedules. The results of the research are presented on the example of the introduction of mini hydroelectric power plants through joint work, the developed approach and the program.

1. Introduction

Currently, technologies based on renewable energy sources (RES) are developing worldwide. And our country is no exception. According to the state energy strategy of Russia until 2035, the main strategic guidelines include the creation of alternative energy sources based on renewable energy sources, which increase energy and environmental safety in the region and save non-renewable energy resources. The location of such a generation near the consumer will increase the capacity in electric networks, reduce the loss of electricity during transmission, and give autonomy of power supply in hard-to-reach areas. This is a positive trend in the development of renewable energy. Such sources of energy will save natural fuel, fully utilize the potential of natural resources.

2. Materials and methods

To conduct a study of the efficiency of the energy system during the implementation of small hydropower plants, a comprehensive approach was used to apply the proposed algorithm for determining the impact on the regional fuel balance of new generating capacities and the developed program in VBA for calculating and scheduling loads and power flows [1].

To accomplish this task, an algorithm is presented for choosing the composition of the included generating equipment when a new consumer is connected to the power system (figure 1). When connecting new capacities, the consumer chooses the method of generating electric energy: its own generation or the country's combined energy system.

In the second case, an application for a technological connection to the grid company is submitted. The question of how to acquire electrical energy is being addressed. All electricity generated at hydroelectric power stations, power plants, thermal power plants, nuclear power plants, as well as generation sources operating on renewable energy sources, is sold by generating companies in the capacity market. Depending on the consumer's reliability category (according to the PUE of Chapter

1.2, paragraphs 1.2.17–1.2.21), the number of independent sources of electricity that will supply the consumer with electricity is selected. Next, modeling is carried out of various options for supplying power to the consumer, taking into account the location of power sources, losses in power lines, fuel consumption for the production of 1 kWh of electric energy, a model with more efficient use of fuel resources is selected.

When calculating the effect of introducing a small generation in the territory under consideration in the proposed algorithm, it is necessary to change the parameters for power sources based on renewable energy sources with a known declared load.

The main actions of the algorithm:

a) selection of power and location of a new source of small generation;
b) determination of the closest consumers of electric energy;
c) creating a scheme of power lines from a source of small generation to a consumer with the calculation of losses in it;
d) the definition of existing lines suitable for the consumer;
e) calculation of losses in power lines from the current source of generation to the consumer with a decrease in power output due to the input of small generation;
f) determination, reduction of losses in electric networks after the introduction of small generation;
g) calculation of the saved fuel at the heat station as a result of a decrease in the power output [2].

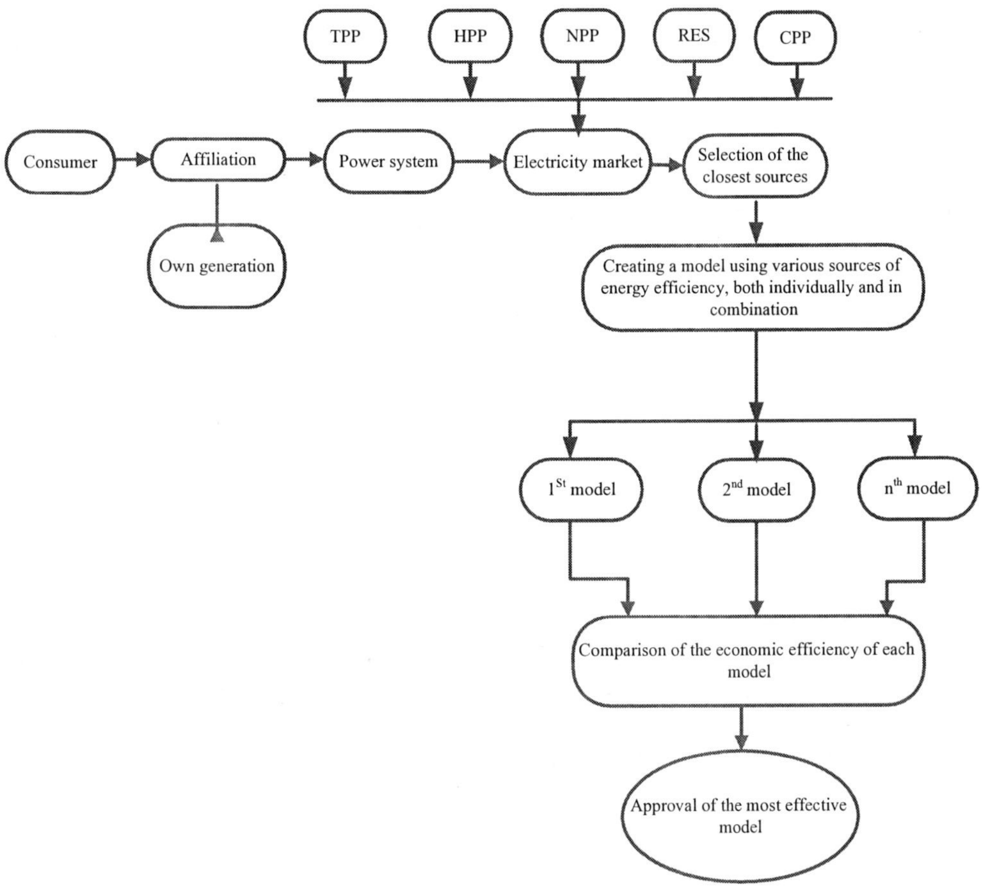

Figure 1. Block diagram of the selection of the composition of the included generating equipment.

IOP Conf. Series: Materials Science and Engineering 1035 (2021) 012012 doi:10.1088/1757-899X/1035/1/012012

In the developed program, the algorithm consists of calculating the main parameters of the graphs of electrical loads. The block diagram shown in figure 2 describes an algorithm for calculating the main parameters of the load graph and its construction for analysis and smoothing of peak loads.

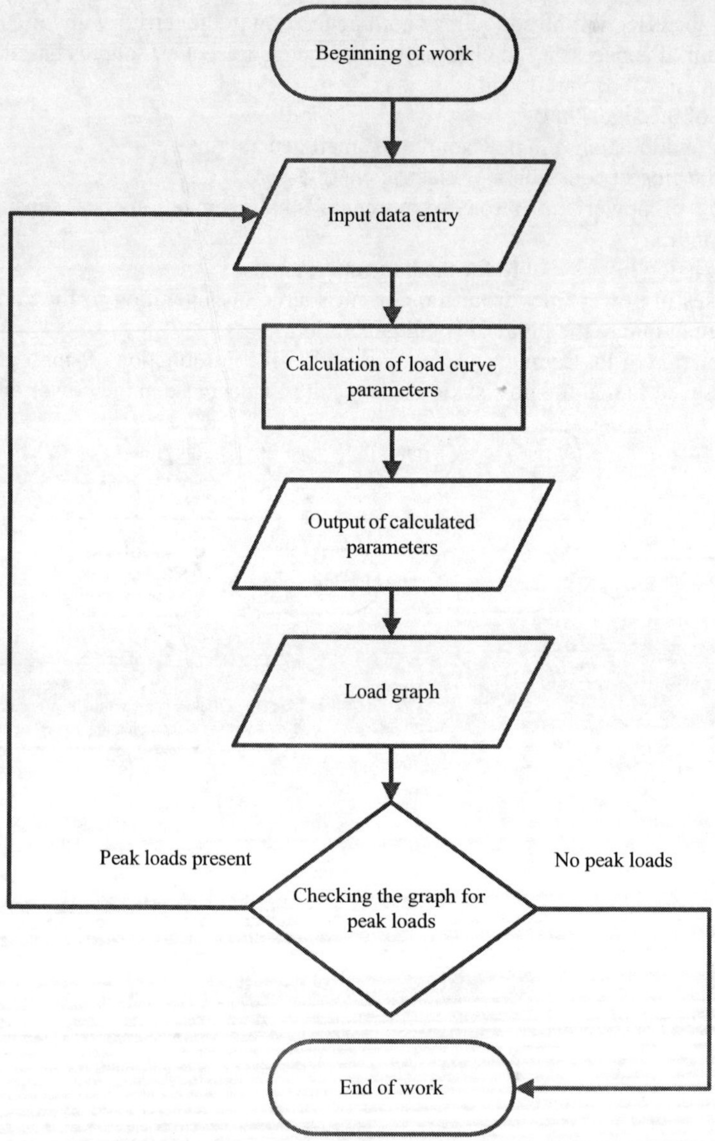

Figure 2. The block diagram of the proposed algorithm of the program.

The main actions of the software algorithm:

a) calculation of maximum, minimum and average daily capacities;

b) the calculation of the fill factors of the graph, the unevenness of the graph and the shape of the graph;

c) the calculation of the annual time of use of the maximum load, the annual time of use of the maximum power loss

d) the calculation of reactive and full power loads;

e) daily load schedules of all capacities are built;

f) measures are taken to regulate the loads (recounting) [3].

In the course of the work, the section of the electric network X of the power system was investigated. Three options for connecting the nth consumer from different generation sources were considered. The following conditions and limitations are indicated: the consumer in question is currently powered by a hydroelectric power station. The power supply to the consumer and the possible reduction in the loss of transmitted power are considered when connected to the CHPP located next to the consumer and the introduction of a small generation, more specifically Mini Hydroelectric Power Station on the River R near the consumer. For the selected source, a small river R and sufficient terrain for its construction are located nearby. The pressure for the station will be 4.4 m, the flow rate is 35 m^3/s, the utilization rate is 0.55, and the guaranteed rated power is 831 kW. This capacity will be considered sufficient for the power supply to the consumer.

Figure 3 shows the consumer power circuit from three-generation sources.

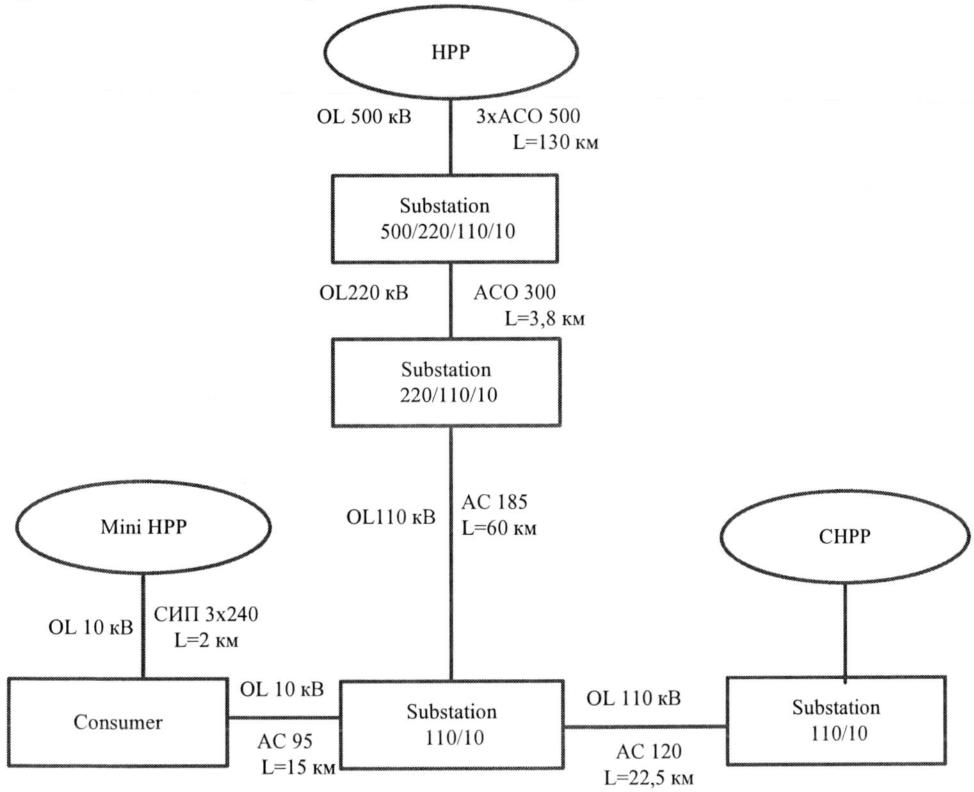

Figure 3. Consumer power scheme from three sources of generation.

We calculate the following consumer power options:
a) from three sources of generation: HPPs, TPPs and Mini HPPs;
b) from two sources of generation: hydroelectric power station and Mini hydroelectric power station;
c) from two sources of generation: TPP and Mini HPP.

3. Results of experimental studies

According to the algorithm, power losses in the network sections were calculated (table 1).

Using the power of Mini Hydroelectric Power Station will reduce losses in overhead lines, as well as reduce the cost of purchasing additional capacity to cover these losses. The calculation results showed that the implementation of Mini HPPs with guaranteed power of 831 kW will reduce grid losses by 17.27 kW, and the power supplied to the HPP will be reduced by 848.3 kW. It was also revealed that the power supplied to the CHPP can be reduced by 847.8 kW.

Table 1. A summary table of the results obtained with the implementation of the Mini HPP.

Circuit number	Plot	Losses, kW	Power generated by the generation source, kW	Power issued by the source, taking into account losses, kW	Amount of fuel saved, t t/year	Loss reduction in power lines, thousand kWh/year
1	Mini HPP - Consumer	2,6	831	833,6	-	23
2	HPP - Consumer	17,27		848,27	-	151
3	CHPP - Consumer	16,83		847,83	911	147

Using the program, we got the following graphs of consumer loading and generation taking into account losses (figures 4).

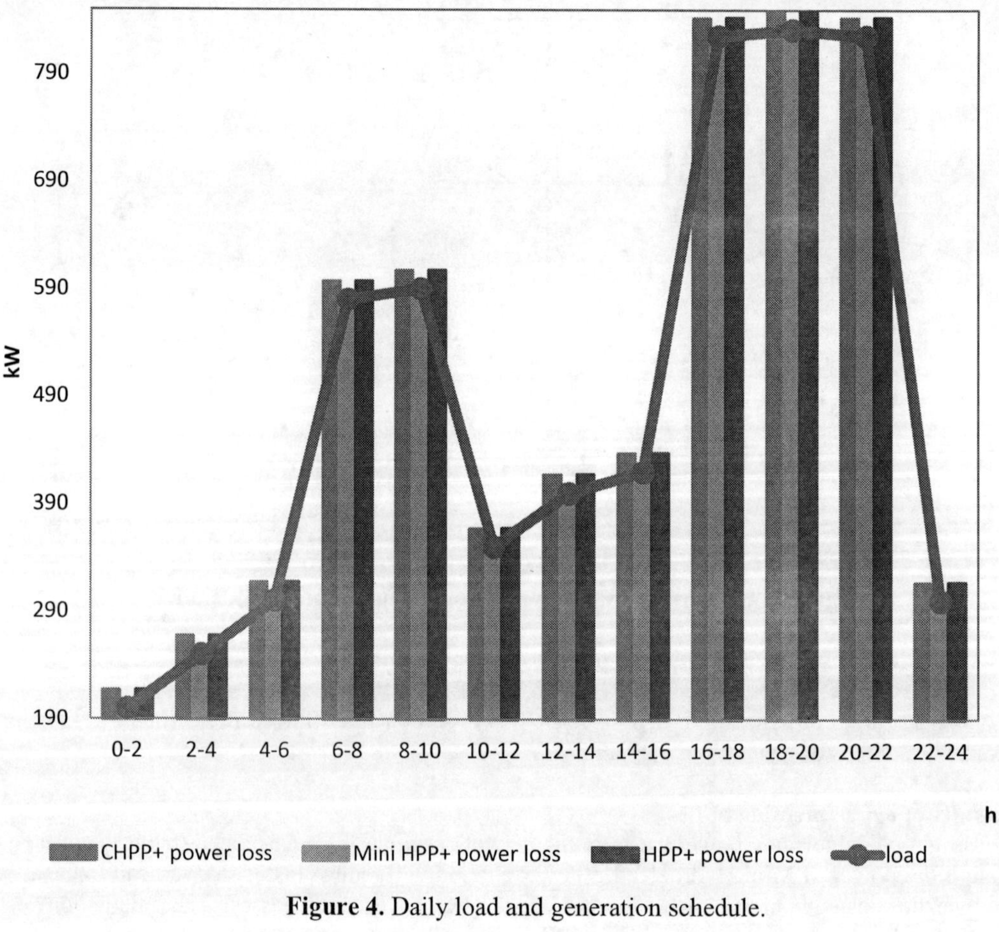

Figure 4. Daily load and generation schedule.

Figure 4 shows that the daily load schedule and the generation schedule of Mini HPPs with losses are practically indistinguishable. When powered by hydropower plants and thermal power plants, losses are approximately equal to 2% of the total power output. When modeling processes in the X power system with the introduction of renewable energy sources, the following results were obtained:

When introducing a Mini HPP, two possible scenarios are possible:

a) losses in the electric grid complex will decrease by 151 thousand kWh per year. The released capacity of hydroelectric power stations can be transferred to other parts of the power system, this is possible both during peak hours of the load and in the base.

b) losses in the electric grid complex will decrease by 147 thousand kWh per year, and fuel economy at the TPP will amount to 911 t. t. per year. The use of a comprehensive assessment of the introduction of the new generation made it possible to assess how much the losses in electric networks will decrease and how much fuel costs will be reduced at thermal power plants when new generation sources are connected to the power system [4].

4. Conclusions

The study showed that when the calculation and modeling methods work together, the accuracy of the selection of generating equipment increases, which contributes to an increase in energy efficiency. The load graphs clearly show the overloaded sections of the network, as well as the underload of consumers. This approach contributes to the modeling of various network load options, both from traditional and alternative energy sources, and the selection of the best option. When applying the latter, an increase in energy and economic efficiency was revealed. In the case of the introduction of small hydropower plants in the power system, it is also possible to unload other generating facilities of the studied area.

References

[1] Sultanov M M, Zhelyaskova O I and Zhelyaskov S S 2019 Modeling the process of analyzing the operation of the energy system at peak loads *The 25th Inter-University Scientific and Practical Conference of Young Scientists and Students* Volzhsky Branch of FSBEI HE "NRU MPEI" ISBN 978-5-94721-145-0 pp 86-88

[2] Sultanov M M, Zhelyaskova O I, Riga I L 2019 *The study of the energy efficiency of generating systems in modern conditions of energy consumption: a training manual* Volzhsky Branch of FSBEI HE "NRU MPEI " 59 p

[3] Sultanov M M, Zhelyaskova O I 2019 T*he program for calculating the parameters and plotting the energy system* Federal Service for Intellectual Property Certificate No. 2019667320 dated 12/23/2019

[4] Riga I L and Sultanov M M 2015 Implementation of the program for the implementation of agreements on the provision of capacity on the basis of modern domestic plants *The 21st Inter-University Scientific and Practical Conference of Young Scientists and Students* Volzhsky Branch of FSBEI HE "NRU MPEI" pp 124-125

The structure and the electrical properties of a multilayer system based on a composite $(Co_{40}Fe_{40}B_{20})_{34}(SiO_2)$ and zinc oxide

O V Zhilova, A V Sitnikov, S Yu Pankov, V A Makagonov, M A Kashirin and I V Babkina

Department of Solid State Physics, Voronezh State Technical University, Moscow Avenue, 14, Voronezh, 394026 Russia

Abstract. A heterogeneous multilayer system $(Co_{40}Fe_{40}B_{20})_{34}(SiO_2)_{66}/ZnO$ was obtained by ion-beam sputtering. X-ray diffraction analysis showed that the obtained films have an amorphous multilayer structure. Transmission electron microscopy analysis (TEM) confirmed the formation of separated $(Co_{40}Fe_{40}B_{20})_{34}(SiO_2)_{66}$ and ZnO layers. The $(Co_{40}Fe_{40}B_{20})_{34}(SiO_2)_{66}$ composite layers are monogranular with a metal particle size of 1 nm. The thin films remain stable after heat treatment at temperatures up to 350 °C. Measurement of the electrical resistance confirms the formation of a multilayer structure and crystallization of films after heating above 350 °C.

1. Introduction

Zinc oxide (ZnO) has a wide range of unique properties such as its wide and direct band gap (≈3.3 eV), large exciton binding energy (60 meV), high transparency in the visible range and low cost [1,2]. Additionally, easy tunability of its morphology for the application type makes it a very suitable material for solar cells, thermoelectric generators, energy hydrogen conversion devices and sensors [3-5]. At present increasing interest attract a variety of ZnO-based nanostructures, such as nanoparticles, nanorods, multilayers and superlattices [6-8].

Such multilayered heterostructures consisting of undoped or doped ZnO layers with thickness variated from 1 to 10 nm are promising materials for alternative energy devices application [7,8]. In most cases, such heterostructures exhibit improved or new physical properties and can combine the properties of two different materials [9]. However, the properties of these materials strongly depend on the many factors (preparation method, thickness and morphology of the layers, chemical interaction and diffusion between the layers). This makes controlling of the multilayer structure stability is an important task.

In this paper, we present the results of this research of the structure and electrical properties of the $(Co_{40}Fe_{40}B_{20})_{34}(SiO_2)_{66}/ZnO$ thin films prepared by ion-beam sputtering.

2. Methods

The $[(Co_{40}Fe_{40}B_{20})_{34}(SiO_2)_{66}/ZnO]_{112}$ thin films were obtained by the ion-beam sputtering method [10]. The deposition of the multilayer system was conducted alternately on ST-50 ceramic substrates using two targets: composite target consisting of a metal alloy $(Co_{40}Fe_{40}B_{20})$ with uniformly distributed SiO_2 plates and ZnO ceramic target. V-shaped screens were placed between the target and the substrate to obtain the different thickness of the layer during the single deposition process. The resulting multilayer system $[(Co_{40}Fe_{40}B_{20})_{34}(SiO_2)_{66}/ZnO]_{112}$ had 112 $(Co_{40}Fe_{40}B_{20})_{34}(SiO_2)_{66}/ZnO$ bilayers.

Sample thicknesses (h) were measured using an MII-4 interferometer. To determine the thickness of $(Co_{40}Fe_{40}B_{20})_{34}(SiO_2)_{66}$ and ZnO interlayers, they were formed by layer-by-layer deposition in separate cycles. The deposition parameters were corresponding to the parameters of the formation of multilayer films. Thin-film layer thickness was calculated from the parameters of the time of one revolution and the time of deposition. The thickness of the multilayer system $[(Co_{40}Fe_{40}B_{20})_{34}(SiO_2)_{66}/ZnO]_{112}$ film and $(Co_{40}Fe_{40}B_{20})_{34}(SiO_2)_{66}$ and ZnO interlayers are shown in table 1.

Table 1. Parameters of the studied thin films.

Film composition	Film thickness (μm)	Layer thickness No. 1 (nm)	Layer thickness No. 2 (nm)
ZnO	0.04 – 0.12	0.63 – 1.68	–
$(Co_{40}Fe_{40}B_{20})_{34}(SiO_2)_{66}$	0.11 – 0.27	1.20 – 2,80	–
$[(Co_{40}Fe_{40}B_{20})_{34}(SiO_2)_{66}/ZnO]_{112}$	0.16 – 0.35	0.73 – 1.57	0.45 – 1.14

The structure was investigated by X-ray diffraction (XRD) on a D2 PHASER diffractometer (BRUKER). To decipher the obtained diffraction patterns a licensed software was used. Transmission electron microscopy (TEM) images of the cross-section and electron diffraction were obtained on a Hitachi HT7700 TEM (acceleration voltage 100 kV, W source). The cross-sectional samples on the sitall substrates for TEM studies were prepared by focused ion-beam (single-beam FIB, Hitachi FB2100) at 40 kV to a thickness of about 40 – 50 nm.

The electrical resistance of the samples was taken by DC two-point probe technique by means of a V7-78/1 multipurpose digital voltmeter with original software. The relative error of the measuring did not exceed 10%.

3. Results and discussion

Figure 1 shows the XRD patterns of $[(Co_{40}Fe_{40}B_{20})_{34}(SiO_2)_{66}/ZnO]_{112}$ thin films with different thickness, measured at average Bragg angles in the range $2\Theta = 10° – 60°$. It can be seen from the figure that the thin films have an amorphous structure characterized by a wide halo. Such a structure can be associated with the presence of several amorphous phases (metal granules of $Co_{40}Fe_{40}B_{20}$ alloy, dielectric phase of SiO_2 and amorphous ZnO).

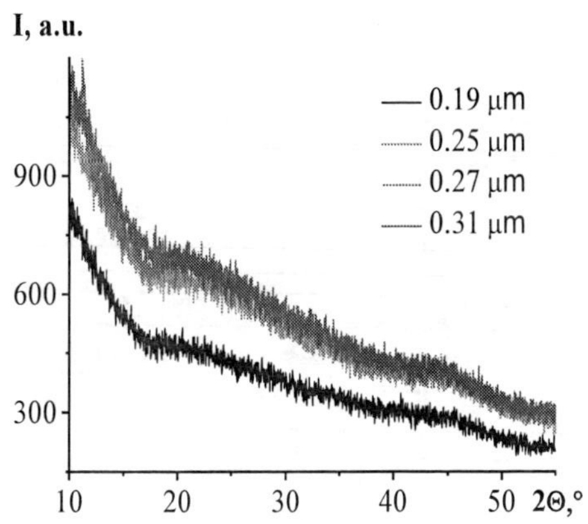

Figure 1. The XRD patterns of the $[(Co_{40}Fe_{40}B_{20})_{34}(SiO_2)_{66}/ZnO]_{112}$ multilayer thin films in the initial state with different total film thickness.

The XRD patterns at small Bragg angles in the range 2Θ = 1° – 10° showed a complex structure (figure 2). The maxima at the diffraction pattern are comparable with the bilayer thickness of the $[(Co_{40}Fe_{40}B_{20})_{34}(SiO_2)_{66}/ZnO]_{112}$ film, which may indicate the multilayer structure.

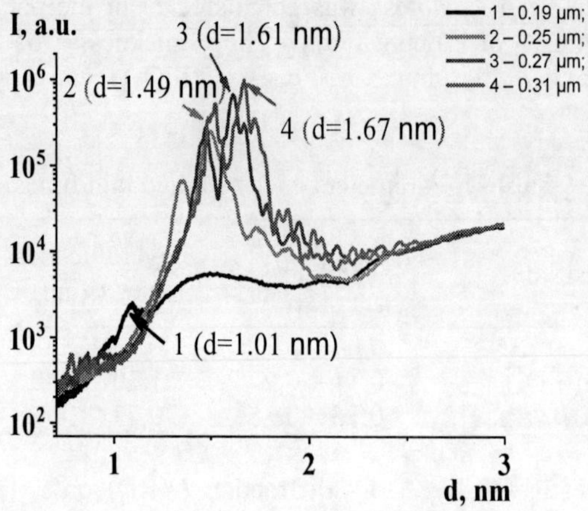

Figure 2. The XRD patterns of the $[(Co_{40}Fe_{40}B_{20})_{34}(SiO_2)_{66}/ZnO]_{112}$ thin films with different total film thickness measured at small Bragg angles (d is the bilayer thickness of the $[(Co_{40}Fe_{40}B_{20})_{34}(SiO_2)_{66}/ZnO]_{112}$ film).

An exception is the sample h = 0.19 μm with a very small thickness of the bilayers (h(ZnO) ≈ 0.61 nm and h(composite) ≈ 1.08 nm). In this case, we can assume the absence of a formed layered structure.

TEM micrographs of $[(Co_{40}Fe_{40}B_{20})_{34}(SiO_2)_{66}/ZnO]_{112}$ thin films confirmed the formation of a multilayer structure with $(Co_{40}Fe_{40}B_{20})_{34}(SiO_2)_{66}$ composite and amorphous ZnO interlayers (Figure 3) (electron diffraction pattern on the inset in figure 3). For $[(Co_{40}Fe_{40}B_{20})_{34}(SiO_2)_{66}/ZnO]_{112}$ multilayer showed in figure 3, the thickness of the $(Co_{40}Fe_{40}B_{20})_{34}(SiO_2)_{66}$ and ZnO layers were 1.1 nm and 1.2 nm respectively. The size of the metal granules is ~ 1 nm which is less than in a bulk composite $(Co_{40}Fe_{40}B_{20})_{34}(SiO_2)_{66}$ (near 2 – 4 nm). Therefore, the obtaining of a multilayer structure allows us to control the size of granules by the thickness of the composite interlayer as part of a multilayer structure in comparison with the equilibrium size of the granule in a bulk composite.

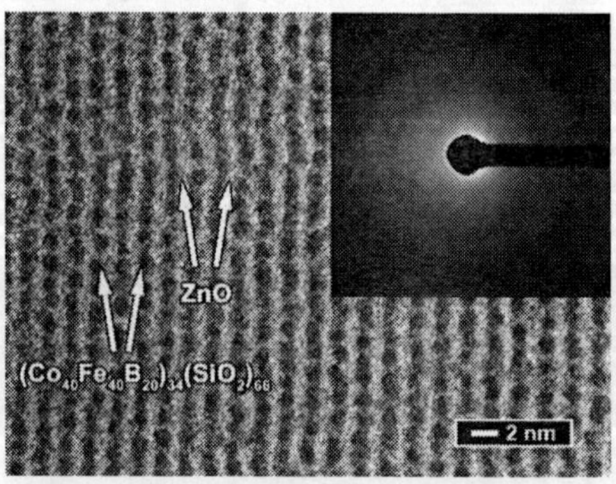

Figure 3. TEM micrographs of the cross section and electron diffraction pattern (inset) for $[(Co_{40}Fe_{40}B_{20})_{34}(SiO_2)_{66}/ZnO]_{112}$ multilayer thin film (h = 0.26 μm).

Figure 4 shows the dependence of the specific electrical resistance film on the thickness of the ZnO layer (ρ(h)) of the $[(Co_{40}Fe_{40}B_{20})_{34}(SiO_2)_{66}/ZnO]_{112}$.

With an increase in the ZnO interlayer thickness from 0.4 to 0.8 nm, a decrease in ρ(h) is observed. If we assume that the main conduction channel is the interface between the wide-gap semiconductor ZnO and the nanogranular composite, then the formation of a quasi-continuous two-dimensional region during the formation of a layered structure leads to a decrease in the electrical resistivity. With an increase in the thickness of the layers per unit thickness of the film, the number of conduction channels decreases; therefore, ρ(h) increases. This assumption is indirect evidence of the presence of multilayers in the films under study and the presence of a critical thickness of the dielectric interlayer allowing the formation of a quasicontinuous multilayer structure.

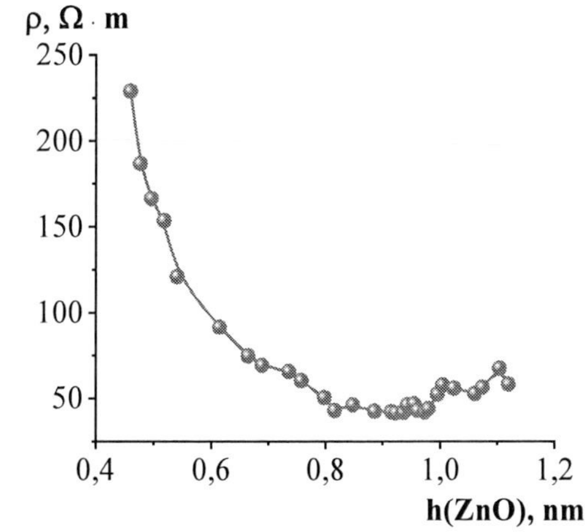

Figure 4. The dependences of the specific electrical resistance of $[(Co_{40}Fe_{40}B_{20})_{34}(SiO_2)_{66}/ZnO]_{112}$ multilayer thin film as a function of ZnO interlayer thickness.

The presence of nanometer thickness interlayers should affect the crystallization processes at heat treatment. If we assume that zinc oxide crystals form in the ZnO layer, then its size should be a few nanometers. Such small crystallites will have excess surface energy, which can lead to chemical interaction between composite and ZnO interlayers at interfaces. It can result in the formation of compounds with complex compositions. An analysis of the diffraction patterns for $[(Co_{40}Fe_{40}B_{20})_{34}(SiO_2)_{66}/ZnO]_{112}$ films annealed at different temperatures for 30 minutes confirmed these assumptions.

Crystallization begins with the formation of the metallic CoFe and iron oxide crystals. Then at 450°C ZnO crystallites appear. At temperatures of 500-600°C more complex compounds $Co_2FeO_2(BO_3)$ and Zn_2SiO_4 are formed. An analysis of the diffractograms taken at small Bragg angles showed that the multilayer structure of the studied films disappears at a temperature of 350°C. With decreasing in the thickness of the interlayers, the crystallization process and the destruction of the multilayer structure begin at lower temperatures. The nonmonotonic behavior of the temperature dependences of the electrical resistivity reflects structural changes in the films under study.

4. Conclusion

In summary, $[(Co_{40}Fe_{40}B_{20})_{34}(SiO_2)_{66}/ZnO]_{112}$ multilayer thin films with layer thicknesses of the order of one nanometer were obtained by layer-by-layer deposition using ion-beam sputtering; the composite layer has a monogranular structure with a metal particle size of the order of 1 nm; the films retain a multilayer structure when exposed to heat up to 350°C for 30 minutes.

References

[1] Benzitouni S, Zaabat M, Mahdjoub A Benaboud A and Boudine B 2018 *Materials science-poland* **36**(3) 427-434
[2] Levy D and Castellyn E 2018 *Transparent Conductive Materials: Materials, Synthesis, Characterization, Applications, First Edition* (Weinheim: Wiley-VCH Verlag GmbH & Co. KGaA) p 392
[3] Dobrozhan O, Vorobiov S, Kurbatov D, Balaz M, Kolesnyk M, Diachenko O, Komanicky V, and Opanasyuk A 2020 *Superlattices and microstructures* **140** 106455
[4] Caglar Y, Zor M, Caglar M and Ilican S 2006 *J. of optoelectronics and advanced materials* **8**(5) 1867-1873
[5] Mahajan S and Jagtap S 2020 *Applied materials today* **18** 100483
[6] Fan Z Y and Lu J G 2005 *Journal of nanoscience and nanotechnology* **5**(10) 1561-157
[7] Shih B, Hsieh W, Shyue J, Tsai F 2020 *Ceramics international* **46**(6) 7122-7130
[8] Yo-Seop Yo *et al* 2019 *Journal of materials chemistry* **7**(6) 1670-1680
[9] Volochaev M N *et al* 2020 *Superlattices and microstructures* **140** 106449
[10] Gridnev S A, Kalinin Yu E, Sitnikov A V and Stognei O V 2012 *Nonlinear Phenomena in Nano-and Microheterogenuous Sistems* (Moscow: BINOM Knowledge Lab.) pp 44–53 (in Russian)

Acknowledgments

This work was supported by the Ministry of Education and Science within the framework of state assignment No. FZGM-2020-0007.

IOP Conf. Series: Materials Science and Engineering 1035 (2021) 012014 doi:10.1088/1757-899X/1035/1/012014

Influence of the separation elements` shape on the process of water-oil emulsion demulsification in the pipe

V E Zinurov[1], A V Dmitriev[1] and I N Madyshev[2]

[1]Kazan State Power Engineering University, Krasnoselskaya street 51, Kazan, 420066, Russia
[2]Kazan National Research Technological University, Karl Marx street 68, Kazan, 420015, Russia

Abstract. In order to intensify the demulsification of water-oil emulsions, various types of separation elements for the settling tanks are proposed, namely, the corrugated plates with different orientation of corrugations, located at an angle of 45°, 135° and the cross-corrugated plates. The experimental studies of water-oil emulsion demulsification by means of a laboratory unit, using the developed separation elements were conducted. The conducted studies showed that the use of various separation elements inside the settling tanks allows intensifying the process of water-oil emulsion demulsification. The separation elements, considered in this paper, allow the performance of demulsification process with minimum 62% efficiency at the average rate of emulsion movement of more than 0.04 m/s. In the course of studies, it was found that the most effective of the considered separation elements are the corrugated plates with an orientation of corrugations, located at an angle of 45°, allowing the process of water-oil emulsion demulsification with an average efficiency of 80.1% at the movement rate within the range of 0.04–0.053 m/s. It is also shown that for the various separation elements, there are the most effective ranges of emulsion rates, at which the maximum values of demulsification efficiency are achieved.

1. Introduction

One of the most important tasks for oil treatment facilities is to improve the quality of oil-water emulsion separation. The relevance of this task is caused by the fact that the oil fields are exploited and the water concentration in the extracted oil increases, while the oil concentration decreases accordingly. Currently, when producing the oil, the concentration of stratum water can reach up to 80–90% [1–3].

The most well-known methods of water-oil emulsion separation are the following: settling, filtration, centrifugation, heat treatment, in-line demulsification, etc. The choice of a particular method depends on the following factors: thermal parameters of oil and water (specific weight, dynamic viscosity, etc.), the concentration of water-oil emulsion components, required efficiency of emulsion separation, etc. [4–9].

It should be noted that in the course of exploitation of oil fields, the parameters of water-oil emulsions may change, for example, due to the changing of original stratum conditions. In this regard, almost all oil treatment facilities have a multi-stage scheme for the purification of water-oil emulsions, the complex of which creates conditions that allow adapting to the changing of physical and chemical properties of the oil, as well as to the component composition. As a rule, the first stage of purification is the settling tank, in which the process of water-oil emulsion separation is based on the difference in water and oil densities. The advantages of settling tanks are the following: operation simplicity, large deposition surface, high reliability of equipment, increased cost-effectiveness since there is no need for periodical replacement of expensive filter elements or the use of demulsifiers, and high operation efficiency [10–14]. The main disadvantages of settling tanks are the following: bulkiness, inability to provide a stable hydraulic regime due to the change in the density of water-oil emulsion flow along the cross-section of apparatus, which is caused by the change of temperature, phase concentration and other parameters, as well as a relatively low rate of water-oil emulsion separation: when the movement rate of emulsions increases, the separation efficiency decreases due to the flow turbulence [15–18].

Therefore, an important task for the petroleum refining industry sector is to increase the efficiency and to speed up the processes of water-oil emulsion separation in settling tanks, since the subsequent stages of purification, including the use of demulsifiers and other methods, are the most expensive from an economical point of view. One of the most promising methods to increase the efficiency and to speed up the processes of water-oil emulsion separation in settling tanks is the use of various separation elements inside them, which allow to speed up the break of film that envelops the oil particles or water drops [19].

2. Purpose of research

The purpose of this research paper is to study the impact of various forms of separation elements by experimental method on the process of water-oil emulsion demulsification. In the course of study, the following types of separation elements were considered: corrugated plates with different orientation of corrugations, located at different angles of 45° or 135° and cross-corrugated plates. For the experimental modeling of water-oil emulsion demulsification process, using the separation elements, the laboratory unit was assembled, which includes the emulsion treatment vessel *1*, the cylindrical settling tank *2*, into which various types of separation elements were inserted: corrugated plates with different orientation of corrugations, located at an angle of 45° or 135° and cross-corrugated plates, pump *3*, a vessel for the sampler *4*, liquid filter *5* and shut-off valve on the emulsion supply line *6* (figure 1).

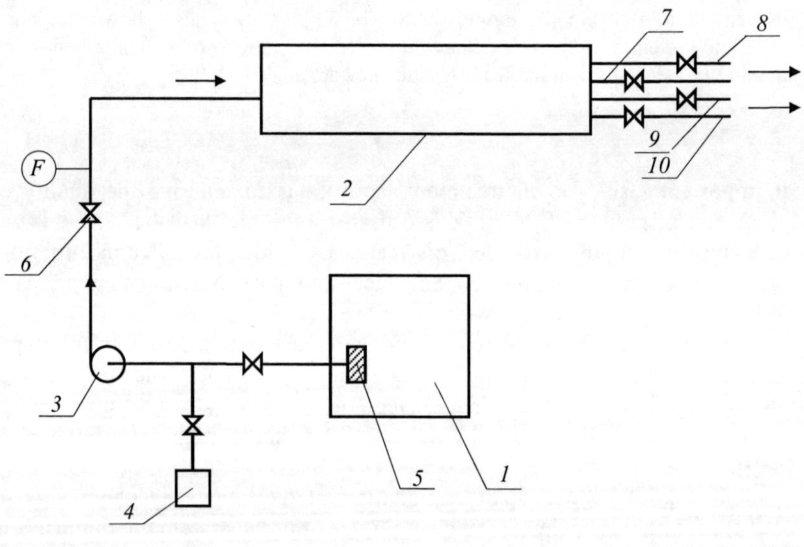

Figure 1. The experimental unit: *1* – vessel with original emulsion; *2* – cylindrical settling tank; *3* – pump; *4* – vessel for the sampler; *5* – liquid filter; *6* – shut-off valves on the emulsion supply line; *7* – the first line for the output of light phase; *8* – the second line for the output of light phase; *9* – the first line for the output of heavy phase; *10* – the second line for the output of heavy phase.

The process of water-oil emulsion demulsification in the laboratory unit can be described as follows: the treated water-oil emulsion was fed to the vessel *1*, after that to the cylindrical settling tank *2* by means of pump *3*; as the emulsion moved into the settling tank, a small part of it was fed to the sampling vessel *4* in order to record the value of oil and water concentration in the water-oil emulsion; the emulsion flow movement along different parts of laboratory unit was provided by opening and closing the shut-off valves *6*; the lines for output of light and heavy phases were arranged at the outlet of the cylindrical settling tank. It should be noted that the light phase was sampled from the first line of output *7* and the second line of output *8*, while the heavy phase was sampled from lines *9* and *10*.

In the course of experimental study, the following values were taken as constant: ambient temperature t_0 = 20°C, water density ρ_{water} = 998.2 kg/m^3 and oil density ρ_{oil} = 880 kg/m^3, dynamic viscosities of water μ_{water} = 0.001004 Pa·s and oil μ_{oil} = 0.0198 Pa·s.

The efficiency of water-oil emulsion demulsification in the pipe was determined by the following formula [20]:

$$E = \frac{\overline{x_D} - \overline{x_F}}{1 - \overline{x_F}}, \quad (1)$$

where $\overline{x_D}$ – mass fraction of oil in the light phase, kg/ kg; $\overline{x_F}$ – mass fraction of oil in the original mixture, kg/kg.

3. Results and discussion

The graphical results of conducted studies are shown in figures 2–4. The use of various separation elements in a round pipe allows intensifying the process of water-oil emulsion demulsification. When the emulsion is in the direct contact with the surface of separation elements, the stratification effect for the oil components from water increases, due to the violation of the integrity of emulsion structure (the film, which covers the oil particles and water drops are broken) and relatively weak vortices near the curving outward, curving inward and other parts of separation elements that increases the probability of coalescence of water and oil drops.

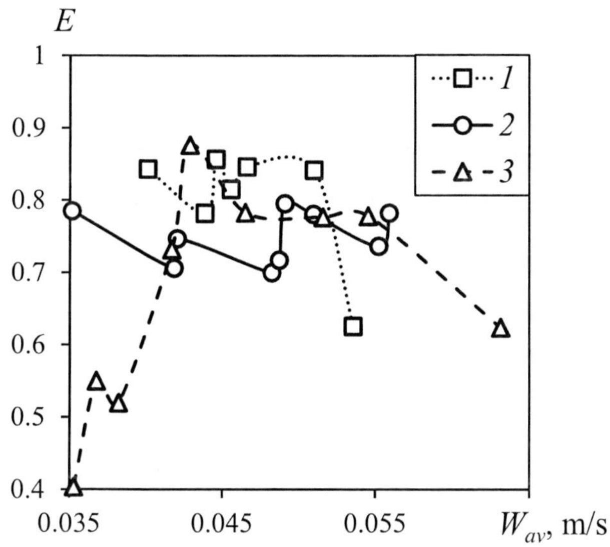

Figure 2. The dependency of change in the efficiency of water-oil emulsion demulsification on its average rate in a cylindrical settling tank with various separation elements: *1* – corrugated plates, with an orientation of corrugations at an angle of 45°, *2* – corrugated plates, with an orientation of corrugations at an angle of 135°, *3* – cross-corrugated plates.

The efficiency of water-oil emulsion demulsification, when using the corrugated plates with different orientation of corrugations, located at an angle of 45°,135° and the cross-corrugated plates, was on average equal to 74.1%. In the course of studies, it was also found that for the various separation elements, there are certain rate ranges W_{av}, at which the maximum values of water-oil emulsion demulsification efficiency are achieved. Thus, for the corrugated plates with different orientation of corrugations, located at an angle of 45°, the most optimal range is 0.04–0.05 m/s. while for the corrugated plates, located at an angle of 135° – 0.05–0.055 m/s and for the cross-corrugated plates – 0.04–0.055 m/s. This is due to the fact that the occurrence of different vortex forces depends not only on the water-oil emulsion rate but also on the shape of separation elements. Therefore, at the same rate, but when using different types of separation elements, different processes can be carried out – additional mixing of water-oil emulsion or coalescence of water and oil drops (figure 2).

According to the results of the study, the highest values of oil removal from the pipe are achieved when it is sampled from the outlet line, which is executed in the device at the highest elevation. When sampling the oil from the line of output *7* (figure 1) and using the corrugated plates with different orientation of corrugations, located at an angle of 45°, 135° and the cross-corrugated plates, its mass fraction in the light phase on average was equal to 89.4% (figure 3).

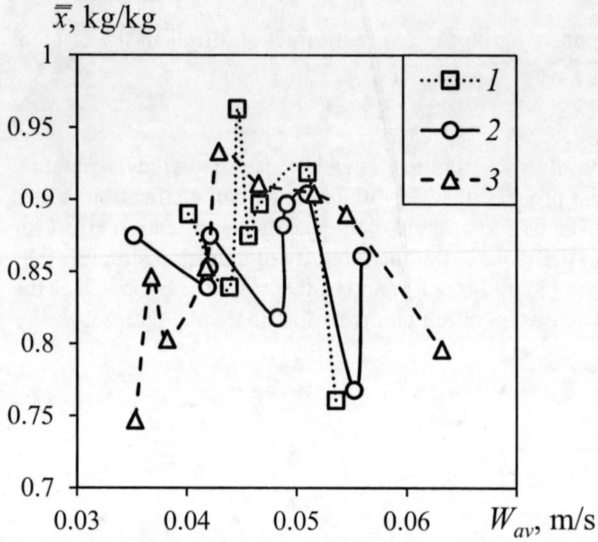

Figure 3. The dependency of oil mass fraction in the light phase, sampled from the first line of output, on the average flow rate when there are various separation elements: *1* – corrugated plates, with an orientation of corrugations at an angle of 45°, *2* – corrugated plates, with an orientation of corrugations at an angle of 135°, *3* – cross-corrugated plates.

When sampling the oil from the line of output *8* (figure 1) and using various separation elements, its mass fraction in the light phase on average was equal to 49.1% (figure 4). The obtained results also show that the required, correct length of pipe was chosen for the oil floating to the emulsion surface inside the pipe, which was separated from the water-oil emulsion when moving along the separation elements by the emulsion.

The efficiency of demulsification in the pipe when using the corrugated plates with different orientation of corrugations, located at different angles of 45° ,135° and the cross-corrugated plates, was on average equal to 80.1, 74.9 and 67.2%, respectively. At the same time, the greatest impact of changes in the average rate of water-oil emulsion in the pipe on the efficiency of its demulsification was recorded when using the cross-corrugated plates, due to a more complex profile of them in comparison with the corrugated plates, located at different angles in the pipe: at an average rate within the range of 0.035–0.039, 0.04–0.054 and 0.055–0.063 m/s, the demulsification efficiency was on average equal to 49.1, 78.9 and 63.1%, respectively. It should be noted that when the rate is less than 0.039 m/s, the low efficiency is caused by the fact that the water-oil emulsion only envelops the cross-corrugated plates, but there is no intensive separation of two phases. When the rate is within the range of 0.04–0.054 m/s, there is an optimal number of vortices, since the emulsion flows around the cross-corrugated surfaces, thereby achieving maximum efficiency values. When the rate is more than 0.055 m/s, the demulsification efficiency decreases due to a large number of vortex points and a greater force than for the rates within the range of 0.05–0.054 m/s, resulting in mixing of two phases (figure 2).

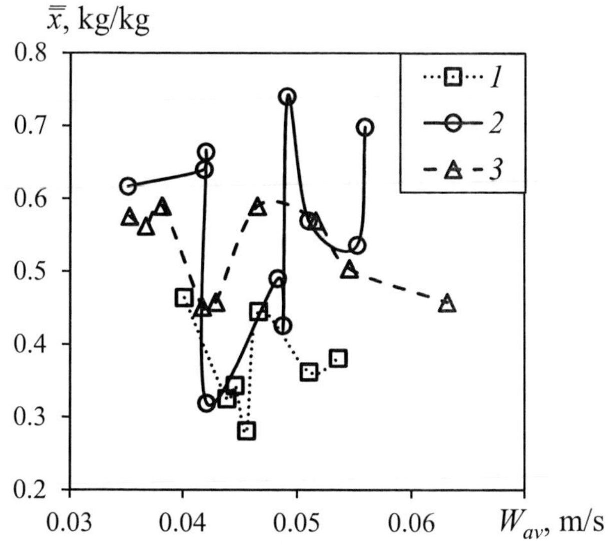

Figure 4. The dependency of oil mass fraction in the light phase, sampled from the second line of output, on the average flow rate when there are various separation elements: *1* – corrugated plates, with an orientation of corrugations at an angle of 45°, *2* – corrugated plates, with an orientation of corrugations at an angle of 135°, *3* – cross-corrugated plates.

When using the corrugated plates with different orientation of corrugations, located at an angle of 45°, 135° and the cross-corrugated plates, the oil mass fraction in the light phase was on average equal to 87.2, 95.2 and 85.3%, respectively. The maximum values of oil in the light phase, sampled from the first line of output *7* (figure 1), when using the corrugated plates with different orientation of corrugations, located at an angle of 45°, 135° and the cross-corrugated plates were recorded at the following rates: 0.044–0.051, 0.042–0.051 and 0.042–0.052 m/s, respectively (figure 3).

When using the corrugated plates with different orientations of corrugations, located at an angle of 45°, 135° and the cross-corrugated plates, the oil mass fraction in the light phase was on average equal to 37.1, 57.1 and 52.8%, respectively. The maximum values of oil in the light phase, sampled from the first line of output *8* (figure 1), when using the corrugated plates with different orientation of corrugations, located at an angle of 45°, 135° and the cross-corrugated plates were recorded at the following rates: 0.047–0.053, 0.049–0.056 and 0.046–0.054 m/s, respectively (figure 4).

4. Conclusions

The conducted studies showed that the use of various separation elements inside the settling tanks allows intensifying the process of water-oil emulsion demulsification. The separation elements, considered in this paper, namely, the corrugated plates with different orientation of corrugations, located at an angle of 45°, 135° and the cross-corrugated plates, allow a performance of demulsification process with minimum 62% efficiency at the average rate of emulsion movement of more than 0.04 m/s. In the course of studies, it was found that the most effective of the considered separation elements are the corrugated plates, located in the apparatus at an angle of 45°, allowing the process of water-oil emulsion demulsification with an average efficiency of 80.1% at the movement rate within the range of 0.04–0.053 m/s. It was also found that for the various separation elements, there are the most effective ranges of emulsion rates, at which the maximum values of water-oil emulsion demulsification efficiency are achieved: for the corrugated plates with different orientation of corrugations, located at an angle of 45°, 135° and for the cross-corrugated plates, these ranges correspond to the rates of 0.04–0.05, 0.05–0.055 and 0.04–0.055 m/s, respectively. The developed separation elements are of practical significance for the petroleum industry since the demulsification process of water-oil emulsion in most of the used settling tanks is ten times slower. Thus, the developed separation elements allow us to speed up the process of demulsification of water-oil emulsions significantly.

References

[1] Ong C, Shi Y, Chang J, Alduraiei F, Wehbe N, Ahmed Z and Wang P 2019 *Sep. Purif. Technol.* **227** 115657
[2] Tseng H-H, Wu J-C, Lin Y-Chen and Zhuang G-L 2018 *J. Membr. Sci.* **559** 148–58
[3] Liu C, Li K, Tian X, Zhao G, Chen Y and Mahlalela B M 2019 *J. Pet. Sci. Eng.* **172** 527–37
[4] Eskin D and Vikhansky A 2019 *Chem. Eng. Res. Des.* **151** 261–9
[5] Zolfaghari R, Fakhru'l-Razi A, Abdullah L, Elnashaie S and Pendashteh A 2016 *Sep. Purif. Technol.* **170** 377–407
[6] Tanudjaja H J, Hejase C A, Tarabara V V, Fane A G and Chew J W 2019 *Water Res.* **156** 347–65
[7] Zeng Q, Wang Z, Wang X, Zhao Y and Guo X 2016 *J. Pet. Sci. Eng.* **145** 83–94
[8] Berman Y and Tamir A 2003 *Chem. Eng. Sci.* **58** 2089–102
[9] Merentsov N A, Balashov V A, Bokhan S A, Nefed'eva E, Tezikov D A and Groshev V V 2019 *IOP Conf. Ser.: Earth Environ. Sci.* **224** 012041
[10] Golovanchikov A B, Balashov V A and Merentsov N A 2017 *Chem. Petrol. Eng.* **53** 10–3
[11] Ochowiak M, Matuszak M, Włodarczak S, Ancukiewicz M and Krupińska A 2017 *J. Environ. Manage.* **189** 22–8
[12] Shah M T, Parmar H B, Rhyne L D, Kalli C, Utikar R P and Pareek V K 2019 *J. Pet. Sci. Eng.* **182** 106352
[13] Zhang H, Liang Y, Yan X, Wang B and Wang N 2017 *J. Pet. Sci. Eng.* **156** 366–72
[14] Jiang W., Chen Y., Chen M., Liu X, Liu Y, Wang T and Yang J 2019 *Sep. Purif. Technol.* **211** 259–68
[15] Dmitriev A V, Dmitrieva O S, Dang S V and Nguen V L 2019 *Chem. Pet. Eng.* **55** 329–35
[16] Li B and Stenstrom M K 2014 *Water Res.* **65** 40–63
[17] Kou J, Chen Y and Wu J 2019 *Chem. Eng. Process.: Process Intensification* 107725
[18] Cambiella A, Benito J M, Pazos C and Coca J 2006 *Chem. Eng. Res. Des.* **84** 69–76
[19] Dmitriev A V, Zinurov V E, Vinh D and Dmitrieva O S 2019 *E3S Web Conf.* **110** 01026
[20] Zinurov V, Sharipov I, Dmitrieva O and Madyshev I 2020 *E3S Web Conf.* **157** 06001

Acknowledgments

The research was conducted with funding from the RF President`s grant project No. MK-616.2020.8.

Analysis of modern environmental management mechanisms for energy providers

M S Ivanitckii, M M Sultanov, E V Kuryanova and A A Konstantinov

Department of Heat Power Engineering and Heat Engineering, Branch of the National Research University "MPEI" in Volzhsky, Lenin Avenue, 69, Volzhsky, 404110, Russia

Abstract. In the paper the essential changes in Federal Law № 219-FZ "About changes in Federal Law "On protection of the environment" and the separate acts of the law of Russian Federation" under 21.07.2014 are concerned, the mechanisms and new energy provider environment management principles are analyzed. The main field of environment protection realization from heat power facilities' negative effects according to the new Russian environmental regulation requirements is presented. The new environmental norming and control principles for industrial facilities based on the best available technologies are determined. Their implementation issues taking into account development financing are analyzed.
Keywords: best available technologies, complex ecological permit, technological norming.

1. Introduction

The heat power source of thermal power plant and energy provider boiler house is fuel consumption. Depending on the fuel type the negative effect on the environment is caused by combustion gases: nitrogen oxides, sulfur dioxide, carbon monoxide and solid particles forming during combustion [1]. The polluting agents' threat is intensified by their interaction with other agents inside the atmosphere.

Therefore, to gradually reduce the negative influence on environment the norms of specific pollutant emissions are established and the technological norming based on the principles of best available technologies to secure the norms is required [2].

According to federal law №219 FZ under 21.07.2014 "About changes in Federal Law "On protection of the environment" and the separate acts of the law of Russian Federation" the new principles of state environment protection policy are developed to mitigate the pollutant negative influence on the environment. These changes provide for the introduction of a technological norming system, based on the basics of best available technologies, grouping all industrial facilities into four categories by pollution concentration, state environmental audit for facility information acquisition and systematization. Depending on equipment power all energy facilities are categories I and II, which have a higher pollutant effect index.

2. Methods

In the new environmental legislation [1-3] there are new details on the importance of environmental management programs for categories I, II and III objects ecological control according to the established regulations. Acquired data should be stored by the facility and represent the necessary measures to reduce pollutant negative influence on the environment.

IOP Conf. Series: Materials Science and Engineering 1035 (2021) 012015 doi:10.1088/1757-899X/1035/1/012015

To apply the environmental legislation all category I industrial objects must get a complex ecological permit. Such a permit that implements an ecological efficiency increase program, include the basics of standard and technological pollution indicators determination [2].

When introducing norming and ecological control of industrial facilities based on the best available technologies category I objects should be equipped with automated pollution continuous control and accounting devices. Applying the devices allows quick and efficient important ecological decisions and reducing the number of over permitted pollutions. Such a control system allows taking measures for following the environment protection legal requirements. However, many industrial facilities are not ready for such financial expenses. The challenge of using such systems to prevent negative effects is the lack of well-qualified personnel.

Depending on the year of power generating installation putting into service and its performance the various pollutant emission reduction measures are taken. Overall there are three installation groups: ones put into service before 1982, in the period from 1982 to 31.12.2000 and after 01.01.2001. For each group, there is a certain range of technological indexes for various types of fuel consumption and installation power. I.e., the state ecological norming with pollution requirement strengthening is introduced.

The indexes proposed in the information and technical reference book is different from the GOST R 50831-95 standard indexes for the worse. Nevertheless, for category I installations, put into service before 31.12.1981 the requirement strengthening is not reasonable. It is firstly because of their deprecation, obsoletion, low energy efficiency and need for capital improvement or decommissioning [1].

3. Results and Discussion

For installation, put into service from 01.01.1982 the limits of the technological parameters for best available technologies particle and nitrogen oxide emissions are set specific emissions, which could be obtained by using the current technologies. The standard emissions are presented in plots 1-5.

Within the framework of industrial ecological control of large combustion installations pollutant, the marker materials should be normed [1].

For gas fuel, thermal power plants the main normed marker materials are nitrogen oxides and carbon oxide.

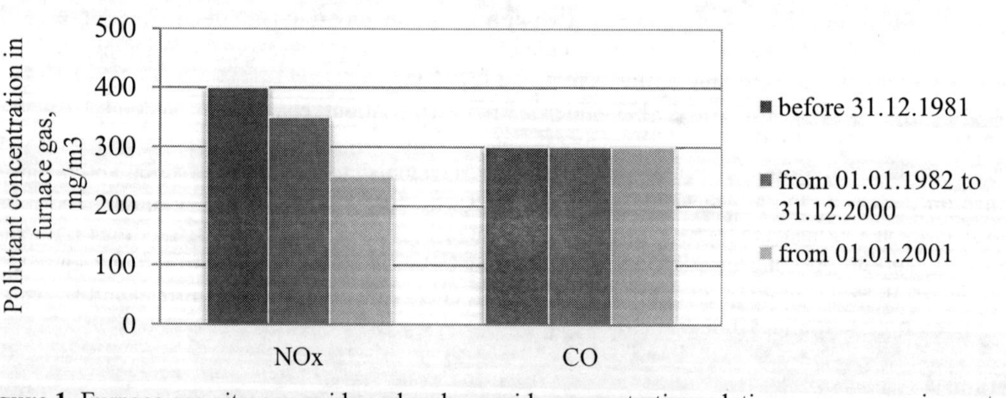

Figure 1. Furnace gas nitrogen oxide and carbon oxide concentration relations on year equipment put into operation for gas fuel consumption.

For liquid fuel thermal power plants, the main normed marker materials are nitrogen oxides, sulfur dioxide, carbon oxide, fuel oil ash (in terms of an equivalent amount of vanadium).

IOP Conf. Series: Materials Science and Engineering 1035 (2021) 012015 doi:10.1088/1757-899X/1035/1/012015

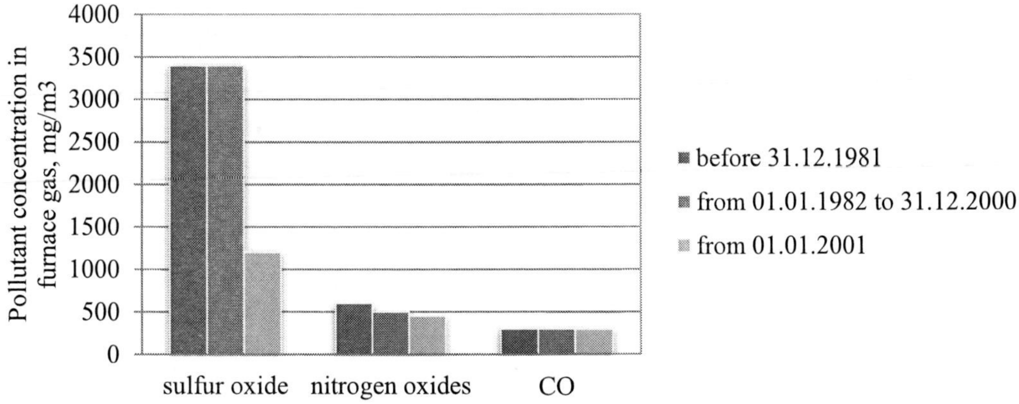

Figure 2. Furnace gas sulfur, nitrogen and carbon oxides concentration relations on year up to 420 t/h capacity equipment put into operation for liquid fuel consumption.

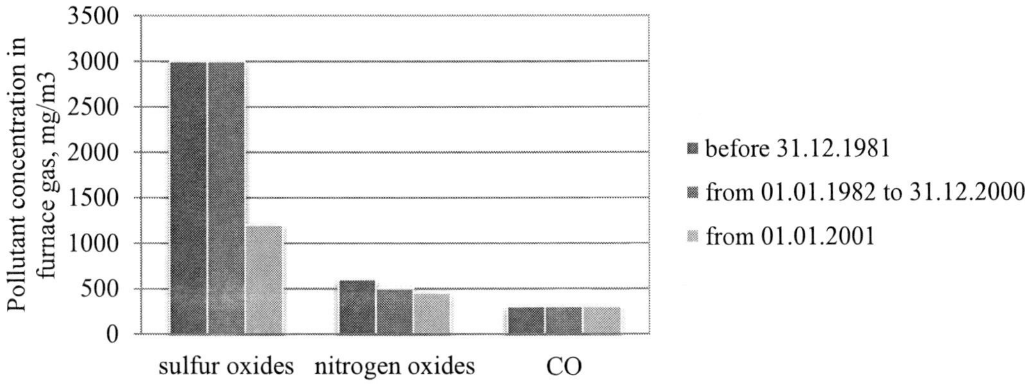

Figure 3. Furnace gas sulfur, nitrogen and carbon oxides concentration relations on year of up to 420 t/h capacity equipment put in operation for liquid fuel consumption.

For solid fuel thermal power plants, the main normed marker materials are nitrogen oxides, sulfur dioxide, carbon oxide, solid fuel ash.

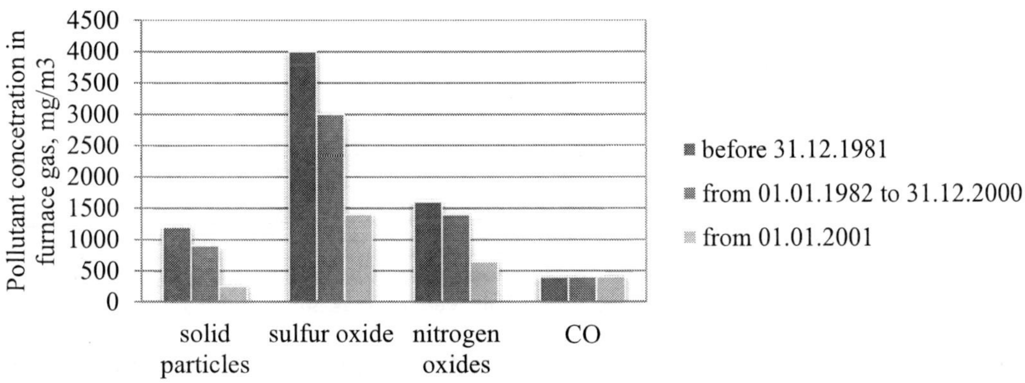

Figure 4. Furnace gas solid particles, sulfur, nitrogen and carbon oxides concentration relations on year up to 420 t/h capacity equipment put into operation for solid fuel consumption.

The analysis of furnace gas pollutants presented in figures 1-5 has shown that depending on the year of putting boiler unit into service, its power and operation mode measures, which do not require sufficient improvements of operating equipment or implementation of a new improved boiler unit, it leads to solid particle, sulfur oxide, nitrogen oxide concentration reduction in furnace gas under the established threshold values of carbon monoxide.

Establishing new specific boiler unit pollutant emission regulations, first of all, is necessary for gradually thermal power plant negative effect reduction and taking into account economic efficiency for the introduction of the best available technologies.

The main purpose of the changes in the regulatory documents firstly is the importance of obsolete equipment modernization at active objects and modern equipment manufacturing meeting all ecological and energy efficiency requirements of energy facilities. To achieve the goal, the best available technologies information and technical reference book have been developed [1].

In the information and technical reference book the structure of thermal power plant generating equipment, which is in operation for more than 30 years, and pollutant ecological indexes are presented. According to the specialists' observations for the last 25 years, thermal power plant gross emissions reduced approximately 3 times and most of the industrial facilities follow the normed indexes of maximum permissible emissions.

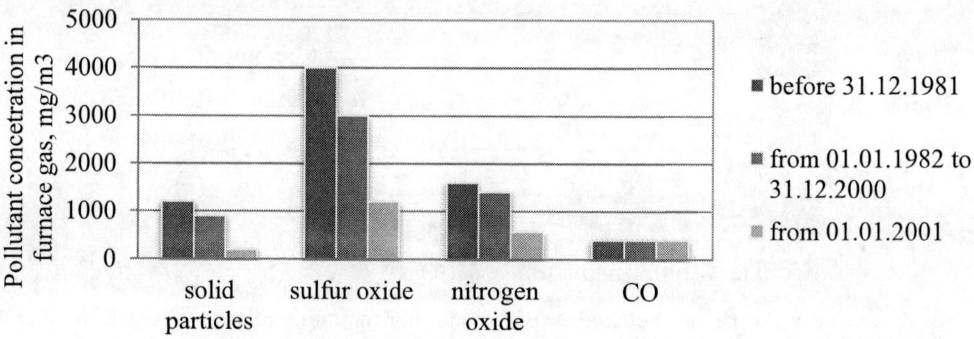

Figure 5. Furnace gas solid particles, sulfur, nitrogen and carbon oxides concentration relations on year up to 420 t/h equipment put into operation for solid fuel consumption.

Energy facility pollutant effect reduction is due to:
- sufficiently increased ratio of combusted gas fuel and decreased consumption of liquid fuel;
- implementation of the efficient measures for zeroing nitrogen and sulfur oxides in thermal power plant boilers;
- implementation of modern dust-collecting equipment on thermal power plants;
- applying combined cycle gas generation equipment.

4. Conclusion

To implement the environment improvement measures in Russian Federation it is necessary to not only develop laws and regulations and design mechanisms for their implementation but also work through the financial support measures for best available technologies implementation.

When implementing such technologies, it is necessary to put together environmental benefits and expenses for the project realization. For energy facilities, there are boiler unit requirements, which consist of plant location, chimney flue height, region climate. Thus there is a situation when one facility does not have a technical capability of best available technologies implementation to reduce pollutants and others, which can afford it, are not motivated [3–6].

Therefore, the changeover to technological norming allows ordering environmental requirements for boiler units and speeding up industry modernization and improvement, which will lead to pollutant reduction of thermal power plants. Thus the planned over the coming years change over to technological pollutant norming makes it possible to increase domestic thermal power plants competitivity versus foreign fuel consumption units.

5. References

[1] ITS 38 2017 *A Technical Reference Guide To The Best Available Technology. Combustion Of Fuels In Large Installations For The Purpose Of Energy Production* M.: Rosstandart

[2] The Law of RF from 21/07/2014 No. 219 federal law 2014 *On Amendments To The Federal Law On Environmental Protection And Separate Legislative Acts Of The Russian Federation Collection Of Legislation Of The Russian Federation* No. 30 4220

[3] PNST 187-2017 *Best Available Technology. Automatic Systems For Continuous Monitoring And Accounting Of Emissions Of Harmful (Polluting) Substances From Thermal Power Plants Into The Atmospheric Air. Primary Requirements* M.: Standartinform

[4] Federal standard R50831-95 *Boiler Installations. Thermal Mechanical Equipment. General Technical Requirements*

[5] 2016 *A Technical Reference Guide To The Best Available Technology. General Principles Of Industrial Environmental Control And Its Metrological Support.* M.: Federal Agency for Technical Regulation of Metrology [electronic version] URL: http:// http://docs.cntd.ru/document/1200143295

[6] SultanovMM, KonstantinovA A, Ivanitckii M S 2017 Environmental aspects of thermal power equipment operation modes optimization *International Journal of Hydrogen Energy* **42** (18) pp 13300-13306

Acknowledgments

The research is funded by Russian Federation public contract No.0720-2020-0025 "Technique development and method analysis for ensuring power system object security and competitiveness based on the digital technologies."

Electrochemical properties of AB_5 type $La_{0.8}Ce_{0.2}Ni_4Co_{0.5}Mn_{0.3}Al_{0.2}$ alloy for unified metal hydride fuel cell

A N Kazakov[1,3], V Yu Bodikov[1,2] and D V Blinov[1,2]

[1]Joint Institute for High Temperatures, Krasnokazarmennaya, 17A, Moscow, 111116, Russia
[2]National Research Institute "MPEI", Krasnokazarmennaya, 14, Moscow, 111250, Russia

[3]Corresponding author: Kazakoffalex09@gmail.com

Abstract. Metal hydride materials are widely applied for electrochemical energy storage. AB_5 type alloys are the main anode material for Ni-MH batteries. Reversible hydrogen uptake and release provides good opportunities to use metal hydrides for alkaline fuel cell applications. Metal hydrides have good activation performance, high hydrogen sorption/desorption rates, catalytic activity in alkaline solution. Current work presents the first experimental results of AB_5 type metal hydride electrodes, measured in a battery mode. The electrochemical properties of the prepared electrodes, such as activation performance, maximum discharge capacity, and high rate dischargeability are investigated in detail.

1. Introduction

Hydrogen is an excellent secondary energy carrier to fit into a global sustainable energy strategy in the next few decades of the 21st century, which withstands the challenges of climate change, depletion of fossil fuels, rising pollution. Hydrogen would play a significant role in medium and long-distance road and rail vehicles; in coastal and international shipping; in air transport; and for longer-term seasonal storage on electricity grids relying mainly on local renewable energy sources and feedstocks [1]. New technologies of electricity/heat production including fuel cells, hydrogen production from renewables and its storage are necessary for the transition to new energy systems based on resource-saving renewable energy technologies. Electrocatalytically produced hydrogen is a perspective way to convert and store renewable energy [2]. Hybrid solar/wind/hydrogen based fuel cell power generation systems can effectively convert solar and wind energy into electric energy for remote areas [3].

Metal hydrides are known as good solid-state hydrogen storage materials with high volumetric capacity, high sorption/desorption kinetics and cyclic stability. They are used in stationary hydrogen storage [4-6], hydrogen compression [7-9], thermal energy storage [10-12] as well as electrochemical energy storage [13-16]. Metal hydrides are widely used in Ni-MH batteries, which have a number of advantages for practical applications, such as high specific energy density, resistance to overcharge/overdischarge, fast charge-discharge kinetics, environmental friendliness and safety. A number of studies [17-19] showed practical applicability of metal hydrides as anode material for alkaline fuel cells. Metal hydride based fuel cells can operate at near ambient conditions and do not require precious catalysts. Their advantage is the possibility to be recharged in two ways: either electrochemically or by hydrogen pressure [19]. Thus, the metal hydride FC is the fuel cell with built-in charge storage, which can operate reversibly as an electrolyzer, and hydrogen is absorbed by the metal hydride alloy, i.e. as unitized regenerative fuel cells.

Metal hydride is formed by reversible interaction of individual metal or multicomponent alloy with hydrogen. Mostly multicomponent intermetallic compounds are used as anode materials. Typically, intermetallic compounds include AB_5 ($LaNi_5$), AB_2 ($ZrMn_2$), AB (TiFe) and super-lattice AB_{3-4} (rare earth-Mg-Ni) alloys. For a binary alloy, metal A is a hydride-forming metal, while metal B is a transition or non-transition metal that does not form stable hydride at normal conditions [20].

Despite the great efforts to develop new hydrogen absorbing materials with enhanced working characteristics [16, 21-29] AB_5 type alloys still stay the most popular materials due to their high surface catalytic activity, easy activation and good cycle stability. Partial replacement of constituent metals can significantly improve the electrochemical properties of metal hydride electrodes. Rare earth metals are hydride forming metals and provide hydrogen uptake, Ni is a key element for catalytic activity, Co improves cyclic stability, Mn increases surface activity, Al and Fe provide corrosion resistance [29-31].
We choose $La_{0.8}Ce_{0.2}Ni_4Co_{0.5}Mn_{0.3}Al_{0.2}$ alloy as a potential metal hydride electrode and investigate it in terms of crystal structure, hydrogen storage and electrochemical properties in a battery mode.

2. Experimental details
The AB_5 type $La_{0.8}Ce_{0.2}Ni_4Co_{0.5}Mn_{0.3}Al_{0.2}$ intermetallic compound is prepared in an arc furnace under argon atmosphere. The stoichiometric amount of initial metals is remelted three times to achieve high homogeneity of the alloy. After melting, the surface cleaned sample is wrapped in a molybdenum foil and sealed in a stainless-steel tube sample holder with titanium powder under the argon atmosphere. The sample is annealed at 1223 K for 20 hours and quenched into ice water.

XRD pattern of the annealed alloy is obtained from X-ray diffractometry. The measurement conditions are as follows: the step size is 0.02° and the exposition time is 1 s. The 2θ angles scanned are in the range from 15 to 80. Crystal structure and phase composition are performed using Jana2006 and Crystal Impact Match software using JCPDS PDF-2 Data Base.

Hydrogen sorption properties are investigated by the volumetric technique in a Sievert's type apparatus. Thermodynamic parameters are calculated from Van't Hoff plot.

In the case of electrochemical studies, metal hydride electrodes are prepared from a powder fraction with particle sizes of 40-60 μm. The electrodes in the form of a tablet with a diameter of 10 mm and a thickness of 1 mm with a total weight of 0.5 g are prepared by cold pressing at a pressure of 25 MPa for 3 minutes. Three electrodes samples named #1, #2 and #3 are manufactured to investigate them in different conditions. The prepared tablets are pressed between the foam nickel plates attached to the current collector.

Electrochemical measurements are carried out in a standard three-electrode cell with a 6M KOH solution as an electrolyte and Hg/HgO reference electrode. A commercial $Ni(OH)_2$ with a larger capacity is used as the counter electrode. The experiments are carried out on a multichannel potentiostat R-20X8. Activation of the metal hydride electrodes includes 10 charge cycles performed at a current density of 100 mA/g for 4.5 hours and a subsequent discharge at the same current density to a cut-off potential -0.6 V for samples #1 and #3. Sample #2 is discharged at current density 60 mA/g. In addition, sample #3 sandwiched between two steel perforated plates to increase the stability of the working electrode. The relaxation time between the charge and the electrode discharge is 0.5 hours. After activation, hydrogen diffusion rate and the high rate discharge ability performance in the range of discharge current densities of 100 - 1000 mA/g are measured.

3. Results and discussion

3.1. Crystal structure and PCT properties

The crystal structure of $La_{0.8}Ce_{0.2}Ni_4Co_{0.5}Mn_{0.3}Al_{0.2}$ alloy is well defined as a single-phase and belongs to the hexagonal $CaCu_5$ type. XRD pattern shows a high homogeneity of the obtained sample after annealing (figure 1a.). Unit cell parameters are defined as a = 0.5032 nm, c = 0.4035 nm. PCT isotherms are measured from 295 to 333 K (figure 1b). Equilibrium pressure and hydrogen content at 295 K are 0.032 MPa and 1.1 wt.% H2. Equilibrium pressure is suitable for use as anode material. Thermodynamic parameters obtained from van't Hoff plot are $\Delta H = 30.1$ kJ/mol and $\Delta S = 88$ J/(mol·K).

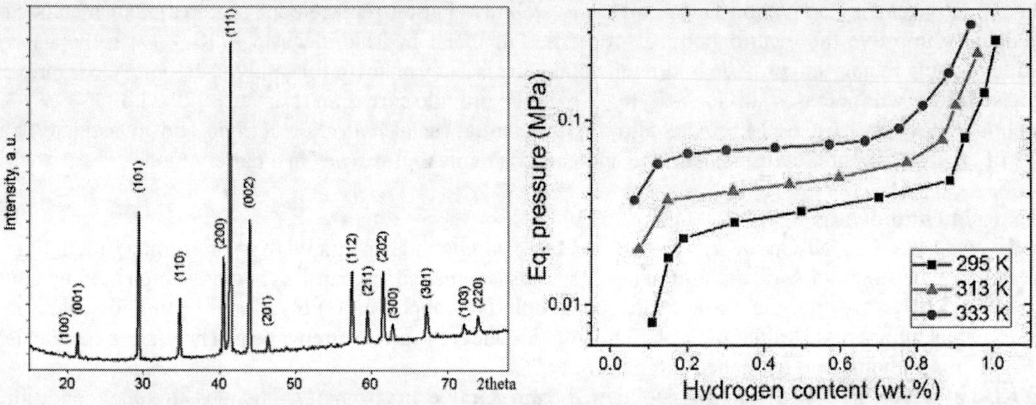

Figure 1. Crystal structure and PCT properties of $La_{0.8}Ce_{0.2}Ni_4Co_{0.5}Mn_{0.3}Al_{0.2}$ alloy.

3.2. Electrochemical activation performance

Different charge/discharge current densities are applied for the activation of metal hydride electrodes [32-34]. We apply two common ways charge/discharge 100/100 mA/g and 100/60 mA/g currents to evaluate the effect of activation current densities on electrochemical properties of metal hydride alloy. Discharge curves at different discharge current rates are shown in figure 2.

Discharge current density effect is observed for first cycles and equilibrium potential. However, there is no significant influence on the maximum discharge capacity. Lower equilibrium potentials for samples 1 and 3 are related to charge transfer limitations at higher current densities. In addition, equilibrium potential and activation performance of sample 3 are strongly affected by applied perforated plates.

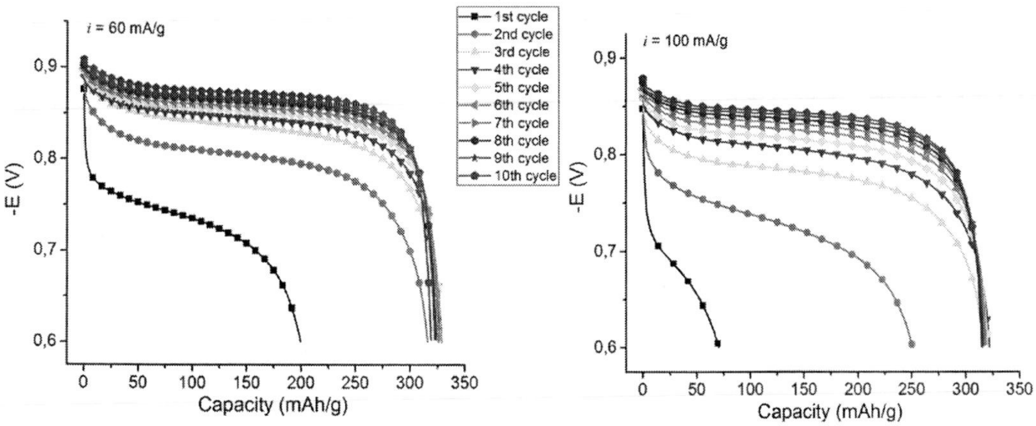

Figure 2. Discharge performance of $La_{0.8}Ce_{0.2}Ni_4Co_{0.5}Mn_{0.3}Al_{0.2}$ metal hydride electrode.

As shown in figure 3, the samples reach the maximum discharge capacity at the third-fifth cycle. Sample #3 is characterized by later activation, but more stable discharge capacity in the next activation cycles. Samples 1 and 2 show a small reduction of discharge capacity during activation, which is likely because of metal hydride particle decrepitation and precipitation on the bottom of the electrochemical cell. It is well known, that AB_5 type alloys are characterized by a significant change (up to 20–25%) of the unit cell volume during hydrogen sorption – desorption. Internal stresses arising from the expansion of particles lead to the gradual destruction of the electrode and the precipitation of metal hydride particles. In this case, sample 3 stabilized by perforated plates shows higher mechanical stability. The highest discharge capacity at the 10th cycle is equal to $C_{max} = 316.5_{\pm 2.5}$ mAh/g for all samples. The equilibrium discharge potentials of the electrodes slightly decrease as higher current density and perforated plates are applied (figure 4).

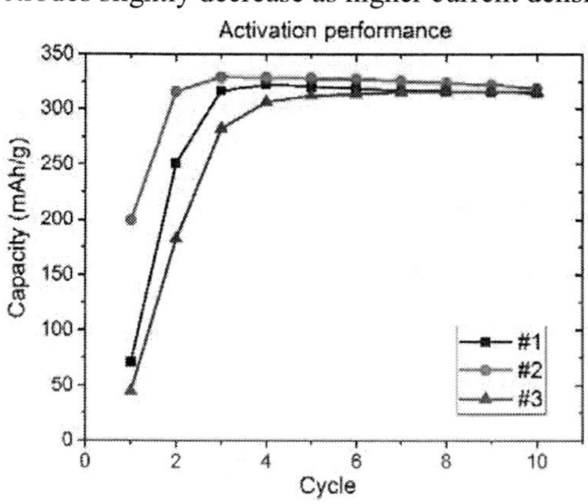

Figure 3. Activation performance of metal hydride electrodes.

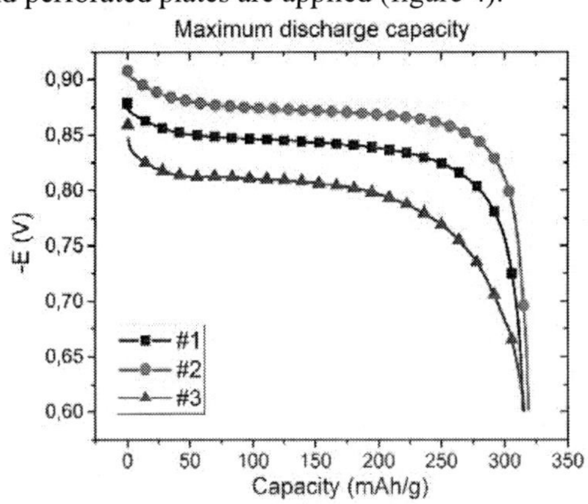

Figure 4. Maximum discharge capacity of metal hydride electrodes.

3.3. Hydrogen diffusion properties

The hydrogen absorption/desorption reactions in alkaline aqueous solutions for metal hydrides involve the following steps: charge transfer step, adsorption and absorption of electroactive species, and mass transfer from the surface to the bulk of metal hydride. In parallel to the hydrogen diffusion reaction, recombination processes described by the Tafel and the Heyrovski reactions are also occurred on the metal hydride surface [37, 38].

The above-mentioned reactions show a complex combination of different processes on the metal hydride electrode during electrochemical reactions. Thus, the estimation of hydrogen diffusion by traditional non-steady state methods is quite difficult. However, diffusion kinetics affect greatly the electrode's rate capability and accurate determination of diffusion constants are great of interest from both practical point of view and understanding of the mechanisms of insertion processes. In comparison with traditional techniques used to estimate hydrogen diffusion in metal hydride electrodes such as constant potential step methods [32, 33, 39-42], electrochemical impedance spectroscopy [38, 43-45], potentiostatic intermittent titration technique has two major advantages: (i) applying of small potential steps makes it possible to linearize the electrochemical system responses to the potential perturbation and (ii) the potential dependence of the diffusion coefficients of intercalated species can be evaluated continuously in the whole potential domain of interest [46].

Current transients at a low amplitude of 20mV are measured in the potential range from -0.80 V to - 0.96 V. Each potential is applied until the constant current is achieved. Experimental curves plotted $It^{1/2}$ vs log t coordinates are time invariant and should provide a horizontal section at short time domains, which corresponds to a typical Cottrellian behavior of diffusion [47]. The horizontal line shows an expected form of $It^{1/2}$ vs log t plot for ideal Cottrellian diffusion (figure 5a).

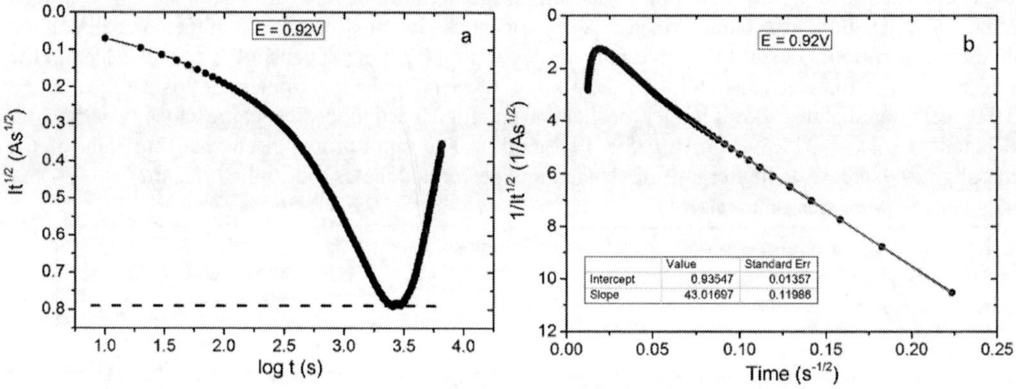

Figure 5. Cottrellian time invariant $It^{1/2}$ vs log t plot (a) and linear fitting analysis of $It^{1/2}$ vs $1/t^{1/2}$ plot (b) at E = 0.92 V for metal hydride electrode.

It is assumed that the deviation of the It1/2 vs log t curves from ideal conditions is because of the action of resistances in the electrochemical system that are not related to diffusion in the solid (for example, charge transfer and surface resistance at the interface), which impede the expected Cottrell behavior at short time domains. However, for small ΔE used in PITT, the measured current I(t) must "obey" Ohm's law (1):

$$I(t) = \Delta E/\left[\left(R_s + R_{sf} + R_{ct}\right) + (DC_{int})^{-1}l(\pi D t)^{1/2}\right] \quad (1)$$

where R_s – solution resistance, R_{sf} – surface resistance, Rc_t – charge transfer resistance, D – diffusion coefficient, C_{int} – differential capacity, l – diffusion path length.

To calculate the effective hydrogen diffusion coefficient, we transform equation (1) to the following form [14, 48]:

$$\frac{1}{I\sqrt{t}} = \frac{R_\Sigma}{\Delta E\sqrt{t}} + \frac{l\sqrt{\pi}}{\Delta Q\sqrt{D}} \quad (2)$$

where R_Σ - the sum of all non-diffusion resistances, ΔQ – integral capacity of the electrode at a given potential.

The linear dependence of experimental data on $It^{1/2}$ vs $1/t^{1/2}$ plot makes it possible to calculate the hydrogen diffusion coefficient for each potential step using equation (2). There is a negligible charge capacity change at low potentials under 0.86 V, which corresponds to less than 1% of the total charge. Highest hydrogen diffusion coefficient values approximately 10^{-10} - 10^{-11} cm²/s are obtained at minimum studied state of charge 5-7% and belong to α-solid solution formation in metal hydride electrodes. As the state of charge increases hydrogen diffusion coefficients become lower decreasing to $1.2\cdot10^{-13}$ - $5.5\cdot10^{-14}$ cm²/s (figure 6), because of the saturation of β-hydride phase. Such behavior of hydrogen diffusion also noted for AB_3 [25, 48] and AB_2 [14] type metal hydride electrodes and assumed to be typical for hydride forming alloys [49]. At the start of measurement, sample 3 shows lower hydrogen diffusion rate because of spatial restrictions caused by perforated plates.

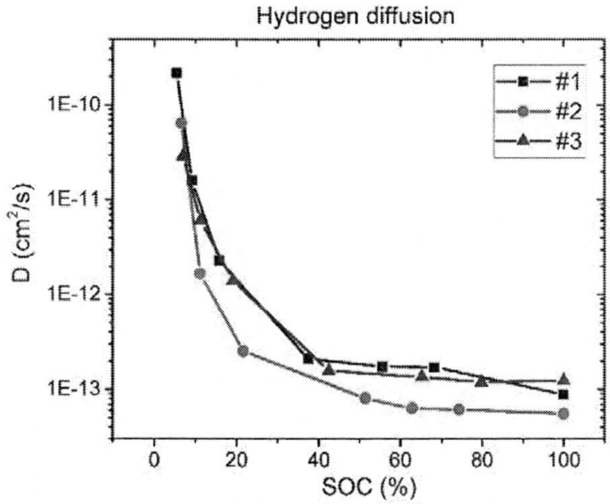

Figure 6. Hydrogen diffusion coefficient vs state of charge.

4. Conclusions

The hydrogen storage and electrochemical properties of metal hydride materials based on $La_{0.8}Ce_{0.2}Ni_4Co_{0.5}Mn_{0.3}Al_{0.2}$ are studied in terms of crystal structure, hydrogen storage properties, activation performance, maximum discharge capacity and hydrogen diffusion. Hydrogen gravimetric capacity in gas-phase reaction reaches 1.1 wt.% H_2, which corresponds to the electrochemical capacity of 316.5 mAh/g. Perforated steel plates provide good mechanical stability of electrode with no significant effect on the hydrogen diffusion and maximum discharge capacity. However, they greatly affect equilibrium potential and may cause side reactions at higher discharge currents. The hydrogen diffusion coefficient is rather low not exceeding $1.2\cdot10^{-13}$ cm²/s in a saturated region. Further improvement of metal hydride composition and its electrochemical properties is necessary for use as anode material.

References

[1] Andrews J and Shabani B 2012 Re-envisioning the role of hydrogen in a sustainable energy economy

International Journal of Hydrogen Energy **37(2)** pp 1184-1203 DOI: 10.1016/j.ijhydene.2011.09.137
[2] Emonts B, et al. 2017 Re-energizing energy supply: Electrolytically-produced hydrogen as a flexible energy storage medium and fuel for road transport *Journal of Power Sources* **342** pp 320-326 DOI: https://doi.org/10.1016/j.jpowsour.2016.12.073
[3] Fathabadi H 2017 Novel standalone hybrid solar/wind/fuel cell power generation system for remote areas *Solar Energy* **146** pp 30-43 DOI: https://doi.org/10.1016/j.solener.2017.01.071
[4] Lototskyy M, et al. 2019 Metal hydride hydrogen storage tank for fuel cell utility vehicles *International Journal of Hydrogen Energy* DOI: https://doi.org/10.1016/j.ijhydene.2019.04.124
[5] Srinivasan S S and Demirocak D E 2017 *Metal Hydrides used for Hydrogen Storage, in Nanostructured Materials for Next-Generation Energy Storage and Conversion: Hydrogen Production, Storage, and Utilization* ed. Y P Chen, S Bashir and J L Liu Springer Berlin Heidelberg pp 225-255
[6] Kubo K, Kawaharazaki Y and Itoh H 2017 Development of large MH tank system for renewable energy storage *International Journal of Hydrogen Energy* **42(35)** pp 22475-22479 DOI: https://doi.org/10.1016/j.ijhydene.2017.06.048
[7] Tarasov B P, et al. 2018 Cycling stability of RNi5 (R = La, La+Ce) hydrides during the operation of metal hydride hydrogen compressor *International Journal of Hydrogen Energy* **43(9)** pp 4415-4427 DOI: https://doi.org/10.1016/j.ijhydene.2018.01.086
[8] Yartys V A, et al. 2016 Metal hydride hydrogen compression: recent advances and future prospects *Applied Physics A* **122(4)** p 415 DOI: 10.1007/s00339-016-9863-7
[9] Lototskyy M V, et al. 2014 Metal hydride hydrogen compressors: A review *International Journal of Hydrogen Energy* **39(11)** p 5818-5851 DOI: http://dx.doi.org/10.1016/j.ijhydene.2014.01.158
[10] Manickam K, et al. 2019 Future perspectives of thermal energy storage with metal hydrides *International Journal of Hydrogen Energy* **44(15)** pp 7738-7745 DOI: https://doi.org/10.1016/j.ijhydene.2018.12.011
[11] Feng P, et al. 2019 Techno-economic analysis of screening metal hydride pairs for a 910 MWhth thermal energy storage system *Applied Energy* **242** pp 148-156 DOI: https://doi.org/10.1016/j.apenergy.2019.03.046
[12] Nyamsi S N, Lototskyy M, and Tolj I 2018 Selection of metal hydrides-based thermal energy storage: Energy storage efficiency and density targets *International Journal of Hydrogen Energy* **43(50)** pp 22568-22583 DOI: https://doi.org/10.1016/j.ijhydene.2018.10.100
[13] Wijayanti I D, et al. 2020 Hydrides of Laves type Ti–Zr alloys with enhanced H storage capacity as advanced metal hydride battery anodes *Journal of Alloys and Compounds* **828** pp 154354 DOI: https://doi.org/10.1016/j.jallcom.2020.154354
[14] Volodin A A, et al. 2019 Study of hydrogen storage and electrochemical properties of AB2-type Ti0.15Zr0.85La0.03Ni1.2Mn0.7V0.12Fe0.12 alloy *Journal of Alloys and Compounds* **793** pp 564-575 DOI: https://doi.org/10.1016/j.jallcom.2019.03.134
[15] Zhang L, et al. 2018 A new AB4-type single-phase superlattice compound for electrochemical hydrogen storage *Journal of Power Sources* **401** pp 102-110 DOI: https://doi.org/10.1016/j.jpowsour.2018.08.080
[16] Ouyang L, et al. 2017 Progress of hydrogen storage alloys for Ni-MH rechargeable power batteries in electric vehicles: A review *Materials Chemistry and Physics* **200** (Supplement C) pp 164-178 DOI: https://doi.org/10.1016/j.matchemphys.2017.07.002
[17] Hu W-K and Noréus D 2009 Metal hydrides as bi-functional catalysts for hydrogen generation and oxidation in reversible MH-air fuel cells *Electrochemistry Communications* **11(11)** pp 2212-2215 DOI: https://doi.org/10.1016/j.elecom.2009.09.033

[18] Chartouni D, et al. 2002 Metal hydride fuel cell with intrinsic capacity *International Journal of Hydrogen Energy* **27(9)** pp 945-952 DOI: http://doi.org/10.1016/S0360-3199(01)00186-0
[19] Folonari C, et al. 1980 Metal hydride fuel cells: A feasibility study and perspectives for vehicular applications *Journal of the Less Common Metals* **74(2)** pp 371-378 DOI: https://doi.org/10.1016/0022-5088(80)90175-7
[20] Hirscher M, et al. 2020 Materials for hydrogen-based energy storage – past, recent progress and future outlook *Journal of Alloys and Compounds* **827** pp 153548 DOI: https://doi.org/10.1016/j.jallcom.2019.153548
[21] Liu Y, et al. 2011 Rare earth–Mg–Ni-based hydrogen storage alloys as negative electrode materials for Ni/MH batteries *Journal of Alloys and Compounds* **509(3)** pp 675-686 DOI: https://doi.org/10.1016/j.jallcom.2010.08.157
[22] Takasaki T, et al. 2013 Cobalt-free nickel–metal hydride battery for industrial applications *Journal of Alloys and Compounds* **580** pp S378-S381 DOI: https://doi.org/10.1016/j.jallcom.2013.01.092
[23] Young K-H, Chang S, and Lin X 2017 C14 Laves Phase Metal Hydride Alloys for Ni/MH Batteries *Applications Batteries* **3(3)** pp 27
[24] Yan H, et al. 2017 Investigations on AB3-, A2B7- and A5B19-type LaYNi system hydrogen storage alloys *International Journal of Hydrogen Energy* **42(4)** pp 2257-2264 DOI: https://doi.org/10.1016/j.ijhydene.2016.09.049
[25] Volodin A A, et al. 2016 Phase-structural transformations in a metal hydride battery anode La1.5Nd0.5MgNi9 alloy and its electrochemical performance *International Journal of Hydrogen Energy* **41(23)** pp 9954-9967 DOI: https://doi.org/10.1016/j.ijhydene.2016.01.089
[26] Verbovyts'kyi Y V and Zavalii I Y 2016 New Metal-Hydride Electrode Materials Based On R1–x Mg x Ni3–4 Alloys for Chemical Current Sources *Materials Science* **51(4)** pp 443-456 DOI: 10.1007/s11003-016-9861-0
[27] Denys R V, Yartys V A, and Webb C J 2012 Hydrogen in La2MgNi9D13: the role of magnesium *Inorg. Chem.* **51(7)** pp 4231-8 DOI: 10.1021/ic202705u
[28] Zhang Y-h, et al. 2012 Electrochemical performances of the as-melt La0.75−xMxMg0.25Ni3.2Co0.2Al0.1 (M = Pr, Zr; x = 0, 0.2) alloys applied to Ni/metal hydride (MH) battery *International Journal of Hydrogen Energy* **37(19)** pp 14590-14597 DOI: https://doi.org/10.1016/j.ijhydene.2012.07.020
[29] Liu Y, et al. 2011 Advanced hydrogen storage alloys for Ni/MH rechargeable batteries *Journal of Materials Chemistry* **21(13)** pp 4743-4755 DOI: 10.1039/C0JM01921F
[30] Zhao X and Ma L 2009 Recent progress in hydrogen storage alloys for nickel/metal hydride secondary batteries *International Journal of Hydrogen Energy* **34(11)** pp 4788-4796 DOI: https://doi.org/10.1016/j.ijhydene.2009.03.023
[31] Feng F, Geng M, and Northwood D O 2001 Electrochemical behaviour of intermetallic-based metal hydrides used in Ni/metal hydride (MH) batteries: a review *International Journal of Hydrogen Energy* **26(7)** pp 725-734 DOI: https://doi.org/10.1016/S0360-3199(00)00127-0
[32] Zhang W, et al. 2009 Study on kinetics and electrochemical properties of low-Co AB5-type alloys for high-power Ni/MH battery *Electrochimica Acta* **54(4)** pp 1383-1387 DOI: https://doi.org/10.1016/j.electacta.2008.09.019
[33] Giza K 2013 Electrochemical studies of LaNi4.3Co0.4Al0.3 hydrogen storage alloy *Intermetallics* **34** (Supplement C) pp 128-131 DOI: https://doi.org/10.1016/j.intermet.2012.11.014
[34] Yao Q R, et al. 2014 Electrochemical properties of the LaNi4.5Co0.25Al0.25 hydrogen storage alloy in wide temperature range *Journal of Alloys and Compounds* **606** (Supplement C) pp 81-85 DOI: https://doi.org/10.1016/j.jallcom.2014.04.026
[35] Tliha M, et al. 2014 Kinetic and thermodynamic studies of hydrogen storage alloys as negative

electrode materials for Ni/MH batteries: a review *Journal of Solid State Electrochemistry* **18(3)** pp 577-593 DOI: 10.1007/s10008-013-2300-3

[36] Borzenko V I, et al. 2019 Hydrogen sorption properties of metal hydride beds: Effect of internal stresses caused by reactor geometry *International Journal of Hydrogen Energy* **44(12)** pp 6086-6092 DOI: https://doi.org/10.1016/j.ijhydene.2019.01.052

[37] Castro B E and Milocco R H 2007 State estimation in Volmer–Heyrovsky reactions coupled with sorption processes: Application to the hydrogen reaction *Journal of Electroanalytical Chemistry* **604(1)** pp 1-8 DOI: https://doi.org/10.1016/j.jelechem.2007.01.002

[38] Zhang B, et al. 2018 Investigations on kinetics properties of hydrogen adsorbing/desorbing reactions for metal hydride electrodes *International Journal of Hydrogen Energy* **43(49)** pp 22427-22437 DOI: https://doi.org/10.1016/j.ijhydene.2018.10.037

[39] Lin J, et al. 2014 High temperature performance of La0.6Ce0.4Ni3.45Co0.75Mn0.7Al0.1 hydrogen storage alloy for nickel/metal hydride batteries *International Journal of Hydrogen Energy* **39(25)** pp 13231-13239 DOI: https://doi.org/10.1016/j.ijhydene.2014.06.112

[40] Liu B, et al. 2012 Microstructures and electrochemical properties of LaNi3.55Co0.2xMn0.35Al0.15Cu0.75(V0.81Fe0.19)x hydrogen storage alloys *International Journal of Electrochemical Science* **7** pp 11966-11977

[41] Zhou W, et al. 2017 Improvement in low-temperature and instantaneous high-rate output performance of Al-free AB5-type hydrogen storage alloy for negative electrode in Ni/MH battery: Effect of thermodynamic and kinetic regulation via partial Mn substituting *Journal of Power Sources* **343** (Supplement C) pp 11-21 DOI: https://doi.org/10.1016/j.jpowsour.2017.01.023

[42] Zhou W, et al. 2017 Low-temperature and instantaneous high-rate output performance of AB5-type hydrogen storage alloy with duplex surface hot-alkali treatment *Journal of Alloys and Compounds* **692** (Supplement C) pp 364-374 DOI: https://doi.org/10.1016/j.jallcom.2016.08.292

[43] Li X, et al. 2012 Electrochemical hydrogen absorbing properties of graphite/AB5 alloy composite electrode *Journal of Alloys and Compounds* **510(1)** pp 114-118 DOI: https://doi.org/10.1016/j.jallcom.2011.09.001

[44] Humana R M, et al. 2012 Electrochemical behavior of metal hydride electrode with different particle size *International Journal of Hydrogen Energy* **37(19)** pp 14966-14971 DOI: https://doi.org/10.1016/j.ijhydene.2011.12.093

[45] Tliha M, et al. 2010 Kinetic behaviour of low-Co AB5-type metal hydride electrodes *Materials Science and Engineering: B* **175(1)** pp 60-64 DOI: https://doi.org/10.1016/j.mseb.2010.07.003

[46] Montella C 2006 Apparent diffusion coefficient of intercalated species measured with PITT: A simple formulation *Electrochimica Acta* **51(15)** pp 3102-3111 DOI: https://doi.org/10.1016/j.electacta.2005.08.046

[47] Aurbach D, Levi M D, and Levi E 2008 A review on the solid-state ionics of electrochemical intercalation processes: How to interpret properly their electrochemical response *Solid State Ionics* **179(21)** pp 742-751 DOI: https://doi.org/10.1016/j.ssi.2007.12.070

[48] Volodin A A, et al. 2015 Hydrogen diffusion in La1.5Nd0.5MgNi9 alloy electrodes of the Ni/MH battery *Journal of Alloys and Compounds* **645** (Supplement 1) pp S288-S291 DOI: https://doi.org/10.1016/j.jallcom.2014.12.201

[49] Young K, et al. 2016 Transmission electron microscope studies in the surface oxide on the La-containing AB2 metal hydride alloy *Journal of Alloys and Compounds* **672** pp 356-365 DOI: https://doi.org/10.1016/j.jallcom.2016.02.182

Hybrid heat supply systems using alternative and renewable energy sources

E G Gasho, A V Govorin and A I Kiseleva

Department of Industrial Heat Power Engineering Systems, National Research University MPEI, Krasnokazarmennaya str., 17, Moscow, Russia

Abstract. The task of creating combined energy sources to cover diverse (thermal, electrical, etc.) loads is not entirely new. In the countries of the European Union (EU), hybrid energy systems for buildings, which combine heating, domestic hot water, as well as air conditioning, have long been widely used. The programs and normative documents adopted in these countries on improving the energy efficiency of heat supply and the use of renewable energy sources (RES) for this have contributed to the widespread introduction of hybrid systems and installations. The study analyzes the potential of non-traditional and renewable sources for energy supply systems of facilities. The world experience of introducing alternative energy sources is considered and the state and implementation of hybrid schemes in Russia is assessed. Hybrid power supply options for facilities for various purposes are proposed. The key problems of active use are examined and ways to solve them are proposed.

1. Introduction

Hybrid energy supply systems are widely used in EU countries. The programs and normative documents adopted in these countries on improving the energy efficiency of heat supply and the use of renewable energy sources (RES) for this have contributed to the widespread introduction of hybrid systems and installations [1]. The share of renewable energy in the EU's energy balance is growing rapidly and should reach 20% by 2020, such a share of renewable energy will be used in the world's energy balance by 2030 (figure 1). Specifically, in the field of cold and heat supply, the share of RES is increasing annually in all countries, without exception.

Although these systems have not yet been in high demand in Russia thermal power plants of various types and capacities were created just to solve such problems, different types of turbines were developed, as well as a set of corresponding circuit solutions. Basically, this concerned large consumers (industrial enterprises, cities), but now the problem of ensuring reliable energy supply to remote and hard-to-reach settlements outside centralized energy supply systems has again become urgent.

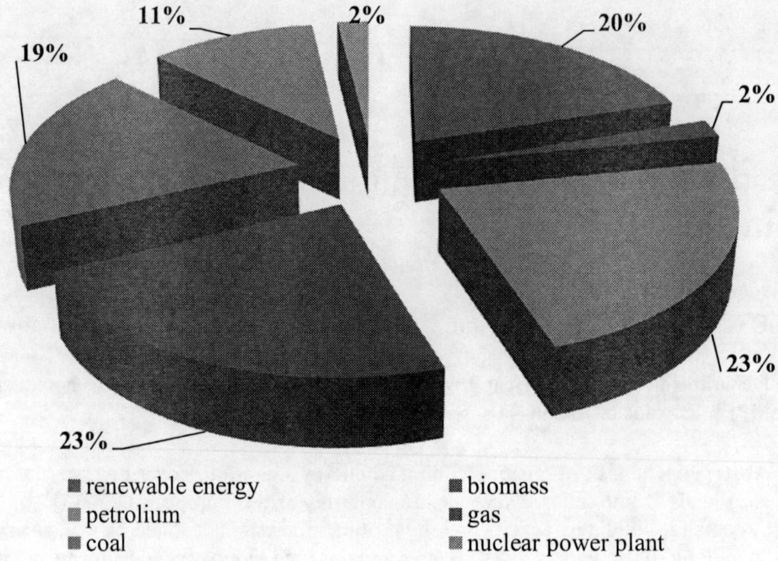

Figure 1. Assessment of the EU energy balance structure by 2030.

For the period 2013-2018, there has been some progress in the development of renewable energy in Russia. The Government of the Russian Federation adopted a number of Resolutions and Orders on the organization of work in the field of renewable energy. However, despite the extensive list of documents designed to ensure the development of renewable energy in Russia, in fact, they are mostly declarative in nature and poorly applied in practice. At the same time, according to Russian experts, the potential for renewable energy in Russia is more than 270 million tons of equivalent fuel equivalent. - about 30% of the total primary energy supply. Unfortunately, the initially established share of renewable energy sources - 4.5% of total electricity production in 2020, was reduced to 2.5%. Today in Russia we are talking about achieving economic growth, the need to develop space with different climatic conditions, corresponding to the construction of large volumes of housing and efficient production capacity [2]. The rational use of combined, hybrid RES-based systems in these cases may well contribute to more efficient solutions leading to various systemic (multiplicative) effects.

The inconsistency of energy output and the mismatch between the places where renewable energy is manifested and the consumed loads (cities) cast doubt on the feasibility of using hybrid energy supply systems. However, at the same time, natural inexhaustibility and environmental friendliness present wide opportunities for the use of heat supply systems based on non-traditional and renewable energy sources.

2. Generation and structure of energy consumption based on non-traditional and renewable energies

China and the USA remain the world leaders in the use of renewable energy sources (table 1). It is worth noting that most countries use the power of non-traditional sources mainly for the generation of electric energy. The largest applications in the energy sector of these countries have been received by solar power plants:
Tengger Desert (China) - 1,500 MW;
Solar Star (USA) - 579 MW;
Topaz (USA) - 550 MW;
Desert Sunlight Solar Farm (USA) - 552 MW.

Solar energy, in the first place, is very popular both in the residential and in the commercial sector [3]. Such demand is explained by rather clear weather in some countries in Europe. For example, according to the Solar Trade Association (STA), the generation of all solar panels in April 2020 reached a peak of 9.68 GW, a record of 32.23 GW was set in Germany. Recent years have shown the high efficiency of solar systems: more often, cases of negative prices for electric energy began to be recorded. However, the high cost of plants and the dependence on weather and time of day are insurmountable barriers to the construction of solar power plants in several countries.

Table 1. Total capacity of generation as of the end of 2017.

Power	1	2	3	4	5
Renewable power capacity (including hydropower)	China	USA	Brazil	Germany	India
Renewable power capacity (not including hydropower)	China	USA	Germany	India	Japan
Bio-power generation	China	USA	Brazil	Germany	Japan
Bio-power capacity	USA	Brazil	China	India	Germany
Hydropower capacity	China	Brazil	Canada	USA	Russia
Hydropower generation	China	Brazil	Canada	USA	Russia
Solar PV capacity	China	USA	Japan	Germany	Italy
Solar PV capacity *per capita*	Germany	Japan	Belgium	Italy	Australia
Concentrating solar thermal power (CSP)	Spain	USA	South Africa	India	Morocco
Wind power capacity	China	USA	Germany	India	Spain
Wind power capacity *per capita*	Denmark	Ireland	Sweden	Germany	Portugal
Heat					
Solar water heating collector capacity	China	USA	Turkey	Germany	Brazil
Solar water heating collector capacity *per capita*	Barbados	Austria	Cyprus	Israel	Greece
Geothermal heat capacity	China	Turkey	Iceland	Japan	Hungary

2.1. Types of renewable energy in Russia

The vast territory and the diversity of climatic zones in Russia provide advantages and favor the use of a wide variety of alternative energy sources. Hydroelectric power plants are the most popular alternative energy source in Russia. Energy production from the existing 200 river hydropower plants reaches 20% of all energy produced in the country.

Solar energy in Russia is used both among the population as the main or alternative source of energy and on an industrial scale [4]. The largest solar plants are SES - "Vladislavovka" (Crym, figure 2), as well as installations in the Samara, Astrakhan, Orenburg regions. The total capacity of all installations is about 400 MW.

Geothermal energy is widespread in Kamchatka, which is facilitated by the abundance of volcanoes. The power of geothermal sources makes it possible to cover about 40% of the consumed power. According to studies, the potential of geothermal sources in Kamchatka is estimated at 5000 MW, but the annual use does not exceed 80 MW.

Figure 2. Solar power station "Vladislavovka".

Other types of alternative energy, such as wind energy, etc., although used in special cases, have not received wide distribution due to lack of investments and gaps in the law.

3. Use of energy supply systems based on res in Russia

Currently, electricity generation in Russia through hybrid systems based on renewable energy sources is less than 1% of the total generation. Despite the low indicators of using the RES power in Russia in comparison with other countries, hybrid energy supply systems are widely used at the following facilities:

- private houses and cottages;
- cottage settlements;
- country estates;
- infrastructure facilities (treatment facilities);
- public institutions (kindergartens, schools);
- public buildings;
- small industrial facilities (workshops);
- small neighborhoods.

For the heat supply of private houses, heat pumps were widely used regardless of the climatic zone. Using a heat pump allows you to heat a house in the cold season, and to cool in the warm season. On the example of an individual residential building (Republic of Buryatia), one can evaluate the effectiveness of using hybrid heat and cold supply systems.

The energy supply scheme of a residential building using a reversible soil heat pump and solar collectors. The source of heat and cold are three wells with a depth of 100 m. Two groups of vacuum solar collectors (figure 3-4): 120 pipes on the roof of the house and 100 pipes in the greenhouse (total 18 m^2). Solar collector has several modes (options) of operation:

1. In the winter (heating period) all 220 pipes work to heat the low circuit of the VT and cover 15-17% of the annual heat energy consumption for heating;

2. From the middle of March (end of the heating season) of the month, 100 tubes installed in the greenhouse switch to heating the soil in the greenhouse and a month later, in mid-April, the garden is planted. At the end of May, all the collectors work to heat the water in the outdoor pool and to heat the wells.

Figure 3. A group of solar collectors for the energy supply of a residential building (location - roof).

Figure 4. A group of solar collectors for the energy supply of a residential building (location - greenhouse).

This combination of energy sources "heat pump and solar collector" is the most common in the Russian Federation. Almost the entire territory of Russia is subject to insolation in one way or another, therefore there are many such examples: a hybrid solar power plant with a capacity of 45 kW in the Krasnodar Territory, a cascade of geothermal pumps in the Omsk Region, a solar hybrid system that provides a high degree of uninterrupted power supply in Rostov-on Don, a geothermal heat pump in the Leningrad region, etc.

4. Main features and problems of active res resources implementation

The successful use of renewable energy sources, heat pumps in Russia still remains a lot of enthusiastic groups, outside of state policy and support, as a result, the country does not realize additional opportunities for the development of the economy and industry.

According to the creators of the above-mentioned projects, the key measures of system support for the development and implementation of renewable energy sources are:
- increase the market by providing subsidies for the purchase of renewable energy technology;
- provision of soft loans for the acquisition and installation of equipment through authorized banks;
- stimulation of the construction of energy-efficient housing with the installation of systems based on renewable energy sources and heat supply units as sources of heat and energy supply;
- applying reduction factors to the local electricity tariff for pilot RES installations (as is done for homes equipped with electric stoves);
- the creation (updating) of domestic regulations, standards, codes of practice, GOSTs, regulating the application of appropriate installations for renewable energy sources in Russian conditions;
- Broad information support for successful projects in various fields and areas.

The key problems that hinder the massive implementation of renewable energy projects are presented in table 2.

Conclusion

Despite the existing obstacles to the active implementation of renewable energy sources, global trends show growth and the possibility of using renewable sources in most countries. Hybrid systems have not yet been appreciated, although their application provides some advantages over traditional systems: a significant reduction in environmental pollution, additional production of electricity in the required volumes in centralized energy supply zones, as well as ample opportunities for foreign investment.

Table 2. Key problems of the active introduction of renewable energy sources and their coordination.

Economic	Technical	Organizational
The high cost of equipment (mainly imported), the absence of any subsidies or benefits for owners who have installed renewable energy systems	The variety of regional climatic, tariff, infrastructural and geological conditions, available sources of renewable energy sources and technical solutions based on different types of renewable energy makes it impossible to create standard projects for the use of renewable energy sources. It is necessary to prepare a thorough and costly feasibility study for each renewable energy project	Insufficient regulatory framework, technical regulations, standards, lack of empirical data on Russian projects, albums of best practices
The relatively low cost of fuel (especially main gas) with relatively high electricity tariffs negatively affects the economics of projects	Restrictions on the use of renewable energy by climatic conditions in the regions of the middle zone and the north of Russia (lower working temperature requires additional calculations and justifications, especially when reconstructing existing heating systems)	Limited domestic production and the lack of Russian components (high-quality compressors, wind turbines, batteries)
High interest rates on loans for expensive equipment	The heterogeneity of the renewable energy potential in the regional context, both in terms of natural (water bodies, soil, geothermal heat), and technogenic sources (drains, ventilation emissions, etc.)	Lack of service support for many manufacturers
Low profitability of renewable energy projects as is, the need for non-standard technical solutions (for heat storage, use of low-grade heat)	The need to ensure high reliability of heat supply from renewable energy sources in case of possible interruptions in the power supply may require both an independent second input of electricity and the installation of redundant backup equipment on fuel, possible restrictions on the capacity of the supply centers and power lines, especially in rural areas	The complexity and opacity of obtaining electric power for the operation of renewable energy in energy retail organizations

References

[1] Gasho E G 2020 Integrated approach to assessment of impact of beyond-design modes on operation of steam supply systems *Safety and Reliability of Power Industry* **12(1)** pp 10-17

[2] Carvalho D, Guardia E, Marangon Lima J. 2019 Technical-economic analysis of the insertion of PV power into a wind-solar hybrid system *Solar Energy* **191** pp 530-539

[3] Eyerer S, Schifflechner C, Hofbauer S, Bauer W, Wieland C, Spliethoff H 2020 Combined heat and power from hydrothermal geothermal resources in Germany: An assessment of the potential renewable and sustainable energy reviews **120** pp 34-49

[4] Li X Min Z, Zhu P 2018 Storage and Recycling of Interfacial Solar Steam Enthalpy Joule 2 pp 2477-248

Acknowledgments

This article was supported by a grant from RSF, project number 16-19-10568.

Heat consumption of apartment buildings during the heating period

D N Kitaev, A V Cheremisin, Z S Hasanov, A A Mikhailov

Voronezh State Technical University, Moscow Avenue, 14, Voronezh, 394026, Russia

Abstract. The work is devoted to the search for methods for calculating the heat consumption of multi-apartment residential buildings for heating and hot water supply, which allows obtaining the closest results to the actual ones. Three current methods of calculating heat consumption and one canceled, but often used in heat supply contracts, were used. Based on a comparison of the actual data obtained by the common-house heat energy metering devices and the calculated values, the possible deviations in the heat consumption of multi-apartment residential buildings of different floors, area and year of construction, ranging from 49.73 to 177.77%, were found. An economic assessment of the consequences of using the most inadequate methods has been carried out. It was found that when using methods with extreme error values, payments for the heat supply service from one house can differ by 3.6 times.

1. Introduction

In the period from 2009 to the present, great importance is given to energy conservation issues. Federal laws were adopted in the field of energy conservation, heat supply, water supply and sanitation, and the "Rules for the provision of public services to consumers" were approved. Much attention from the administrations of the constituent entities of the Russian Federation is paid to the quality control of heat supply services to consumers [1,2]. An important issue is the streamlining of the system of payments for consumed thermal energy [3,5].

Currently, in most cases, the calculation of the consumer of thermal energy with the supplier is carried out by metering devices (ODPU). But there are a number of situations in which the calculation method is used. Heat supply agreements usually indicate a specific calculation method. The applied technique should have the property of adequacy, i.e. to obtain results comparable with the instrument [4,6]. The choice of the most appropriate methodology for calculating the consumed thermal energy is an important task, the solution of which allows balancing the interests of the consumer and supplier.

2. Materials and methods

As a calculation method, several techniques can be used. In this paper, the following calculation methods were considered: method 1 - CR 373.1325800.2018; Method 2 - Resolution of the Government of the Russian Federation No. 306 as amended on September 29, 2017; Method 3 - Order of the Department of Housing and Utilities and Energy of the Voronezh Region No. 239 dated 12.12.2014; Method 3 - MDS 41.4.2000. These techniques are often used in the implementation of contractual relations between the consumer and the supplier of thermal energy in the territory of the Voronezh region. The fundamental similarity of all the considered methods is the use of specific values of the heat flux per unit area or volume. Some methods use corrective (correction) coefficients for the outdoor temperature, infiltration, heat loss of the home heating systems. In all methods, design climatological data (CR 131.13330.2018) and actual, as well as the duration of the heating period, can be used. Obviously, using different combinations of data will lead to different results.

According to the methods presented above, calculations were carried out and heat consumption values were obtained for the heating period 2018-2019 and for the year for 58 residential buildings of various floors located in the city of Voronezh.

3. Results

Table 1 presents the results of calculations of heat consumption for the needs of hot water supply for a building having the following characteristics: commissioning year - 2008; the number of storeys - 16; total area - 12503.7 m^2; The design heat consumption for heating according to the technical passport is 0.966 Hcal/h. Figure 1 shows the annual heat consumption of the building for the needs of hot water.

Table 1. Heat consumption of a residential building for the needs of hot water supply, Hcal.

period of the year	Month	Methodology 3		Methodology 1				Actual readings	Corrected methodology
		CR 131.13330	fact. temp.	CR 131.13330		fact. temperature			
				calculation	Enlarged standard	calculation	Enlarged standard		
unheated	April	5,293	4,058	4,992	6,758	3,981	5,011	8,019	7,909
	May	20,215	20,215	16,733	19,035	17,205	18,917	36,873	29,877
	June	17,362	17,362	16,07	18,826	14,967	18,011	37,885	29,628
	July	10,014	10,014	8,341	10,551	8,514	10,56	20,144	17,008
	August	20,006	20,006	16,578	18,209	16,057	18,991	30,996	32,15
	September	18,478	18,478	15,547	18,315	16,48	19,002	37,287	29,489
	October	7,408	3,087	6,346	7,873	2,644	3,474	5,229	5,002
heating	October	18,328	25,080	15,699	18,474	21,482	25,281	35,421	32,513
	November	29,903	29,903	25,613	30,142	25,613	30,142	43,522	38,766
	December	28,938	28,938	24,787	29,170	24,787	29,170	48,026	37,516
	January	29,903	29,903	25,613	30,142	25,613	30,142	48,122	38,766
	February	27,009	27,009	23,135	27,225	23,135	27,225	36,442	35,015
	March	29,903	29,903	25,613	30,142	25,613	30,142	42,166	38,766
	April	19,292	21,221	16,525	19,447	18,177	21,391	29,470	27,511
Total	unheated	98,776	93,220	84,607	99,567	79,848	93,966	176,433	151,063
	heated	183,276	191,957	156,985	184,744	164,422	193,495	283,169	248,853

The data presented in table 1 allow us to conclude that when using the calculation method in various combinations of design and actual data, the results are underestimated, on average, only 58% of the actual. The use of the regional standard for the consumption of hot water and the actual climatological data on the number of days in the heating period make it possible to obtain underestimated values on average by 37.9%.

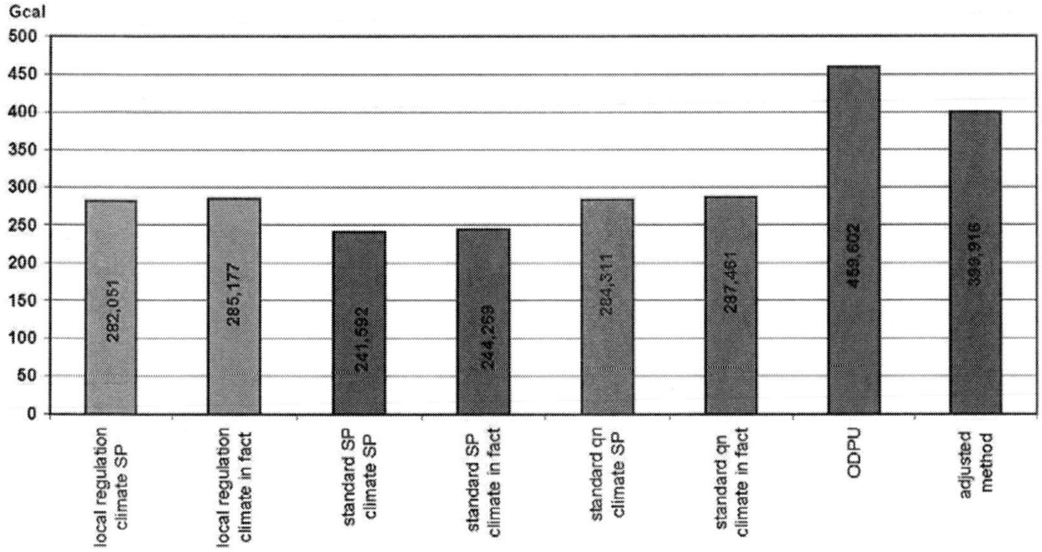

Figure 1. Heat consumption for domestic hot water according to all standards.

It can be seen from the calculation results that the data using the local standard and actual climatological data are almost identical to those obtained using the specific aggregated indicator of the average heat consumption for hot water supply per consumer and the climatological data for the joint venture (the difference is less than 1 Hcal for the whole year).

The formula for calculating the average heat consumption for the hot water supply of the design procedure for autonomous heat supply sources of the latest edition (2018) contains the old requirement for hot water temperature of 55 ° C. The current value is 65 ° C. Using the old value leads to underestimated results of the calculation of heat consumption. The heat loss by the risers and heated towel rails of the hot water supply system is estimated by a factor of 1.2 and does not depend on the type of system, place of installation, type of insulation. The previously used methods took into account the "degree of improvement" and the coefficient was taken in the range of 1.1-1.35. The maximum value of the coefficient was related to the case of the presence of hot water networks after the central heating station on the consumer's balance, which is practically not found in modern systems. The last column of table 1 presents the heat consumption values obtained by the adjusted method taking into account the temperature of 65 °C and the coefficient value of 1.3. Using the proposed recommendations to improve the calculation accuracy leads to a decrease in the difference for the year on average from 37.9 % to 13.08 %. However, the actual value is higher than the estimated value, and the difference is significant.

Table 2 presents the results of calculating the heat consumption of the building under consideration for the needs of heating and hot water supply for the heating period. Figure 2 in the form of a diagram shows the total values of heat consumption for heating and hot water supply.

From figure 2 and the data in table 2, it follows that the closest to the actual results are observed when using method 3 with climatology data adopted by CR. In this case, the difference is - 30.501 Hcal for the heating period. The calculation results when using method 3 in any combination of climatological data are not significantly different: with actual climatological data (difference -96.327 Hcal) and taking into account full months (difference - 192.646 Hcal). The largest deviations are observed when using method 2 with actual data on climatology (the difference is 708,092 Hcal, i.e. more than two times less) and method 4 with climate data on the joint venture (the difference is 1,605.01 Hcal, more than twice as much).

Table 2. Heat consumption of a residential building for heating needs, Hcal.

Month	Methodology 1		Methodology 2				Methodology 3			Methodology 4		Actual readings
			CR 131.13330	actual temperatures	design load		CR 131.13330	actual temperatures	for full months	CR 131.13330	actual temperatures	
	CR 131.13330	actual temperatures			CR 131.13330	actual temperatures						
10	56,849	34,365	45,586	27,556	120,612	72,908	137,944	188,766	225,067	173,209	104,702	125,93
11	145,987	157,279	117,064	126,119	309,727	333,684	225,067	225,067	225,067	444,794	479,197	204,54
12	181,085	176,870	145,209	141,829	384,192	375,250	225,067	225,067	225,067	551,731	538,889	259,04
1	205,672	201,075	164,925	161,238	436,356	426,602	225,067	225,067	225,067	626,643	612,635	264,07
2	183,583	154,079	147,212	123,553	389,491	326,894	225,067	225,067	225,067	559,341	469,447	220,63
3	156,472	115,419	125,472	92,552	331,973	244,873	225,067	225,067	225,067	476,740	351,657	216,06
4	50,995	2,347	40,892	1,882	108,192	4,979	150,044	165,049	225,067	155,373	7,150	92,55

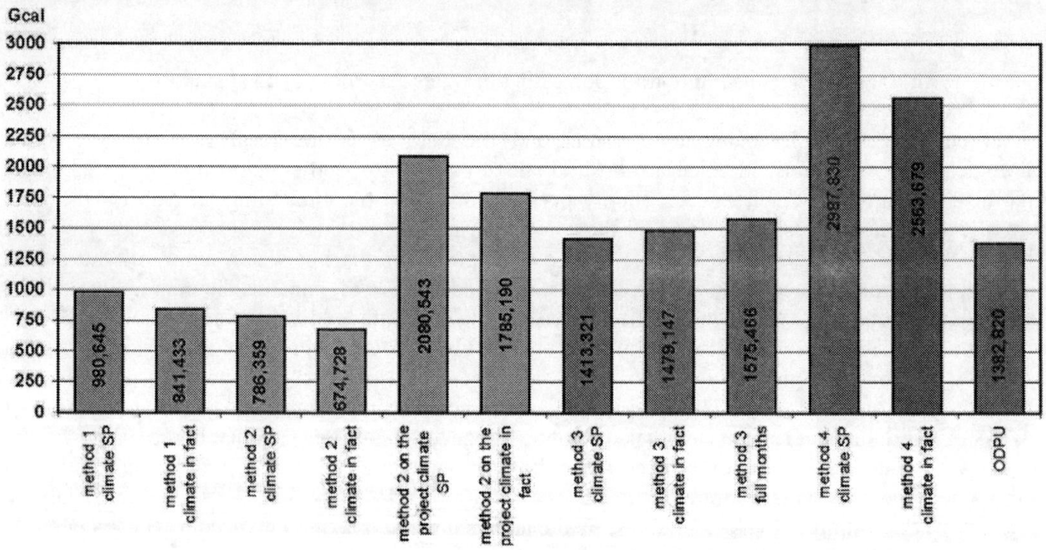

Figure 2. Heat consumption for heating.

According to the results of calculations for all considered buildings, the following can be indicated. Method 1 in any combination of climatological data gives underestimated results, and significantly. Method 4 (currently canceled) gives, as a rule, significantly overestimated results.

The maximum discrepancies ranged from 0.452 to 2.16. When using different methods, the calculation results in comparison with the actual heat consumption can differ by more than two times, both smaller and larger.

Of the currently existing methods, 3, in combination with climatology data for CR, gives in most cases the most adequate results.

Table 3 presents the results of calculating the cost of heat consumed for a building with the maximum deviations of the calculation methods for heating and hot water supply for the heating period 2018-2019 and for the year, taking into account the heat energy tariff for extreme values of the possible ranges. The values of the difference between the received value of each method and the actual value are also given.

Table 3. Cost of consumed thermal energy, rub.

min	methodology 3	max	Metering device	difference, million rubles		
				min	methodology 3	max
for the heating period						
1,408963	2,704715	5,389323	2,822268	1,413	0,118	-2,567
in a year						
1,552292	2,872045	5,548507	3,121155	1,569	0,249	-2,427

The usc of inadequate methods for calculating heat consumption can lead to significant losses for both the consumer and the supplier in monetary terms. Using an example of a building with a maximum difference in results (see table 3) obtained by calculation methods, it can be seen that using the minimum heat consumption will result in the supplier losing 1.191 million rubles, and using a method that receives the most overstated results will result in an overpayment of 2.567 million rubles for one heating period. Using the most appropriate methodology, as follows from table 4, can also lead to a shortfall in profit by the supplier.

4. Conclusions

Existing methods for determining the heat consumption of residential buildings for heating and hot water supply using the calculation method are considered. The calculations of the total heat consumption of multi-unit residential buildings for heating and hot water supply for the heating period 2018-2019 were carried out. and for the whole year. Based on a comparison of the actual data obtained from common-use heat energy metering devices and calculated values, the possible deviations in the heat consumption of multi-unit residential buildings of different floors, area and year of construction, ranging from 49.73 to 177.77%, were found. It has been established that the methodology for calculating the consumed heat, taking into account local specific standards for heat consumption for heating and hot water supply in combination with climatology data according to CR 131.13330.2018, allows you to get the closest values to the actual ones, but the results are underestimated by an average of 11.8%. The calculation of the possible economic consequences of errors in the values of heat consumption obtained by the calculation method is carried out. When using a methodology that obtains minimal results, it is possible to receive payments from one residential building in the amount of 1.191 million rubles. For the heating period, and the maximum - an overpayment of 2.567 million rubles.

References

[1] Chicherin S V 2017 Planning the magnitude of the loads of district heating systems, taking into account the features of modern heat consumption *Bulletin of the Irkutsk State Technical University* **21(6)** pp 103-10

[2] Kitaev D N 2013 The error in calculating the temperature graph of the heating network when using indicators of heating devices *Industrial energy* **7** 34-7

[3] Kitaev D N, Schukina T 2017 Calculation of actual heat consumption of the industrial site *Urban planning. Infrastructure. Communications* **3(8)** 40-5

[4] Kitaev D N, Filshina I U 2019 Temperature graphs of heating networks in the conditions of Voronezh *High technology in the construction complex* **1** 78-83

[5] Frolov Yu V, Filshina Iu, Nedobezhkin D O and Bogdanov V M 2019 Methods for calculating heat consumption for heating multi-apartment residential buildings *Urban planning. Infrastructure. Communications* **1(14)** 34-9
[6] Belykh S V, Potapov A A 2019 Improving the accuracy of calculating the heat consumption of hot water in autonomous heat supply systems *High technology in the construction complex* **2** 44-9
[7] Blinkova E S, Ryzhkov O AND 2019 Heat consumption for hot water supply of a residential building *Urban planning. Infrastructure. Communications* **3(16)** 12-5

IOP Conf. Series: Materials Science and Engineering 1035 (2021) 012019 doi:10.1088/1757-899X/1035/1/012019

Virtual simulator of a boiler room

E V Kocharyan, V V Shaposhnikov[1] and A S Macharashvili

Kuban State Technological University, Moskovskaya Str., 2, Krasnodar, 350072, Russia

[1]Corresponding author: shaposhnikov.valentin@gmail.com

Abstract. The use of virtual simulators, computer models of equipment and production processes, is already a necessity in the modern world. Teachers of KubSTU have developed and tested a virtual simulator of a hot-water boiler for theoretical training and development of skills in the work of both future and current thermal power engineers. This simulator is not only an effective learning tool but also the basis for creating a remote control system for a complex production facility, which is a boiler room.

1. Introduction

Currently, there are many computer simulators for training boiler-house personnel and students for working with heat-power equipment, including training in emergency situations. However, most of these simulators are presented in the form of thermal diagrams or flat schematic images of boilers and other auxiliary equipment. Such programs can improve the skills of students, as well as gain an understanding of the relationships between the various elements of the boiler room (or the boiler itself) and their influence on each other. However, without moving to the most realistic visual display of the equipment under study using three-dimensional models, it will not be possible to achieve truly effective training in accordance with modern requirements. Moreover, the use of such simulators becomes obligatory for a number of industrial enterprises.

Teachers of the Kuban State Technological University have developed a specialized virtual simulator for a hot water boiler (VSHWB) that meets modern requirements [1]. The VSHWB training complex is designed to develop skills for working with the equipment of a hot water boiler room, as well as studying individual elements and equipment. The visual image and placement of elements of the boiler room are identical to real solutions (figures 1, 2).

This makes it possible to simplify and accelerate the transition from a theoretical study of boiler plants and auxiliary equipment to practical experience. The calculation of the main parameters of the heat power equipment and of the boiler room thermal circuit is based on mathematical dependencies and fundamentally coincides with the real characteristics of the facility, but it uses certain assumptions and limitations. These assumptions allow us to speed up the system operation. This happens while maintaining the correspondence between the data obtained in the model and during the operation of the existing facility in real conditions.

At the same time, the acquisition of primary operating experience takes place in virtual mode, saving the working time of the personnel, ensuring absolute safety when committing incorrect or erroneous actions that are unacceptable at a real facility.

Figure 1. Photo of a real boiler house.

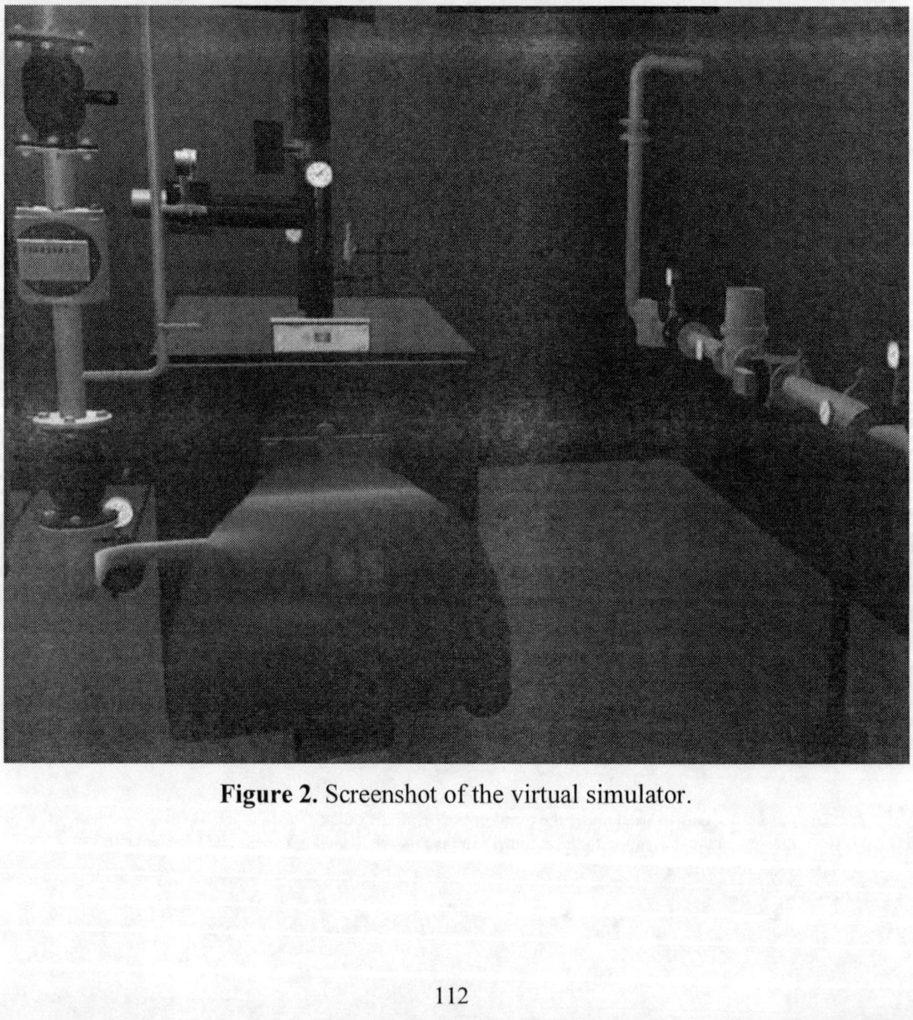

Figure 2. Screenshot of the virtual simulator.

IOP Conf. Series: Materials Science and Engineering 1035 (2021) 012019 doi:10.1088/1757-899X/1035/1/012019

2. Research object description

The object of modeling is a boiler room. The boiler house has two main boilers with a capacity of 418 kW and one summer boiler with a capacity of 154 kW. The heat load is set at 600 kW for heating and 100 kW for hot water supply (HWS). The main fuel is natural gas with an inlet pressure of 25 kPa. The reserve fuel is diesel. It is fed by gravity from a tank installed outside the boiler room. Smoke removal is carried out due to stack draft through a chimney which is 15m high. A chemical water treatment system is provided at the boiler room due to an automatic system based on Na cation exchange. The circulation of the coolant in the supply and return piping is provided by a group of network pumps.

3. Methodology

The mathematical model that establishes the relationship between the control members, control objects, as well as indicating devices, in general, is as follows:

$$F_{ij}=(\sum K_i)\cdot S_j, \qquad (1)$$

where F_{ij} is the desired function, which can be either instrument readings or certain operations, for example, turning on the boiler or opening the valve;

K_i, S_j are input parameters, which are used as calculated values (temperature, pressure, etc.), and logical coefficients (1 - “on”; 0 - “off”).

It is obvious, that the model (1) contains both additive and multiplicative components, which allows one to take into account both quantitative and qualitative characteristics of the described processes. The proposed model is effective due to its simplicity and accuracy acceptable for practical calculations. Calibration of the model and its verification was carried out on the basis of a real-life hot-water boiler, which served as a prototype when creating the simulator. The calculated characteristics showed good convergence with real data, and the calculation error is permissible for educational purposes.

4. Simulator functionality

Functionally designed virtual simulator consists of the following sections:

1. Theoretical study of the following documents: hot water and steam boiler designs, designs of main and auxiliary equipment of the boiler room, instructions for the operators of the boiler room, safety rules for the operation of thermal power plants.
2. Practical study of structural elements of a boiler room taking into account their location in the boiler room. The effectiveness of training is provided by visual identity to the real equipment used.
3. Virtual study of the start and stop modes of the boiler room. Testing the knowledge of the start and stop mode of the boiler room.
4. Study of the personnel actions in emergency situations. Testing of knowledge of emergency procedures.

In the theory section, the user is provided with the necessary theoretical information for the successful study of the basic regulatory documents and safety rules. This knowledge is required for the personnel working with thermal power plants. In addition, brief information on the basic designs of boilers and auxiliary equipment is given [2].

After this, we provide the possibility of intermediate control of the studied material. Control is carried out using theoretical tests for each subsection.

In the start and stop simulator there are two modes - “Study” and “Exam”.

In the “Study” mode, the sequence of operations is strictly regulated and is accompanied by different prompts. These prompts appear in the upper corner of the screen when it is necessary to perform any actions. The learner's task is to remember the sequence of operations in order to start and stop the boiler.

The sequence of operations depends on the initial state. The simulator provides the following initial conditions:

1. Initial start-up of the boiler room after installation. In this mode, all pipelines require filling. It is necessary to start the “summer” boiler and the corresponding “summer” network pump.
2. Starting a gas-fired boiler room in summer. In this mode, all heating mains are filled with water, the gas pipeline is empty. It is necessary to start the “summer” boiler and the corresponding “summer” network pump.
3. Start-up of a boiler room in summer on reserve fuel. In this mode, all heating mains are filled with water, the fuel line with diesel fuel is empty. It is necessary to start the “summer” boiler and the corresponding “summer” network pump.
4. Start-up of a gas-fired boiler room in winter. In this mode, all heating mains are filled with water, the gas pipeline is empty. Both large boilers and associated mains pumps must be started.
5. Start-up of a boiler room in winter with reserve fuel. In this mode, all heating mains are filled with water, the fuel line with diesel fuel is empty. Both large boilers and associated mains pumps must be started.
6. Stop of a boiler room in winter on gas fuel. In this mode, all heating mains are filled with water; the gas pipeline is under pressure. Both boilers are in operation. It is necessary to stop the boilers and accessories, as well as to blow out the gas line.

In the section “Emergencies” the following typical accidents are currently provided:
- There is no voltage at the input;
- The accident of boiler number 1;
- The accident of boiler number 2;
- The accident of boiler number 3.

The system also provides for the “Study” of emergency situations and the “Exam” mode.

The algorithm of operations before working in the “Study” mode is displayed on the screen in the form of a notepad with the tasks recorded in it. These tasks are crossed out as they are performed.

In the “Exam” mode, there are no prompts in the form of a notebook. Only three errors are allowed during the course of a particular procedure. After completing all the necessary tasks, an assessment of the learner's actions is displayed on the screen.

The simulator allows us to acquire skills while working with the heat and power equipment of the boiler room, simulating the operation of a real object. There is a multi-user mode, in which two or more people can train at the same time. This greatly expands training opportunities. The integration of this virtual simulator into the training system for heat supplying organizations will significantly reduce the risk of accidents associated with errors of personnel, and will also help develop an action plan for employees in case of emergency situations.

The developed virtual simulator of a hot-water boiler room has already been introduced into the educational process when preparing bachelors and masters in the fields of heat power engineering and heat engineering at the Kuban State Technological University and is demonstrating its effectiveness. This software product is not the first one developed by teachers of the department of heat power engineering and heat engineering. A series of programs have been created and are actively used both in the educational process and in scientific research to determine the thermodynamic properties of air, water and water vapor, as well as to calculate the thermal circuits of gas turbine and combined cycle plants online [3, 4].

In addition, it is planned that the mathematical model implemented in a computer simulator will serve as the basis for creating a remote control system for a boiler room, which will be the so-called digital double.

5. Conclusion

Teachers of KubSTU have developed and tested a virtual simulator of a hot-water boiler for theoretical training and development of skills in the work of both future and current thermal power engineers. This simulator is not only an effective learning tool, but also the basis for creating a remote control system for a complex production facility, which is a boiler room.

IOP Conf. Series: Materials Science and Engineering 1035 (2021) 012019 doi:10.1088/1757-899X/1035/1/012019

Reference

[1] Kocharyan E V and Shaposhnikov V V and Macharashvili A.S. 2019 Virtual simulator of a water boiler Certificate of registration of a computer program of the Russian Federation No 2019664785 appl. 31.10.2019, publ. 13.116.2019

[2] Biryukov B V 2012 *Boiler Plants And Steam Generators: A Training Manual* (Krasnodar: Publ. KubSTU) p 357

[3] Biryukov B V and Shaposhnikov V V 2016 Program complex "System for the automated calculation of gas turbine plants and combined-cycle gas mixing plants" *Industrial energy* **4** 41-45.

[4] Biryukov B V and Shaposhnikov V V 2017 Increasing efficiency of CCP-based TPP with injection of dry saturated steam from recovery boiler into regenerator *Journal of physics: conference series* **891** 012185

Acknowledgements

The reported study was funded by Innovation assistance fund according to the research project №139ГУЦЭС8-D3/56294.

IOP Conf. Series: Materials Science and Engineering 1035 (2021) 012020 doi:10.1088/1757-899X/1035/1/012020

Combined use of the nearest vector selecting technique and low-frequency quarter-wave symmetric space vector PWM in control of three-phase multilevel voltage source inverter

N N Lopatkin

Department of Mathematics, Physics and Informatics, Shukshin Altai State University for Humanities and Pedagogy, Korolenko str., 53, Biysk, 659333, Russia

E-mail: nikolay_lopatkin@mail.ru

Abstract. The paper offers the problem's solution of the three-phase multilevel voltage source inverter (MLVSI) control with low dynamic losses in power semiconductor switches. The proposed technique is based on a single approach that operates with values of the integer and fractional parts of the relative values of the reference delta voltages, and it allows combining the use of the space-vector control, choosing the nearest vector (SVC), with the low-frequency space-vector pulse-width modulation with the quarter-wave symmetry of the output voltages (QWS-SVPWM). Depending on the impedance kind of a generalized load, the integrated voltage harmonics factor (weighted THD) of the appropriate order is being minimized by applying one of the two mentioned modulation methods (SVC or QWS-SVPWM) for a current range of voltage amplitude modulation index. Like its components, the combined algorithm is suitable for MLVSI of any circuit configuration and any number of equal output voltage levels.

1. Introduction

In addition to existing for several decades circuitry solutions of multilevel converters [1], this concept brings new and yet more effective results, leading, as expected, to the wide industrial application [2], including applications in photovoltaics power conversions with their grid integration [3]. Modified control techniques are constantly being developed to improve performances of both existing and new converters, and in particular inverters [4].

With an increase in the number of levels of multilevel voltage source inverters (MLVSI), that is precipitated by the necessary increase in their output voltages and powers, the high quality of the voltages and currents is achievable without increasing the frequency of power switches' switching. The possible solutions are both the space vector control with the choice of the nearest vector (SVC) [5], and the low-frequency space vector pulse-width modulation (SVPWM).

The combined use of two alternatives, SVC and SVPWM, is possible within the framework of a single approach, operating with the values of the integer and fractional parts of the samples of the relative values of the reference delta voltages and realizing the two-component formation of the two output delta voltages. This is exactly how the SVPWM with quarter-wave output voltage symmetry (QWS-SVPWM) [6] and the new algorithm for realizing the choice of the closest vector without storing data in lookup tables [7] were implemented separately.

This paper describes both the main principles of the combined use of the techniques and the approach to the specified MLVSI output variable THD minimizing by means of consideration of the integrated voltage harmonics factor [8] (weighted THD) of the appropriate order, depending on the

IOP Conf. Series: Materials Science and Engineering 1035 (2021) 012020 doi:10.1088/1757-899X/1035/1/012020

impedance kind of a generalized load. The ranges of using SVC and QWS-SVPWM modulation methods are revealed for the cases of the generalized load equivalent circuit, having numbers of the orders from zero to three (by the order of a corresponding differential equation and the relevant to it weighted THD).

2. Space vector's oblique-angled coordinates and barycentric coordinates on triangle of nearest vectors

Despite the space vector PWM algorithms, using the orthogonal decomposition has been well-established ones [9], the techniques operating in the context of oblique-angled coordinates' systems have proved to be a more simple and fast [10].

The so-called g- and h-coordinates in [10] have turned out to be the coordinates of the delta voltages relative values u^*_{ab} and u^*_{bc}. Here and after

$$u^*_{\mathrm{xy}} = u_{\mathrm{xy}} / U_{\mathrm{d}} , \tag{1}$$

where U_{d} is the direct voltage of the unit (base) level, i.e. the delta voltages' step value.

So, one of the available and convenient representation for a complex-plane form $\dot{u}$ of some voltage space vector (VSV) is [11]

$$\dot{u} = \frac{2}{3} U_{\mathrm{d}} \cdot \left(u^*_{\mathrm{ab}} + u^*_{\mathrm{bc}} \cdot e^{j\frac{\pi}{3}} \right). \tag{2}$$

MLVSI voltage space vectors are shown in figure 1. Figure 1,a demonstrates the vectors' diagram in the oblique-angled 60-degree coordinates with the space vectors designated as $\left(u^*_{\mathrm{ab}}, u^*_{\mathrm{bc}}\right)$.

If some VSV is hitting a point R inside a certain triangle of the nearest vectors, its decomposition can be presented in accordance with figure 1,b. Thanks to a property of barycentric coordinates on the upward-pointing and downward-pointing elementary triangles, the VSV $\dot{OR}$ is described, respectively, as follows [11]:

$$\dot{O_\Delta R} = b_{\mathrm{ab}} \cdot \dot{O_\Delta D} + b_{\mathrm{bc}} \cdot \dot{O_\Delta E} + b_{\mathrm{ca}} \cdot \dot{O_\Delta F} \text{ and } \dot{u} = \dot{OR} = b_{\mathrm{ab}} \cdot \dot{OD} + b_{\mathrm{bc}} \cdot \dot{OE} + b_{\mathrm{ca}} \cdot \dot{OF} , \tag{3}$$

$$\dot{O_\nabla R} = b_{\mathrm{ba}} \cdot \dot{O_\nabla E} + b_{\mathrm{cb}} \cdot \dot{O_\nabla D} + b_{\mathrm{ac}} \cdot \dot{O_\nabla G} \text{ and } \dot{u} = \dot{OR} = b_{\mathrm{ba}} \cdot \dot{OE} + b_{\mathrm{cb}} \cdot \dot{OD} + b_{\mathrm{ac}} \cdot \dot{OG} , \tag{4}$$

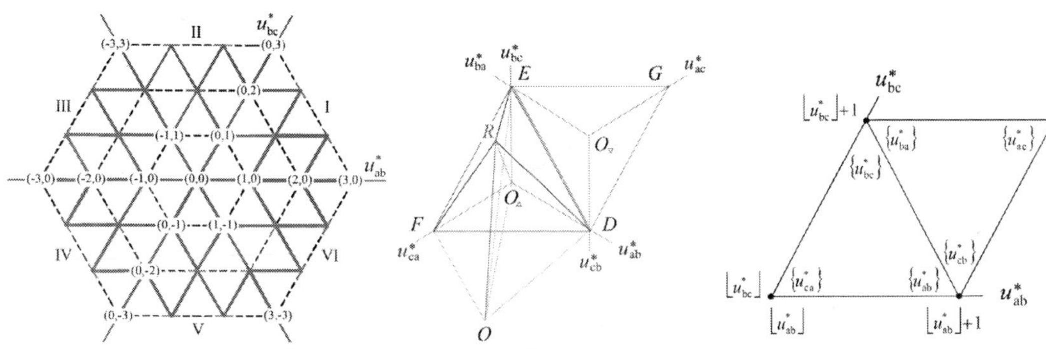

Figure 1. Voltage space vectors of three-phase multilevel voltage source inverters: a – vectors' diagram in 60-degree coordinates; b – elementary triangles of nearest vectors with six delta voltages' axes; c – parts of delta voltages' relative values as coordinates of specified voltage space vector.

where point O is the zero vector $(0,0)$ position, points O_Δ and O_∇ are the barycentres of triangles ΔDEF and ∇EDG, b_{xy} and b_{yx} are VSV barycentric coordinates on these two triangles, respectively,

$$b_{xy} = u^*_{xy} - \lfloor u^*_{xy} \rfloor = \{u^*_{xy}\},\ b_{yx} = u^*_{yx} - \lfloor u^*_{yx} \rfloor = \{u^*_{yx}\}, \quad (5)$$

here $\lfloor x \rfloor$ and $\{x\}$ are the integer and fractional parts of x,

$$b_{ab} + b_{bc} + b_{ca} = 1,\ b_{ba} + b_{cb} + b_{ac} = 1. \quad (6)$$

Thus, as can be seen in figure 1,c, the integer parts of the two delta voltages' relative values are the oblique-angled coordinates of the base vector of the three vectors which are the nearest vectors to the specified VSV, and the fractional parts of the three delta voltages' relative values are the barycentric coordinates of the specified VSV on these vectors' triangle. The three direct delta voltages should be used if the VSV's endpoint is in some upward-pointing elementary triangle, otherwise, the three inverse delta voltages should be considered (see figure 1,b).

The same quantities serve as affine coordinates on the upward-pointing and downward-pointing elementary triangles in following VSV decompositions, respectively:

$$\dot{u} = \dot{OR} = \dot{OF} + \dot{FR} = \dot{OF} + b_{ab} \cdot \dot{FD} + b_{bc} \cdot \dot{FE}, \quad (7)$$

$$\dot{OF} = \frac{2}{3} U_d \cdot \left(\lfloor u^*_{ab} \rfloor + \lfloor u^*_{bc} \rfloor \cdot e^{j\frac{\pi}{3}} \right),\ \dot{FR} = \frac{2}{3} U_d \cdot \left(\{u^*_{ab}\} + \{u^*_{bc}\} \cdot e^{j\frac{\pi}{3}} \right); \quad (8)$$

$$\dot{u} = \dot{OR} = \dot{OG} + \dot{GR} = \dot{OG} + b_{ba} \cdot \dot{GE} + b_{cb} \cdot \dot{GD}, \quad (9)$$

$$\dot{OG} = \frac{2}{3} U_d \cdot \left(-\lfloor u^*_{ba} \rfloor + \lfloor u^*_{cb} \rfloor \cdot e^{-j\frac{2\pi}{3}} \right),\ \dot{GR} = \frac{2}{3} U_d \cdot \left(-\{u^*_{ba}\} + \{u^*_{cb}\} \cdot e^{-j\frac{2\pi}{3}} \right). \quad (10)$$

3. Space vector pulse-width modulation with quarter-wave output voltage symmetry, new nearest vector selecting technique in space vector control and two-component formation of two output delta voltages

The executed output voltages here mean the voltages with waveforms, that are generated at MLVSI output by switching available space vectors as vertices of vectors' diagram in figure 1,a. They also can be treated as the coordinates of the MLVSI executed voltage space vector (EVSV), the latter is understood in the same sense.

If the reference voltage space vector (RVSV) of SVPWM is sampled inside a certain elementary triangle of the nearest vectors, called as modulating triangle, its sampled coordinates are being reproduced at the related PWM clock period (sampling period) in EVSV as average values of the output coordinates. Using the delta voltages as the reference voltages is especially convenient for SVPWM, since the duty-cycles of the three modulating vectors are exactly equal to the values of the above mentioned barycentric coordinates of the RVSV, i.e. the fractional parts of the relative values of the MLVSI reference delta voltages [11].

Thanks to (6) and other simple relationships for the rhombus in figure 1,c, transition from three to two VSV barycentric coordinates is easily accomplished. So, all the calculations for both kinds of the elementary triangles of the rhombus have been reduced to treating the only two direct delta voltages relative values instead of the values of six related delta voltages. In particular, direct using descriptions of the downward-pointing triangle, which are like (10), is no longer needed.

SVPWM with quarter-wave output voltage symmetry provides the absence of the even-order harmonics and the absence of the cosine component of the remaining harmonics in Fourier decomposition of the output voltage' waveform. As a result, we have reached considerable progress in reducing low-order higher harmonics contents. QWS-SVPWM and this supplementary advantage have been achieved due to applying a new switching sequence algorithm with a three-segment switching sequence scheme [6].

The used here scheme of the nearest vector selecting space-vector control SVC [7] applies no special operations and generates the same output voltage waveforms as the traditional nearest vector selecting space-vector control [5]. But it utilizes the conventional delta voltages SVPWM attributes, namely the duty-cycles of the three vectors of the modulating triangle (i.e. of the three vectors, which are nearest to the reference voltage space vector RVSV). As have been shown above, these values are also the barycentric coordinates on the corresponding modulating triangle for the reference voltage space vector RVSV endpoint. So, the closer to unity the fractional value of certain relative delta voltage is, the closer to the related vertex the RVSV endpoint is. Moreover, the highest value of the three barycentric coordinates on the modulating triangle makes it possible to recognize the appropriate vertex as the closest to the reference point, and such a way one can choose the space vector nearest to the vector RVSV endpoint on each sampling period [7].

This approach to space vector control, which is choosing the nearest vector, needs no preliminary finding of anything coefficients and holding them in look-up tables. Similar to the presented SVPWM technique, it is suitable for any arbitrary MLVSI circuit with any arbitrary number of the equal feeding DC voltage levels. All the related logical functions and Matlab-Simulink implementation of the controller model have been offered in [7].

The two-component formation of two independent output delta voltages is another one base of the general technique for combining SVC and QWS-SVPWM modulation methods. Specifically, in accordance with the main principle of separated treating the integer and fractional parts of the samples of the relative values of the reference delta voltages, each of the two executed output delta voltages is generated as follows:

$$u^*_{\mathrm{EXExy}}(t) = \lfloor u^*_{\mathrm{sREFxy}}(t) \rfloor + f_{\mathrm{sxy}}(t), \tag{11}$$

where $u^*_{\mathrm{sREFxy}}(t)$ is a respective relative delta voltage reference signal, which is sampled and held in each sampling period with a duration of T_s, and being a function of continuous model time, namely, we have two such reference signals, that are shifted to be corresponding to the middle point of the sampling period,

$$u^*_{\mathrm{sREFab}}(t) = m_{\mathrm{aD}} \sin\left(\left(2\pi k(t) - \pi\right)/m_{\mathrm{sf}} + 2\pi/3\right),\ u^*_{\mathrm{sREFbc}}(t) = m_{\mathrm{aD}} \sin\left(\left(2\pi k(t) - \pi\right)/m_{\mathrm{sf}}\right), \tag{12}$$

m_{aD} is the delta voltage amplitude modulation index,

$$m_{\mathrm{aD}} = \sqrt{3}\, m_{\mathrm{aY}} = \sqrt{3}\, U/U_{\mathrm{d}} = \sqrt{3}\, U^* = U^*_{\Delta\mathrm{m}},\ m_{\mathrm{aY}} = U/U_{\mathrm{d}} = U^*, \tag{13}$$

m_{aY} is the phase voltage (wye voltage) amplitude modulation index, U and U^* are the value and relative value of RVSV magnitude, $U^*_{\Delta\mathrm{m}}$ is the amplitude relative value of the reference delta voltages, $k(t)$ is the clock-cycle number function and it runs the integer values from 1 to m_{sf} within each MLVSI output voltage's period with a duration of T,

$$k(t) = \left\lfloor \left(t/T - \lfloor t/T \rfloor\right) \cdot m_{\mathrm{sf}} \right\rfloor + 1, \tag{14}$$

m_{sf} is the sampling frequency index, it is similar to the frequency modulation index in PWM,

$$m_{\mathrm{sf}} = f_s/f = T/T_s, \tag{15}$$

here f_s and f are, respectively, the sampling frequency (sampling rate) and modulating (output) frequency;

$f_{sxy}(t)$ is a pulse function, possessing the values 0 and 1, and exactly a difference in formation of this function in SVC and QWS-SVPWM defines an overall difference in generated output voltages.

While in QWS-SVPWM technique a sector's number, a type of a modulating vectors' triangle and values of the barycentric coordinates of the RVSV specifies positions and durations of the high-potential pulses on each clock cycle period [6], SVC sets the only level per the sampling period in accordance with a maximum barycentric coordinate on a related triangle [7]. In contrast to PWM, SVC at some enough sampling frequency value generates the same voltages' waveforms as for "instantaneous SVC" with a vanishingly small sampling period, and the switching frequency of power switches in MLVSI is therefore absolutely independent of the sampling frequency value.

Thus, to have in the offered generalized approach a low level of the switching losses in semiconductor switching devices (IGBT, MOSFET, etc), QWS-SVPWM should be applied at enough low values of the sampling frequency and switches' commutation frequency, while the SVC, as a priori a low-frequency method with regard to MLVSI power circuits, should be used with enough high sampling frequency value.

4. Integrated voltage harmonics factors of inverter voltage waveform as indicators of power quality of a generalized load

The simplified equivalent circuits of the MLVSI for the cases of four various kinds of its load are shown in figure 2, where $u_{ag}(t)$ and $u_{an}(t)$ are the phase-to-ground ("g") and phase-to-neutral ("n") generated voltages of the phase "a", respectively; $\underline{z}_a$ is the phase "a" impedance, L_a and R_a are the prospective phase "a" load inductance and resistance, L_f and C_f are the elements of phase "a" MLVSI output LC filter; i_a is the phase "a" load current.

The MLVSI load current's THD, designated here as the factor of current's harmonics $K_{h/a}$, can be calculated as follows:

$$K_{h/a} = I_{a(hh)} / I_{a(1)}, \tag{16}$$

where $I_{a(1)}$ and $I_{a(hh)}$ are the RMS values of the current's fundamental harmonic and higher harmonics component, respectively.

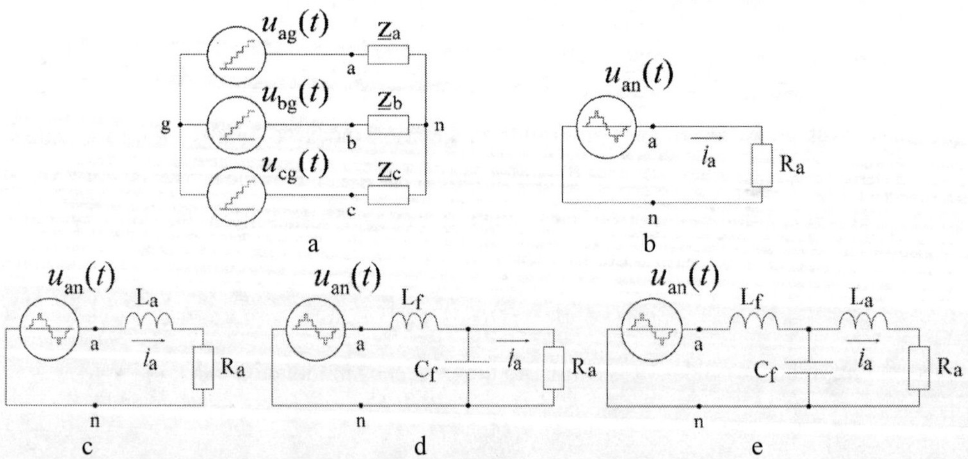

Figure 2. Simplified equivalent circuits of three-phase multilevel voltage source inverter and its load: a – generalized three-phase three-wire circuit; b – per-phase R load circuit; c – per-phase RL load circuit; d – per-phase R load circuit with inverter output LC filter; e – per-phase RL load circuit with inverter output LC filter.

Instead of considering the full three-phase equivalent circuit (figure 2,a), for some three-phase three-wire circuit with symmetrically-controlled MLVSI and some balanced load one of the simplified per-phase equivalent circuits may be studied.

The trivial case is presented in figure 2, b for cases when a generalized load, including some output circuit of the MLVSI itself, can be treated as the active resistance. The respective equation is a degenerate one and lead to THD simple and obvious relation:

$$\mathrm{R}_a i_a = u_{an}, \; K_{hia} = K_{hu} = \overline{K}_{hu}^{(0)}, \tag{17}$$

where K_{hu} is THD of MLVSI phase-to-neutral voltage u_{an}, also treated as the zero-order integrated voltage harmonics factor $\overline{K}_{hu}^{(0)}$ (here and after index "a" is omitted from subscript of such the factors).

The simplest of the more realistic MLVSI phase load circuits in figure 2,c is described by the first order differential equation and reduced integral equation derived from it:

$$\mathrm{L}_a \frac{di_a}{dt} + \mathrm{R}_a i_a = u_{an}, \; i_a + \frac{\mathrm{R}_a}{\mathrm{L}_a} \overline{i}_a^{(1)} = \frac{1}{\mathrm{L}_a} \overline{u}_{an}^{(1)}, \tag{18}$$

here and after $\overline{x}^{(n)}$ designates the result of the taking of n-fold indefinite integral of some instantaneous AC variable x (can have any subscript).

To assess the load current's higher harmonics component and THD the method, developed by Prof. G.S. Zinoviev (NSTU, Novosibirsk), namely the differential equations' algebraization (ADE) method [8] is used here. The ADE method is intended for direct obtaining expressions of RMS voltages and currents values in a closed analytical form from source differential equations without their solving, and ADE is asymptotically approximate method.

After the ADE procedure applying to (18), written for the higher harmonics component $i_{a(hh)}$, the phase "a" load current i_a THD value can be defined as asymptotic quantity, being approximately calculated as the sum of the finite series with alternating members' signs:

$$K_{hia} \approx \left[\left((\mathrm{R}_a/\mathrm{L}_a)^2 + \omega^2 \right) \cdot \sum_{n=1}^{N_f} \left((-1)^{n-1} \cdot \left(\frac{(\mathrm{R}_a/\mathrm{L}_a)^{n-1}}{\omega^n} \cdot \overline{K}_{hu}^{(n)} \right)^2 \right) \right]^{1/2}, \tag{19}$$

where N_f is the approximation level, related to the "filter hypothesis" [8], [12], namely $\overline{i}_{a(hh)}^{(n \geq N_f)} \equiv 0$, $\overline{K}_{hu}^{(n)}$ is the n-order weighted THD (WTHD) coefficient of the MLVSI output voltage $u_{an}(t)$, which ADE method's author called n-order integrated voltage harmonics factor (IHF).

So, the first approximation $N_f = 1$ corresponds to

$$K_{hia} \approx \overline{K}_{hu}^{(1)} \cdot \left[1 + \left(\mathrm{R}_a / (\omega \mathrm{L}_a) \right)^2 \right]^{1/2}. \tag{20}$$

Since the load current's THD, with only one exception (see (17)), does not correlate directly with the output voltage's THD, the latter cannot be a criterion for completely sufficient quality comparative evaluation of some voltage source output voltage. It shows only a quantity ("weight") of higher harmonics in relation to a fundamental one and does not talk about the distribution of harmonics in a frequency range of a voltage spectrum, in particular, about their proximity to the voltage fundamental component, which is important, for instance, in some designing process, which is taking into account filtering properties of both load itself and specially installed circuits. The criteria that take into account both the "weight" of high-frequency harmonics and their position in the spectrum are the IHF factors, i.e. WTHD, of various orders. They make weighted (by the harmonic order number) summation of harmonics, as a matter of fact, modeling the effect of the action of the amplitude-frequency characteristic of the idealized electric circuit of the corresponding order. Such the ideal filter reduces harmonics' magnitudes by the number of times equal to the harmonic order number raised to the filter order power. That is exactly what we can see in the following equation for the n-order integrated voltage harmonics factor [8]:

$$\overline{K}_{hu}^{(n)} = WTHD^{(n)} = \frac{\overline{U}_{(hh)}^{(n)}}{\overline{U}_{(1)}^{(n)}} = \frac{\overline{U}_{(hh)}^{(n)} \cdot \omega^n}{U_{(1)}} = \left[\sum_{k=2}^{\infty} \left(\frac{U_{(k)}}{k^n \cdot U_{(1)}} \right)^2 \right]^{1/2}, \tag{21}$$

where for estimated voltage $u_{an}(t)$ values $U_{(k)}$, $\overline{U}_{(1)}^{(n)}$, $\overline{U}_{(hh)}^{(n)}$ are the RMS values of the k-order harmonic component $u_{an(k)}$, and of the n times integrated values of the fundamental harmonic $\overline{u}_{an(1)}^{(n)}$ and the higher harmonics component $\overline{u}_{an(hh)}^{(n)}$.

The MLVSI phase load circuit with a resistive load and some inverter output LC filter (figure 2,d) produces the second order reduced integral equation:

$$i_a + \frac{1}{R_a C_f} \overline{i}_a^{(1)} + \frac{1}{L_f C_f} \overline{i}_a^{(2)} = \frac{1}{L_f C_f R_a} \overline{u}_{an}^{(2)}, \tag{22}$$

Here the first level approximation current's THD estimate is

$$K_{hia} \approx \overline{K}_{hu}^{(2)} \cdot \left[1 + \left(\frac{1}{(R_a C_f)^2} - \frac{2}{L_f C_f} \right) \frac{1}{\omega^2} + \frac{1}{(L_f C_f)^2 \omega^4} \right]^{1/2}, \tag{23}$$

At last, the MLVSI phase RL-load circuit with LC filter (figure 2,e) leads to the following third order reduced integral equation and current's THD first level estimate:

$$i_a + a_1 \cdot \overline{i}_a^{(1)} + a_2 \cdot \overline{i}_a^{(2)} + a_3 \cdot \overline{i}_a^{(3)} = b_3 \cdot \overline{u}_{an}^{(3)}, \tag{24}$$

$$a_1 = \frac{R_a}{L_a}, \; a_2 = \frac{1}{C_f} \left(\frac{1}{L_f} + \frac{1}{L_a} \right), \; a_3 = \frac{R_a}{L_f C_f L_a}, \; b_3 = \frac{1}{L_f C_f L_a},$$

$$K_{hia} \approx \overline{K}_{hu}^{(3)} \cdot \left[1 + \frac{a_1^2 - 2a_2}{\omega^2} + \frac{a_2^2 - 2a_1 a_3}{\omega^4} + \frac{a_3^2}{\omega^6} \right]^{1/2}. \tag{25}$$

As can be seen, each estimate of load current's THD correlates to the voltage IHF of the order (from zero to three), equal to the order of a differential equation corresponding to the considered equivalent circuit of a generalized load.

The similar THD assessing can be performed for any element of any equivalent circuit regarding current in this element or voltage across it. In particular, equations for voltage u_R across the resistance R_a in circuits of figures 2,b…e are, respectively, as follows:

$$u_R = u_{an}; \; u_R + \frac{R_a}{L_a} \overline{u}_R^{(1)} = \frac{R_a}{L_a} \overline{u}_{an}^{(1)}; \; u_R + \frac{1}{R_a C_f} \overline{u}_R^{(1)} + \frac{1}{L_f C_f} \overline{u}_R^{(2)} = \frac{1}{L_f C_f} \overline{u}_{an}^{(2)};$$

$$u_R + a_1 \cdot \overline{u}_R^{(1)} + a_2 \cdot \overline{u}_R^{(2)} + a_3 \cdot \overline{u}_R^{(3)} = a_3 \cdot \overline{u}_{an}^{(3)}. \tag{26}$$

Their compositions repeat the compositions of equations (17), (18), (22) and (24) with the difference in coefficients of the variables only. Thus, one can obtain first level estimates of voltage u_R

THD, similar to formulae (17), (20), (23) and (25), and check that they are proportional to IHF $\overline{K}_{hu}^{(n)}$ of the appropriate orders. The approach to the required MLVSI output variable THD minimizing should be therefore based on consideration and minimization of the integrated voltage harmonics factor (weighted THD) of the appropriate order, depending on the impedance kind of a generalized load.

5. Results of computer simulation and delineation of modes and ranges of amplitude modulation index between two techniques

Regarding the choice of QWS-SVPWM frequency modulation index (here the sampling frequency index), we should abide by the principle of reasonable sufficiency in the sampling frequency increasing and the related improvement of MLVSI output voltages' higher harmonics content because of the higher frequency of power switches' commutations, the higher related power losses in them and the lower efficiency of MLVSI. So, the quarter-wave symmetric space vector PWM must remain a low-frequency and have a comparable with SVC technique number of switches' commutations per output voltages' period.

We shall confine ourselves here to considering the case of the QWS-SVPWM sampling frequency index value of 30 ($m_{sf} = 30$), i.e. for output voltages' frequency value of 50 Hz the sampling frequency value is 1500 Hz.

SVC simulation corresponds here to a continuous survey of the reference sinusoidal signals of the delta voltages. The same resulting voltage waveforms and the related values of the weighted THD (IHF) take place at enough high sampling frequency index value (likely, above 500). However, as have been mentioned above, transitions between available space vectors and changes in voltages values are comparatively rare.

The computer simulation for the two techniques was performed in LabVIEW, where the controllers' models have been implemented. The resulting waveforms of the delta and phase MLVSI output voltages for SVC and QWS-SVPWM at $m_{aY} = 3.14$ are shown in figure 3.

The curves of the dependences of the integrated voltage harmonics factors on the phase-voltage amplitude modulation index m_{aY} are presented in figure 4. They make it possible to delineate modes and to reveal ranges of using SVC and QWS-SVPWM techniques.

It should be noted that the sub-range $m_{aY} = U^* < 1/3$ is SVC zero space vector proximity zone, i.e. the dead band of SVC, therefore this is the sub-range of QWS-SVPWM technique, without distinction as to the kinds of the generalized load.

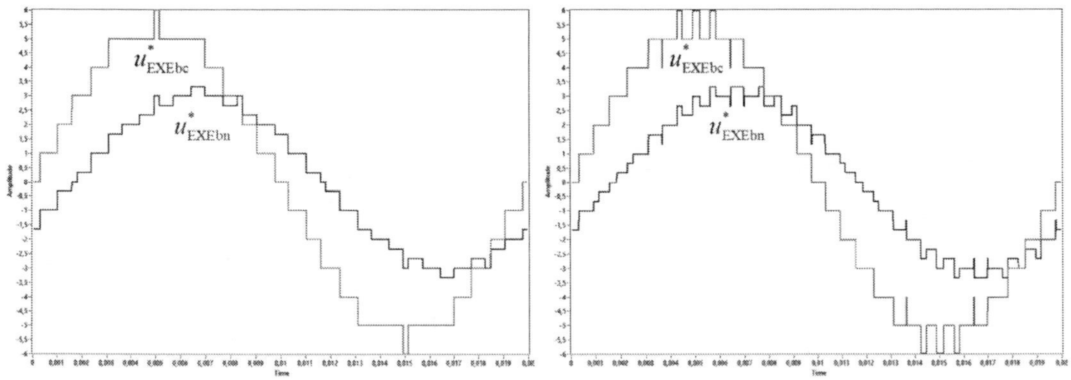

a b

Figure 3. Examples of LabVIEW-simulated delta and phase voltages' waveforms of two techniques: a – space vector control executed voltages idealized signals (pu); b – quarter-wave symmetric space vector pulse-width modulation executed voltages idealized signals (pu).

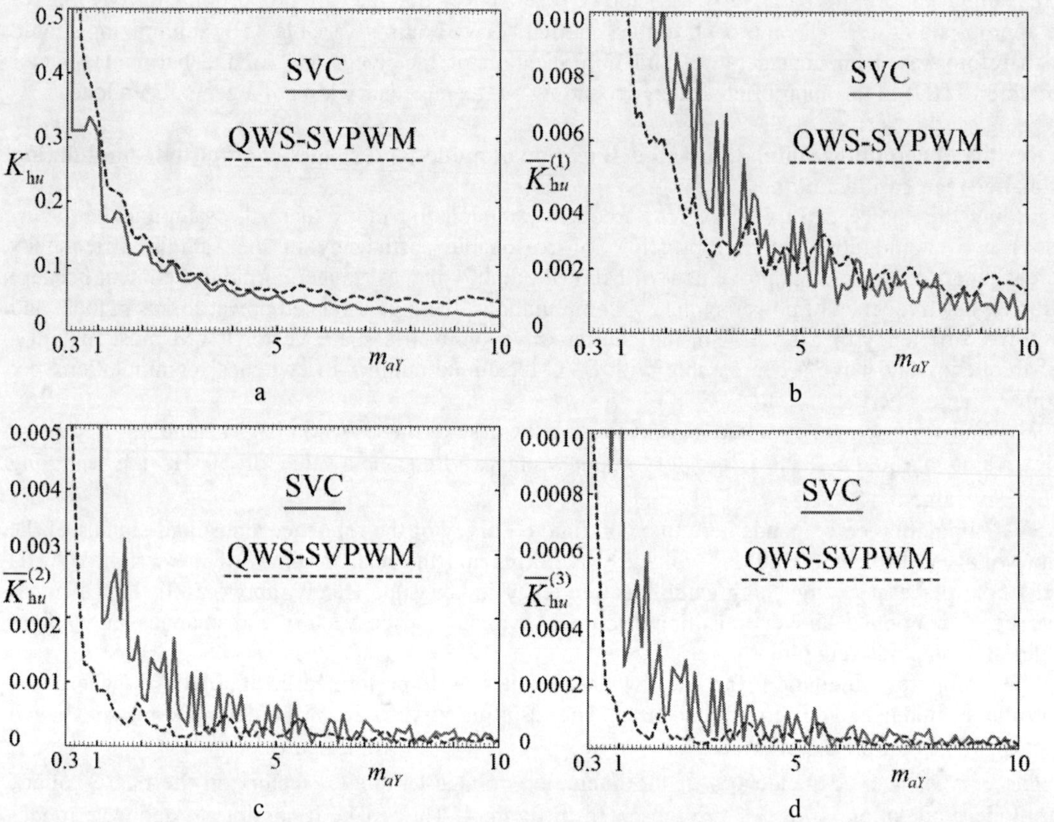

Figure 4. Integrated voltage harmonics factors (IHF, weighted THD) versus phase voltage amplitude modulation index: a – zero-order IHF (THD); b – first-order IHF; c – second-order IHF; d – third-order IHF.

The delineation of the main range $m_{aY} > 1/3$ is as follows. If the generalized load for some being examined MLVSI application can be treated as a pure resistive circuit, the SVC is being employed, as having lower THD values in accordance with figure 4,a.

As can be seen from figure 4,c and d, for the both of the considered circuits with inverter output LC filter (figure 2,d and e), QWS-SVPWM with $m_{sf} = 30$ should be applied in the full range of the phase voltage amplitude modulation index as having an advantage in IHF of the second and third orders. This advantage is clear at low values of the amplitude modulation index. With the increase of output voltage magnitude, some competition from SVC takes place, but the related sub-ranges are too meager to be taken into account.

At last, some generalized load, representable in the form of a resistive inductive series circuit (figure 2,c), must of necessity combine both the considered control algorithms to minimize values of the voltage IHF of the first order, see figure 4,b. This case is very common, it occurs, for instance when some transformerless-by-output MLVSI circuit is loaded by some three-phase symmetrical squirrel-cage induction motor without any filter, and its equivalent per-phase circuit can be simplified and accordingly reduced to this low-order RL-form. Here SVC is losing the competition against QWS-SVPWM under low phase-voltage amplitude modulation index values, namely, when $m_{aY} \leq 5.1$.

Thus, the offered concept can easily be implemented with a guaranteed positive effect on the quality of voltages, generated by an arbitrary circuit of a three-phase multilevel inverter loaded by a specified generalized consumer circuit.

6. Conclusion

The combined control algorithm for a three-phase multilevel voltage source inverter with any circuit configuration and any number of equal output voltage levels is developed to improve the quality of its output power, namely THD of some specified output variable. The approach is oriented to minimization of an integrated voltage harmonics factor (a weighted THD) with an order, dependent on the kind of an equivalent circuit of a generalized load. The algorithm employs the space-vector control, which is choosing the nearest vector to the reference one, and the recently offered space vector pulse-width modulation technique with quarter-wave output voltage symmetry to keep low-frequency nature of power switches' commutations and low level of related losses in them. The combined algorithm is based on processing values of the integer and fractional parts of the samples of the relative values of the reference delta voltages and realizing the two-component formation of the two output delta voltages. The obtained curves of the dependences of the integrated voltage harmonics factors on the phase-voltage amplitude modulation index made it possible to delineate modes and to reveal ranges of applying of the two alternatives.

References

[1] Lai J S and Peng F Z 1996 Multilevel converters - a new breed of power converters *IEEE Trans. on Industry App.* **32(3)** pp 509–517

[2] Gonzalez S A, Verne S A and Valla M I 2013 *Multilevel Converters for Industrial Applications* (Boca Raton: CRC Press) p 241

[3] Maswood A I and Tafti H D (Editors) 2018 *Advanced Multilevel Converters and Applications in Grid Integration* (Hoboken: JohnWiley & Sons) p 496

[4] Leon J I, Kouro S, Rodriguez J and Wu B 2016 The essential role and the continuous evolution of modulation techniques for voltage-source inverters in the past, present, and future power electronics *IEEE Trans. on Industr. Electron.* **63(5)** pp 2688–2701

[5] Rodriguez J, Moran L, Correa P and Silva C 2002 A vector control technique for medium-voltage multilevel inverters *IEEE Trans. on Industr. Electron.* **49(4)** pp 882–888

[6] Lopatkin N N 2017 Simple space vector PWM scheme with quarter-wave symmetric output voltage waveform for three-phase multilevel inverter *Proc. Int. Multi-Conf. on Engineering, Computer and Information Sciences (Novosibirsk)* IEEE pp 433–438

[7] Lopatkin N N 2018 New implementation of nearest vector selecting space vector control for three-phase multilevel voltage source inverter *Proc. Int. Conf. on Industrial Engineering, Applications and Manufacturing (Moscow)* IEEE (pp 1–7)

[8] Zinoviev G S 2012 *Power Electronics* (Moscow: Jurajt) p 667

[9] Holmes D G and Lipo T A 2003 *Pulse Width Modulation for Power Converters* (Hoboken, Piscataway: Wiley-IEEE Press) p 744

[10] Celanovic N and Boroyevich D 2001 A fast space-vector modulation algorithm for multilevel three-phase converters *IEEE Trans. on Industry App.* **37(2)** pp 637–641

[11] Lopatkin N N 2015 Some new representations of the multilevel inverter voltage space vector in the complex plane *Proc. Int. Siberian Conf. on Control and Communications (Omsk)* IEEE (pp 1–11)

[12] Zinoviev G S 1990 *Direct Methods of Calculation of Power Indicators of Valve Converters* (Novosibirsk: NSU Publishing House) p 220

Distributed energy technology development

O G Losev, D A Melnik and A S Grigoriev

National Research Center "Kurchatov Institute", Akademika Kurchatova sq., 1, Moscow, 123182, Russia

Abstract. This article presents preliminary results of work on the development of advanced technologies for the development of distributed energy. In particular, the results of the first stage of creating control algorithms in a distributed power supply network and creating a simple scheme of interaction between two consumers with their own generation sources are shown. Calculations and testing of the selected principles of interaction in the MATLAB/Simulink program showed the efficiency of the decisions made.

1. Introduction

The large-scale tasks of the technical modernization of the electric power industry in Russia require the development of new efficient and environmentally friendly distributed energy technologies that can operate in a wide range of capacities in various climatic conditions [1].

To solve these problems, new structural schemes of distributed energy systems based on advanced technologies for generating, converting and storing electricity are being developed (see figure 1).

The development and debugging of systematic mathematical tools and computer programs are underway, which allows using the methods of computational experiments to model, evaluate the efficiency and optimize the parameters of distributed power systems [2, 3].

2. Energy sources in distributed energy

There are several technologies that can supply a residential or commercial property with the electricity they need. Among them are solar panels, wind generators, gas reciprocating plants and microturbines operating on gaseous hydrocarbons, micro hydroelectric power stations, and fuel cells operating on pure hydrogen or other hydrocarbons [4, 5]. These technologies are used to create distributed energy generation complexes. At the present stage of development, so far these technologies have received rather limited development due to their high cost, the lack of agreed standards for interaction with the central electric grid and the lack of a full-fledged legislative base [6].

The experience of distributed generation technologies operating shows that energy storage devices are needed for the energy complexes built on their basis. These energy storage facilities can supply the consumer with the necessary energy in case of energy supply failure from the main source of electricity. Storage devices in conjunction with power generation devices can, in addition to providing the consumer, also supply electricity to the utility network. In many countries, federal government agencies, as well as private utilities, encourage the transfer of surplus electricity by users with their own generation sources to the central network.

In recent years, the international community has placed great emphasis on the development of environmentally friendly energy sources. This gave a pulse to the development of distributed generation technologies, which led to an increase in production capacities and increased market competitiveness for both power generation devices and storage devices. In Russia, meanwhile, according to a number of experts, the relatively low cost of electricity offered by the central electric grid is a barrier to mass adoption of distributed energy technologies.

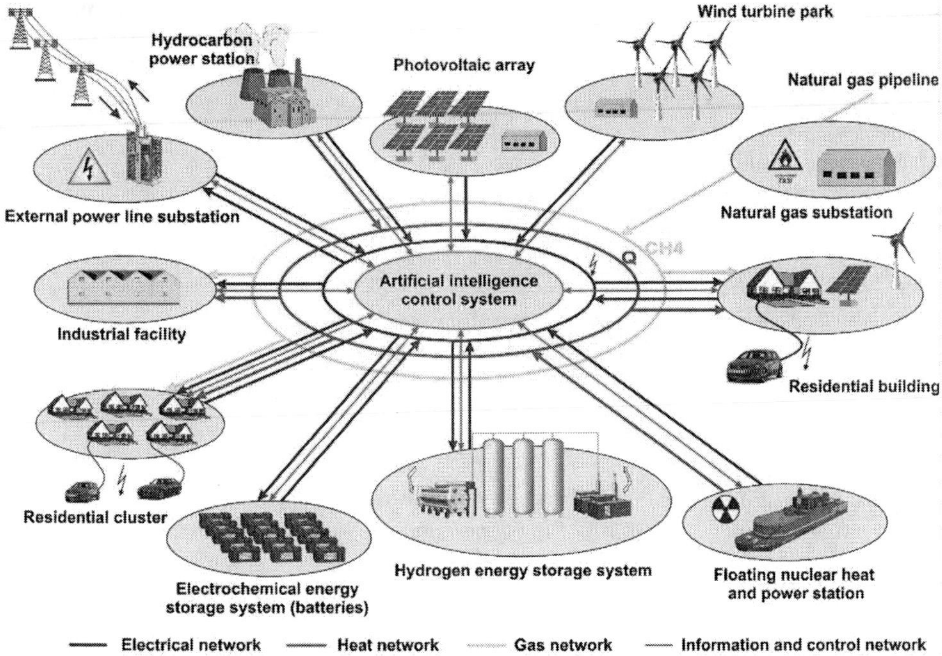

Figure 1. Network architecture of power grid with hydrogen-based energy storage system.

However, one of the most promising segments, where distributed energy technologies can be actively developed in Russia, are isolated energy systems that do not have a connection to the central power grid [7]. Formally, the energy supply of these regions is carried out according to the centralized principle, with only one difference, that generation facilities have a small capacity, which increases the number of such generators. They, in turn, require constant and costly maintenance, which naturally turns the energy supply in this region into subsidized.

The implementation of distributed energy technologies and energy storage technologies takes place in conjunction with the development of advanced measuring infrastructure. The measuring infrastructure of the energy complex consists of modern electronic/digital equipment and software. These systems make it possible to obtain detailed indications about the operation of the energy complex and collect and transmit the received information to various parties [8].

The measuring infrastructure includes meters at the customer's site, communication networks between the customer and the service provider, such as electric, gas, or water networks, as well as data reception and management systems that provide information to the service provider. Using the measuring infrastructure, utilities can manage installed devices in consumers' homes, selectively turn off energy-consuming devices (such as hot water heaters or air conditioners), thereby controlling peak consumption. In addition, utilities may in some cases remotely activate and aggregate distributed sources of energy and energy supply to increase the supply of electricity to limited parts of the electric network.

So, the key elements of distributed energy are:

1. Power generation devices.
2. Energy storage devices.
3. Measuring infrastructure.
4. Management body.

3. Requirements for design of distributed network nodes

It is necessary to form the needed voltage for the consumer inside the house/building from the Battery Pack (Battery), both with an increase in voltage (for example, to get 220 V for powering typical devices and devices) and with a decrease in voltage (for example, to get 12 V for powering LED lamps). It is necessary to connect the batteries of all network participants into a common network in such a way as to compensate for the lack of energy in some facilities, excess energy in others, which in essence is voltage management of all parts in the entire network.

It is necessary to install bi-directional energy meters between objects to account for energy production and consumption.

The work of the body ("Management") that monitors and controls the consumption of electricity for an individual house includes the following functions:

1. General dispatching and control of common network energy sources.
2. Management and control of energy generation by network participants.
3. Billing for electric energy for the network as a whole and for participants – house holdings individually.
4. Management of full or partial restriction of network electricity consumption.
5. Control of electric energy quality.
6. Accounting and budgeting of expenses for losses, operation maintenance, planned equipment change.

For the residential village, lighting control of the general village load is also added.

To obtain reliable information about the real energy consumption, intelligent DC and AC electric meters are used with the function of transmitting data via Ethernet and the ability to remotely control switching units.

In order to create a distributed energy network from the very beginning, it is necessary to start with an extremely generalized equivalent circuit that can be used to work out the principles of interaction of key elements of a distributed energy supply system. As an example, we offer two users with their own sources of energy generation and storage, which are networked and have the ability to both independently provide their own load with their own generation sources, and exchange energy with each other (figure 2). In order to explore the features of power generation and power consumption, it is proposed to create two users with a set of equipment of different nominal values, i.e. user No. 1 will have a 48V battery and the corresponding equipment, which should ensure the operability of this system and user No. 2 with a 220V battery and the equipment corresponding to this nominal.

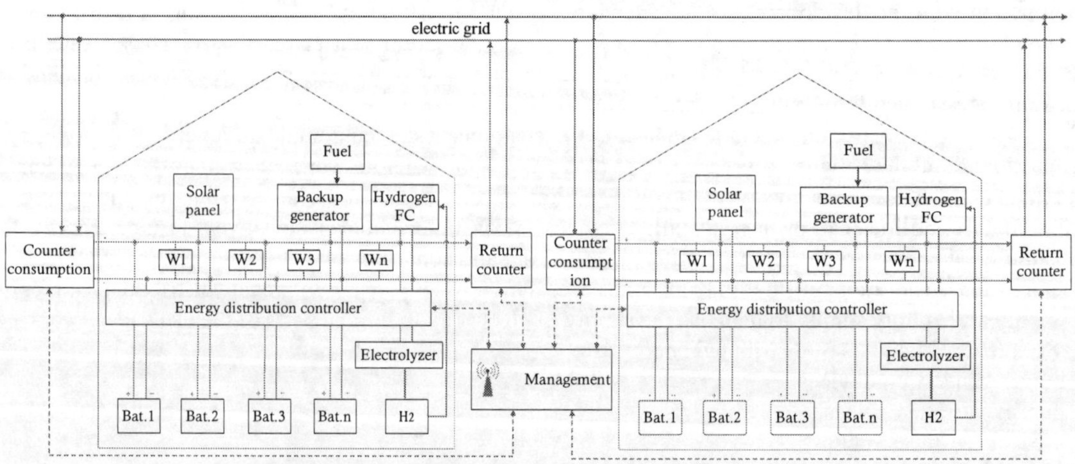

Figure 2. Two objects with their own generation sources.

Each element of the system that is associated with energy consumption and/or generation has its own controller; whose function is to coordinate the operation of the controlled element of the system with the entire system as a whole. With the help of this controller, the Central energy distribution controller receives information about the state of charge of energy storage devices, if they are present in this system element. It also receives information about the amount of fuel that is needed to run backup generators, information about the battery status (age, number of cycles, temperature). Up-to-date information is also transmitted by generation sources, such as the remaining engine life, the number of on/off cycles, the type of fuel, and data on combustion product emissions. All this information is necessary to assign a particular element of the system priority for the return of energy in the event of a request for energy from another element of the system.

The electrical functional diagram below shows two users, similar to figure 2. These users have a set of identical loads and their own sources of energy generation and storage. The load in households is a standard element, powered by a 220V network. The difference between these two users is that the batteries are of different voltage. For user # 1, the battery voltage is 48 V, for user # 2, it is 220V. Thus, for the first user standard charge controllers are required, they will allow using energy from primary energy sources (sun, fuel cell, motor generator) to charge 48 V battery, in the second case, a special charge controller is required that can charge the 220 V battery. This selection of parameters is done in order to determine all the pros and cons of these power supply schemes.

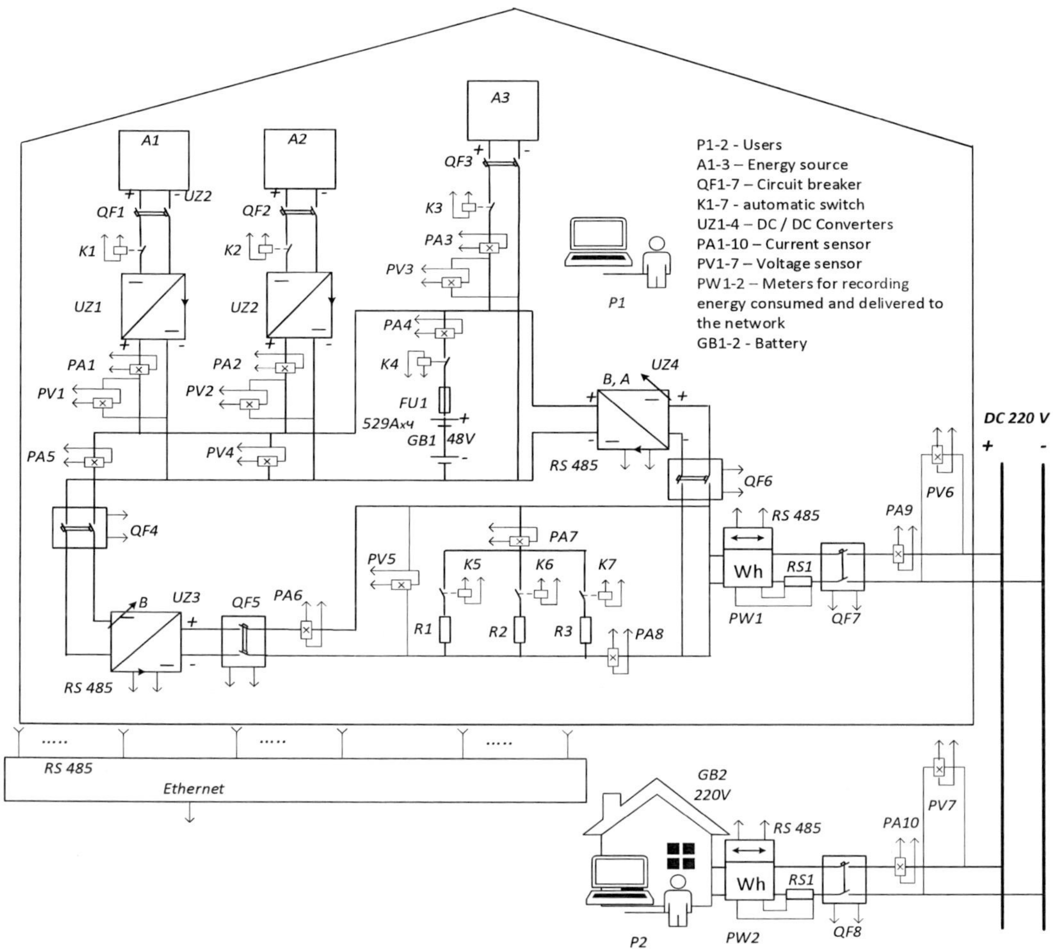

Figure 3. Electrical functional diagram.

The specific feature of this research and the proposed scheme of the power supply is that all elements of the system are connected by direct current lines. Also, the study of the interaction details of the elements of the storage and generation system configured for a nominal voltage of 220 V is necessary in order to create a DC power plant that could supply electric energy with a minimum number of conversions to a standard house load, which was previously powered by a 220 V AC network.

4. Preliminary results of the modeling

In order to make the necessary calculations, a simulation model was created in the MATLAB Simulink environment, where the above scheme was created using standard blocks.

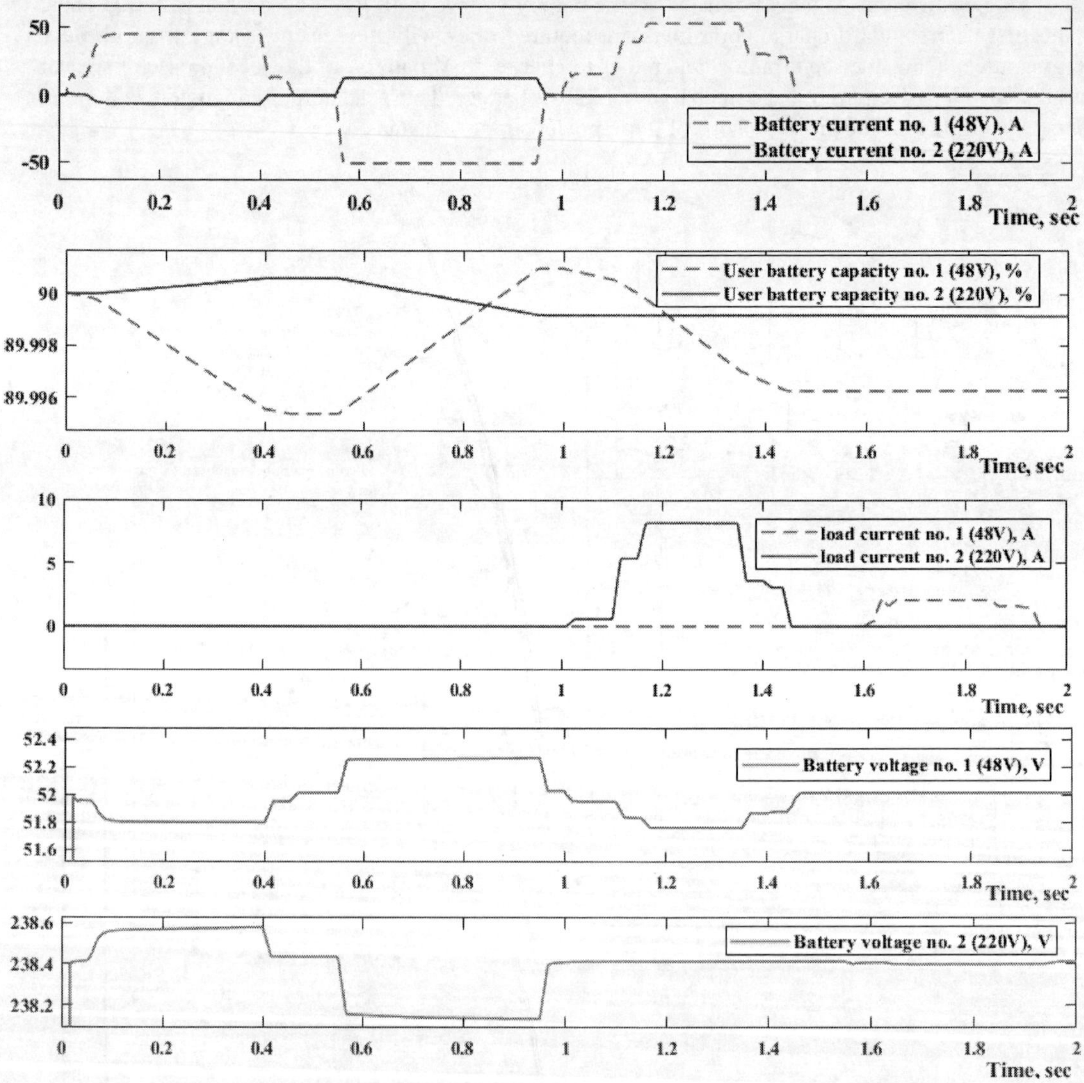

Figure 4. Battery and load parameters for two users.

During the simulation, the following conditions and parameters were set and debugged:
1. Actuation of controlled switching elements in accordance with commands from the microcontroller for controlling nodes of the distributed power system.

2. The presence of current flowing from the 48V node to the 220V node.
3. The presence of current flowing from the 220V node to the 48V node.
4. The presence of the load current of the 220V node's own consumers during the power supply from the node 48V.
5. The presence of the load current of the 48V node's own consumption during the power supply from the node 220V.
6. The value of the charge current for 220 V battery powered from DC 220 V, which is converted from the node 48V.
7. The value of the charge current for 48V battery powered by DC 48V, which is converted from the 220V node.

This is achieved by using special DC / DC converters.

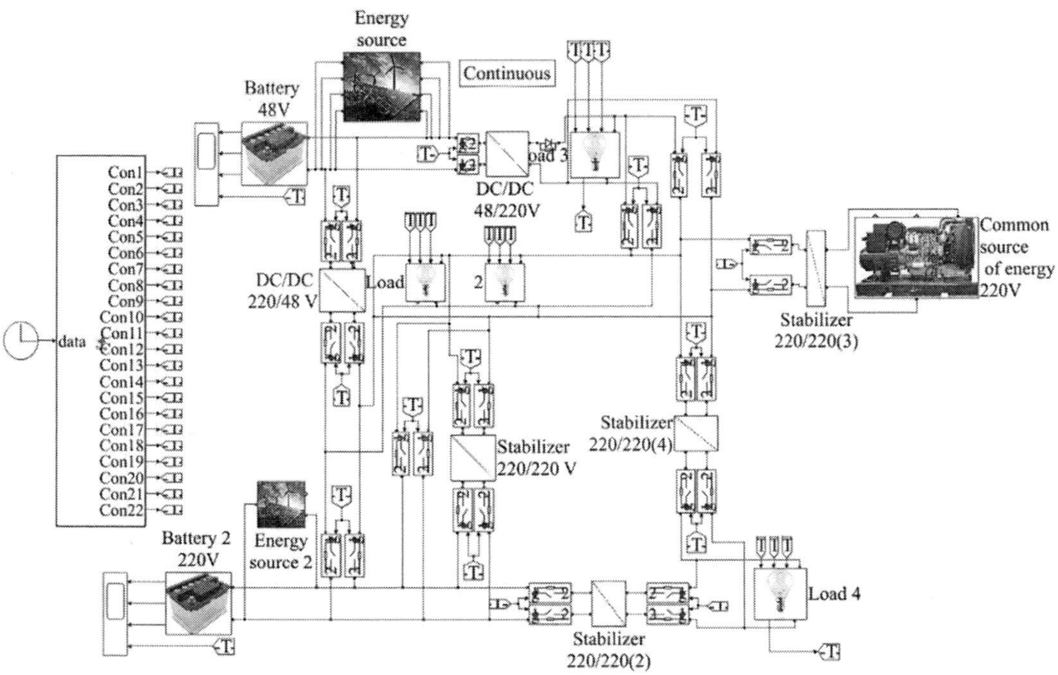

Figure 5. Simulation model in the Simulink.

This model is arranged on the approach widely used, which consists of using standardized mathematical descriptions in calculations. Using the Matlab/Simulink tool for calculations, one gets access to the source database for performing the calculation analysis. Each block used in the model has a mathematical description and is already implemented in the information module.

5. Conclusion

Simulation has revealed some features of the DC network operation. Because the two storage systems are tied to a particular voltage, in this case to the nominal battery voltage, which in one case is 48V and the other is 220V, the load is set to the nominal value of 220V, in one case, this voltage must be converted from 48V through step-up DC/DC Converter, and in another case, this voltage must be stabilized with the help of DC/DC Converter, because the voltage of the 220 V battery charged is 238 V. Hence, it was found that the regulation of the load supply at a stable voltage can only be carried out through current regulation. And for the case when the customer voltage is 220V and the load current is 4A, the electric power will be 880 Wt, discharge current the 220 V battery will be the same 4A. In the case when the load of 220V must be supplied from a system of the nominal voltage of 48V with the conversion to 220V, and the electric power of 880Wt will require the discharge current of the battery of nominal voltage of 48V be of 18.3 A. The obtained values must be taken into account for designing or selection of DC/DC boost converters and other equipment.

References

[1] Electronic resource: http://www.nrcki.ru/catalog/index.shtml?g_show=41353

[2] Grigoriev A, Grigoriev S and Skorlygin V 2013 Sarov modeling of stationary processes in the heat supply system based on renewable energy sources *Alternative Energy and Ecology* **128** pp 10-17

[3] Sokolov P 2013 Mathematical model for optimizing the structure of distribution networks in distributed generation *News of Irkutsk State Economics Academy* **6** pp 4-16

[4] Melnik D, Filimonov M, Losev O, Pechak V, Grigoriev A, and Grigoriev S 2017 *Power distribution Controller for an Autonomous power plant* Patent No 179979

[5] Losev O, Grigoriev A, Melnik D and Grigoriev S 2020 Charging station for electric transport based on renewable energy sources *Electrochemistry* **56(2)** pp 175-182

[6] Losev O, Maruseva I, Pushkarev A, Grigoriev S and Grigoriev A 2013 Fuelling station based on renewable energy sources for hydrogen and electric vehicles *Alternative Fuel Transport* **3** p 15

[7] Boute A 2016 Off-Grid renewable energy in remote arctic areas: an analysis of the Russian Far East *Renewable and Sustainable Energy Reviews* **59** pp 1029–1037

[8] Grigoriev A, Skorlygin V, Grigoriev S, Melnik D and Filimonov M 2018 Hybrid power plant based on renewable energy sources and electrochemical energy storage and generation systems for decentralized power supply in the Northern territories *Int. Electrochemical J. Sci.* **13** p 1822

[9] Electronic resource: http://www.rtsoft.ru/project-cards/sgt/resheniya/

Acknowledgments

This work was carried out with the financial support from the Ministry of Science and Higher Education of the Russian Federation within the framework of the agreement on the provision of grants from the federal budget in the form of subsidies in accordance with paragraph 4 of Article 78.1 of the Budgetary Code of the Russian Federation No. 075-15-2019-1847 dated 04.12.2019 (unique project identifier RFMEFI60419X0243).

Analysis of technical condition and development of practical recommendations to improve the reliability of steam turbine control system

M M Sultanov[1], V S Lunenko[2] and Y M Chubko[1]

[1] Department of Energy Engineering of Heat Technology, Branch of the National Research University "MPEI", Lenin Avenue, 69, Volzhsky, 404110, Russia
[2] Department of Automated Control Systems, Branch of the National Research University "MPEI", Lenin Avenue, 69, Volzhsky, 404110, Russia

Abstract. Questions of ensuring reliability and safety of operation of complex technical systems are relevant during the operation of power equipment of thermal plant (TPP). The paper presents a generalized statistical analysis of defects and damages over the past 15 years in the control systems of steam turbines of types ET (extraction turbine) and IET (industrial extraction turbine), with an installed capacity of 50 to 135 MW. As a result of the calculation study, the main cause-and-effect relationships of the resulting defects that lead to an emergency stop of the turbines were identified. This paper presents the results of the analysis of defects in steam turbine control systems and provides recommendations for eliminating and preventing unplanned downtime due to an emergency shutdown of turbine units.

1. Introduction

Reliability theory is developed to select optimal technical solutions related to the need to preserve the main technical characteristics of equipment and its elements for the required period of time in certain operating conditions and avoid failures. The main reasons for the failure of technical systems such as steam turbines are sudden (accidental) failures, failures due to deterioration of the characteristics of elements (aging, wear), as well as hidden manufacturing defects characteristic of violation of operating conditions. Aging of equipment is accompanied by a decrease in its reliability and an increase in operating costs [1]. An increase in the failure rate of technical systems is usually associated with stricter conditions for their functioning (operation) and insufficient qualification of service personnel [2].

In modern conditions, when most of the equipment of the domestic heat and power complex has exceeded the standard service life, it is important to ensure the safe and efficient operation of generating enterprises. When managing operation and repair modes, it is necessary to take into account the actual level of reliability of power equipment, taking into account the impact of real operational factors [3]. In this work, generalized statistical analysis of damages in the control systems of steam turbines is performed based on the data of operation statistics for a 13-year period.

2. Methods

For the quantitative assessment of equipment reliability in the heat power industry, a number of complex indicators are used (table 1) [4].

IOP Conf. Series: Materials Science and Engineering 1035 (2021) 012022 doi:10.1088/1757-899X/1035/1/012022

Table 1. Analytical dependencies of reliability indicators calculation.

Criterion	Label	Dimension	Equation
Average load	Nal	MW	$N_{al} = \sum \frac{Pr}{T_P} \cdot 100$
Production	Pr	million kWh	-
Runtime	Tp	h	-
Coefficient of working hours	Kp	%	$K_p = \sum \frac{T_p}{T_\kappa} \cdot 100$
Calendar time in the reporting year (8760 or 8784 hours in leap)	T_k	h	-
Coefficient of technical use	K_{tec}	%	$K_{tec} = \frac{\sum T_p + \sum T_{res} + \sum T_{dep}}{\sum T_K} \cdot 100$
Duration of downtime in forced (emergency) repair	T_f	h	-
Duration of idle time in reserve	T_{res}	h	-
Duration of dependent downtime	T_{dep}	h	-
Coefficient readiness	K_r	h	$K_r = \frac{\sum T_P}{\sum T_P + \sum T_f} \cdot 100$
Coefficient of scheduled downtimes	$K_{sch.d.}$	%	$K_{sch.d.} = \frac{\sum T_{over} + \sum T_{av.r} + \sum T_{cur.r} + \sum T_{res}}{\sum T_K} \cdot 100$
Duration of downtime in overhaul	T_{over}	h	-
Duration of downtime in average repair	$T_{av.r}$	h	-
Duration of idle time in current repair	$T_{cur.r}$	h	-
Coefficient of not scheduled downtimes	$K_{n.sh}$	%	$K_{n.sh} = \frac{\sum T_f + \sum T_{dep}}{\sum T_K} \cdot 100$

Average time to failure	T_f^{av}	ч	$T_f^{av} = \frac{\sum T_p}{\sum n_f}$
Number of stops in forced (emergency repairs)	n_f	-	-
Fault flow parameter	w	blackouts/ 1000h	$w = \frac{\sum n_f}{\sum T_p} \cdot 100$
Average recovery time	T_{rec}^{av}	h	$T_{rec}^{av} = \frac{\sum T_f}{\sum n_f}$

The complex of quantitative indicators allows estimating and analyzing the reliability of equipment and determining the expediency of measures aimed at increasing the reliability, including the expediency of repair [2] of equipment.

For estimation of probable damage from failures/accidents and for forecasting of investments, capital expenses, insurance volumes the analysis of the risk of operation of the equipment according to requirements of GOST 51.901.1-2002 is carried out. By means of collection and processing of information on the reliability of equipment the following problems are solved:

1. Determination of failure/accident causes
2. Identification of those parts and components of the equipment that limit its reliability
3. Identification of operating conditions and modes that affect reliability
4. Determination of economic efficiency of reliability improvement measures

The analysis of rejections is based on the investigation of rejections and other reporting forms. In order to increase the reliability and comparability of the performed assessments, the risk analysis should be performed using the same methodology [5], the same working group, and the same source of source data.

In this paper, we consider control systems with the following types of speed controllers:

Table 2. Brief description of station control systems.

Type (brand) of the turbine	PT-61(65)-115(130)/13	T-48(50)-115(130)	T-97(100)-115(130)	PT-133(135)-115(130)/15
Manufacturer	Leningrad metallurgical plant	Ural metallurgical plant	Ural metallurgical plant	Ural metallurgical plant
Regulatory system	hydrodynamic	hydrodynamic	hydrodynamic	hydrodynamic
Turbine speed regulator	SR-3000-6	hydraulic membrane - tape SR	hydraulic membrane - tape SR	hydraulic membrane - tape SR

IOP Conf. Series: Materials Science and Engineering 1035 (2021) 012022 doi:10.1088/1757-899X/1035/1/012022

Figure 1 shows a T type turbine control system with a single control diaphragm and hydraulic connections (simplified) [5].

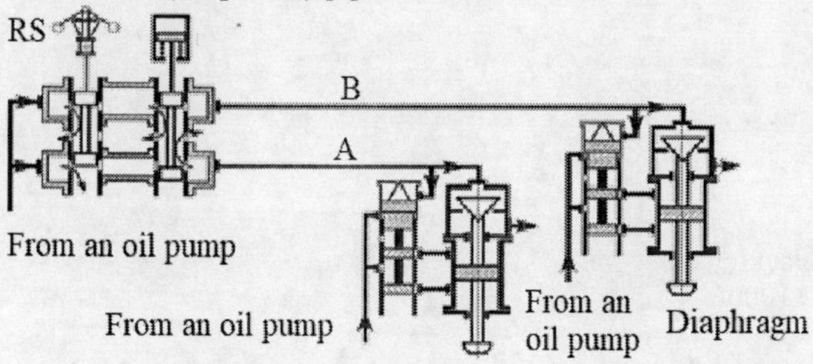

Figure 1. Turbine control system of T type with one control diaphragm and hydraulic connections (simplified).

In accordance with GOST, single and complex reliability indicators are used to quantify the reliability of power equipment. Individual indicators of reliability are parameter flow of failures, the average time between failures, mean time to repair; as complex indicators are used - the coefficient of readiness, coefficient of technical use, utilization of installed capacity, the ratio of unplanned failures and more. Figure 2 shows the control scheme of a PT-type turbine with hydraulic connections (simplified) [6].

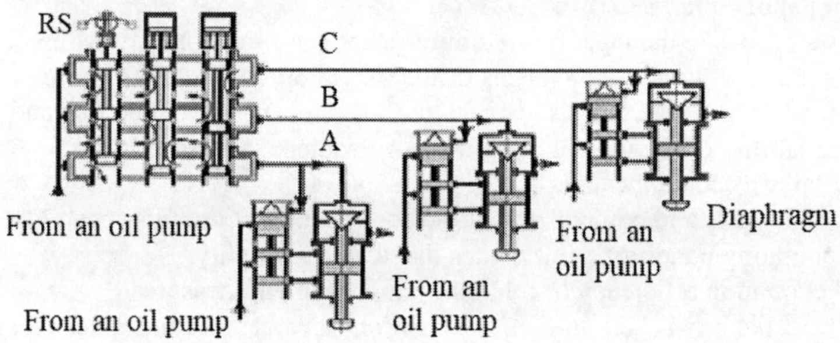

Figure 2. Turbine control diagram of PT type with hydraulic connections (simplified).

For calculations in research dot estimates of parameters of the most applied in the theory of reliability laws of distribution are used and the corresponding methods of finding approximate trust estimates are applied. The structural scheme of reliability of the regulating system is represented in the form of a set of consistently connected subsystems, assembly units, units, mechanisms, etc., failure of each of which leads to failure of the system as a whole. Each subsystem of the structural scheme, in turn, includes a set of elements, which have their own intensities of failures [7-9].

When element-by-element assessing the reliability of calculated point estimates:
- time between failures of i-th component (i-th element)
- failure rate of i-th element
- time between failures of the system in operation or testing
- time between failures point estimation variance of i-th system component
- time between failures point estimation variance of the system
- time between failures point estimation of system ensemble and its variance

Complex indicators of system reliability are also calculated - availability, technical use factors and assessment of system failure rate as a whole is calculated [10].

The paper presents a generalized statistical analysis of defects and damages over the past 13 years in the control systems of steam turbines of types T (extraction turbine) and ПТ (industrial extraction turbine), with an installed capacity of 50 to 135 MW.

3. Results and Discussion

Figures 3 and 4 show statistical data on defects that occurred in the control systems of steam turbines manufactured at LMP and UMP, respectively.

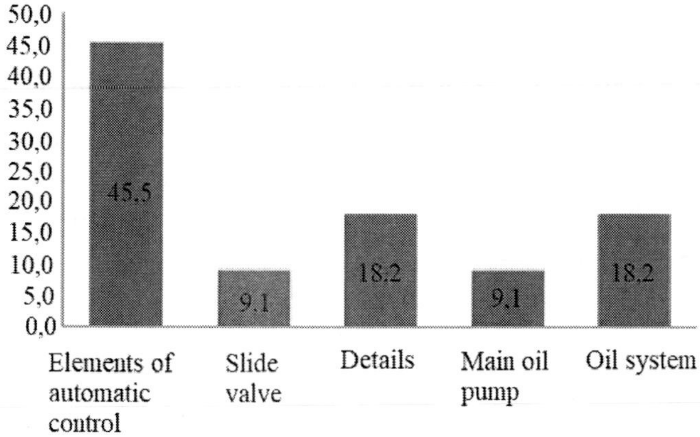

Figure 3. Statistics of defects in the regulatory system of LMP turbines.

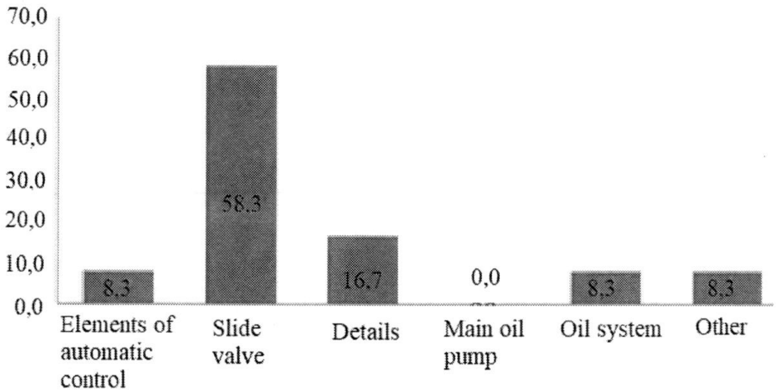

Figure 4. Statistics of defects in the regulatory system of UMP turbines.

As a result of the design study, the main causal relationships of the resulting defects leading to the emergency shutdown of turbines of types PT-61 (65) -115 (130)/13, T-48 (50) -115 (130), T-97 (100) -115 (130) and PT-133 (135) -115 (130)/15 were identified. It has been established that in the vast majority of cases, faults arise as a result of violations during equipment operation, installation defects after completion of current and major repairs, as well as physical and moral wear of units and assemblies.

Table 3. List of some of the main emerging defects in steam turbine control systems and practical recommendations for their elimination.

Failure	Solution
Industrial side steam out of control	Side steam regulator takedown, screw joint jam fixing, stem making and stem gear joint restoration
Industrial and heat side steams out of panel control. Pressure regulator actuator shaft and its manual control arm shaft malalignment	Pressure regulator actuator shaft and manual control arm shaft alignment

Industrial side steam out of control. Cumulative valve doesn't rotate in full swing	Restoring bellow connection rod moving according to the turbine data sheet, making an M6 retention screw
Heat side steam not working (doesn't maintain pressure), steam emission near pressure regulators. Seal failure of heat side steam pressure regulator steam section (bellow damage)	Takedown, bellow replacement, pressure testing
Feedback cone connecting rod oil leak in lower pressure actuator. Bronze bushing wearing	Takedown, blushing to make with gaps according to the turbine datasheet, assembly
Unstable turbine generating unit operation under load and idle motion	Takedown, replacement with newly manufactured ones. Linear dimension adjustment according to technical drawing

4. Conclusion

In this paper, we analyze the evaluation of reliability indicators of technical systems, including indicators of reliability of TPP equipment. The risk analysis of equipment operation is given. The principles of operation of control systems for steam turbines of thermal power plants are considered. Automatic control systems for T type turbines and PT type turbines are described. The analysis of methods for collecting and processing information used to assess the reliability of equipment of existing thermal power plants under operating conditions is performed. The generalized statistical analysis of damages in the systems of steam turbine regulation based on the statistical data for the 13-year period was made.

According to the results of the study, it was determined that the main defects arise due to contamination of the working fluid and failure of the control system elements (bellows, slide valves). To eliminate situations that occur due to clogging of the liners and the oil pipeline, it is recommended to flush the system every time the turbine unit is put out of operation, as well as carefully monitor the quality of the oil and fill the oil filters. Stops of the turbine unit associated with deficiencies in the operating parameters of the control system and operational characteristics are associated with the resource output of the control system elements beyond the design limits. It is recommended to replace the hydraulic part of the system with an electronic one in order to completely eliminate the defects that occur with the turbine control systems in use. These measures will help to reduce the number of nodes that fail during operation, as well as increase the accuracy of setting and triggering the control system. The modern electro-hydraulic control system meets the requirements of reliability, safety and trouble-free operation of generating facilities.

The study found that the failure of the regulation system can not only lead to delays and accidents but in the current job affect the mode of operation of the turbines, the quality of issuance of electric and thermal power, industrial and cogeneration turbines, stability of pressure and temperature, work on the market of electric energy and power according to the set schedule of the load, which greatly complicates the work of operational staff and impact on technical and economic indicators of the station.

References

[1] Brodov Y M, Murmansky B E and Aronson R T 2017 Improving the reliability of technological subsystems equipment for steam turbine unit in operation *Journal of Physics: Conference Series* 891(1)

[2] Sultanov M M, Balandin O I, Skopova E M 2018 *Increasing The Efficiency Of Power Generating System Equipment Operation: Training Manual* Volzhskiy: Branch of National Research University "MPEI" 85 p

[3] Federal Standard 27.002-89 1990 *Reliability In The Equipment. Basic Concepts. Terms And Definitions* M.: Standard 37 p

[4] Federal Standard 27625-88 1998 *Power Units for Thermal Power Plants. Requirements to Reliability, Maneuverability and Economy* M.: Standard 11 p
[5] STO 70238424.27.100.011-2008 2008 *Thermal Power Stations. Methods of Assessment of Main Equipment Condition* 61 p
[6] Trukhanov V M, Tarnayeva A G 2016 *Reliability and Diagnostics of Complex Systems: Textbook* M.: ID "Spectrum" 175 p
[7] Trukhanov V M 1996 *Reliability of Products Mechanical Engineering: Theory and Practice* Moscow: Mechanical Engineering 336 p
[8] Trukhanov V M 1995 *Methods of Ensuring Reliability of Mechanical Engineering Products* Moscow: Mechanical Engineering 304 p
[9] Trukhanov V M 1999 *Reliability in Engineering* Moscow: Mechanical Engineering 598 p
[10] Sultanov M M 2016 *Evaluation of Reliability, Extension of Service Life and Optimization of Repair of Tpp Equipment and Power Systems: Tutorial* Volzhskiy: Branch of National Research University "MPEI" 100 p

Acknowledgments
The research is funded by Russian Federation public contract №0720-2020-0025 "Technique development and method analysis for ensuring power system object security and competitiveness based on the digital technologies."

Application of digital twins of equipment for managing complex renewable energy

A A Smirnov, V S Lunenko and I A Boldyrev

Department of Automated Control Systems, Branch of the National Research University "MPEI", Lenin Avenue, 69, Volzhsky, 404110, Russia

Abstract. The paper describes a control system for renewable energy complexes with optimization of operating modes based on digital twins of equipment. The application of digital twins allowed us to develop a simulator that simulates the main processes occurring in the ground heat pump system. The simulator allows: to perform computer experiments to study different modes of Polygon operation; to demonstrate the physical essence of the ongoing processes to form the qualifications and skills of students in the field of non-traditional and renewable sources of electric and thermal energy. The use of digital twins also allows the implementation of a renewable energy combination management system to improve the energy efficiency of hybrid power complexes. The algorithm developed by the authors for selecting combinations of renewable energy sources to increase the energy efficiency of the complexes takes into account the current values of the operating parameters of the equipment, the forecast of energy consumption, as well as the forecast of power generation by each of the components of the complex.

1. Introduction

Currently, the use of renewable energy sources for generating heat and electricity is widely developed. One of the important aspects of using a combination of different types of renewable energy sources is the real-time optimization of the combinations of equipment involved.

The Polygon of renewable energy sources [1] located on the territory of the branch of "MPEI" in Volzhsky one of the functions of which is to develop algorithms for managing renewable energy complexes and identify parameters for describing digital models of equipment.

When you expanding the hardware model description elements that lead to digital twins creation are virtual prototypes of a real object, a group of objects, or processes.

2. Methods

The technology of creating and using digital twins in the energy sector is becoming more widespread. Leading companies use this technology in their activities [2]. According to [3], the use of digital twins can significantly increase efficiency at all stages of the creation and operation of objects. For example, according to [5], the use of digital twins of operating wind turbines can enable asset owners to virtually and remotely monitor, test, and predict the performance and condition of turbines and subcomponents and carry out performance and reliability tests in a virtual environment.

The Polygon of renewable energy sources [1] located on the territory of the branch of "MPEI" in Volzhsky is used to create digital twins. The following equipment was considered: heat pumps of the "heat carrier-water" and "air-water" types, solar vacuum collectors, circulation pumps, electric heaters, storage tanks, fancoils, heating radiators.

The digital twin stores information about the technical parameters of the equipment used, the date of its commissioning, information about operating hours, repairs, failures, dynamic characteristics, data from measuring devices, experimental data, etc. When using the equipment, the current values of technological parameters, reports, and other data can be synchronized with the digital twin. Based on the collected data, forecasts and estimates are formed that can be used to improve the operation and maintenance of a real object.

When describing renewable energy equipment, you can select duplicate data fields, as well as individual data fields. For example, the first group of parameters can include name, model, date of manufacture, date of commissioning, geometric dimensions, 3-D model, etc. The second group includes characteristics that are individual for a specific type of equipment, for example, for a heat pump, you can distinguish: operating temperature range, flow characteristics, and dynamic characteristics.

An example of the description of digital twins is shown in table 1.

Table 1. The parameters of the equipment of renewable energy.

Parameter	Heat pump	Circulation pump	Electricheater
Title	MammothJ036WHEBLA/ CS	WILO Star 25/8	EVAN -5.6
Company manufacturer	MammothInc (USA)	WILO	EVAN
Productiondate	2010	2010	2010
Commissioning	2011	2011	2011
Size, mm	820 x 660 x 530		500x200x160
Power consumption	2,29/2,93 kW	176.0 W	6 kW
Permissible temperature range	15-55 ° C (heating cycle)	-10° C / +110° C	30-75°C
Dynamic characteristics	$W(p) = \frac{1}{183 \cdot p + 1}$	-	$W(p) = \frac{1}{3100 \cdot p}$

The use of digital twins allowed us to develop a simulator that simulates the main processes occurring in the ground heat pump system [6].

Using the ground heat pump simulator allows you to: perform computer experiments to study various modes of operation of the Landfill; show the physical nature of the processes, their mutual dependence, as well as a number of significant subtleties; to form the qualifications and skills of students in the field of non-traditional and renewable sources of electric and thermal energy (heat pump type "Ground-water").

3. Results and Discussion

The basis for the construction of the simulator is a mathematical model of heat pump control based on existing equipment in the branch of the MPEI in Volzhsky, obtained using digital twins.

Based on the literature data, as well as analytical calculations and experimental data, models of equipment components that are part of the simulated system were obtained. The simulator's mnemonic diagramis shown in figure 1.

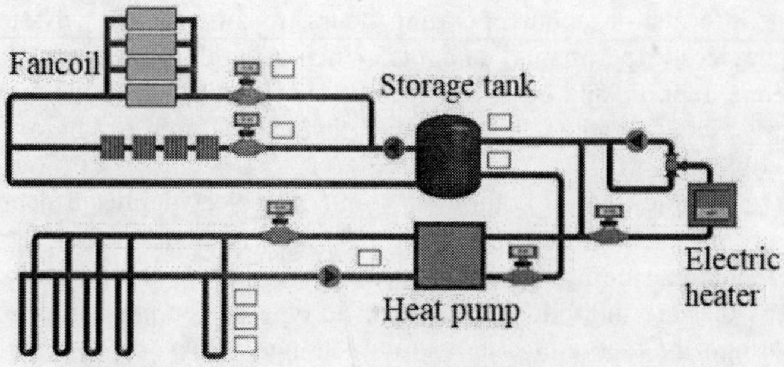

Figure 1. The mnemonic "control of a heat pump."

Also, based on digital twins, you can use the hardware database in the training process, which allows you to select the desired model, export data to CAD for design and calculations.

The use of digital twins also makes it possible to implement a system for managing combinations of renewable energy sources to improve the energy efficiency of hybrid power complexes. A hybrid energy complex is a technical system that combines within a single technological process generator of electric, thermal and other types of energy of various types, accumulators, means of switching, transmitting and converting energy into a form suitable for use by consumers.

The authors developed the algorithm for selecting combinations of renewable energy sources to improve energy efficiency of the complexes takes into account the current values of the parameters of the equipment, the forecast of energy consumption and forecast energy production by each component of the complex. Generation forecasting is based on the developed models of digital twins of renewable energy equipment [6].

Implemented on the basis of the considered ACS algorithm, it has a three-level structure, including the level of field equipment and devices, the level of local control of the complex, as well as the cloud level of implementation of optimization tasks.

The combination of modern technologies (IoT, SmartGrid, digital twins) will significantly increase the efficiency of energy facilities. Similar projects have already been implemented in many countries [8,9,10].

Based on the collected data, rather complex algorithms are used using machine learning methods, fuzzy logic, etc. [11,12].

When controlling hybrid power complexes, it is important to be able to predict energy production and consumption. Implementation of such functions is proposed for control systems of hybrid power complexes [13-19].

The authors used the technology of digital twins in describing algorithms for predicting heat energy consumers. For example, figure 2 shows a thermal energy forecast graph using digital twins. Characteristic points are highlighted during construction:

t_{start} – The time when heat consumption starts.

t_{max} – Time of maximum heat consumption.

t_{end} – The end point of heat consumption.

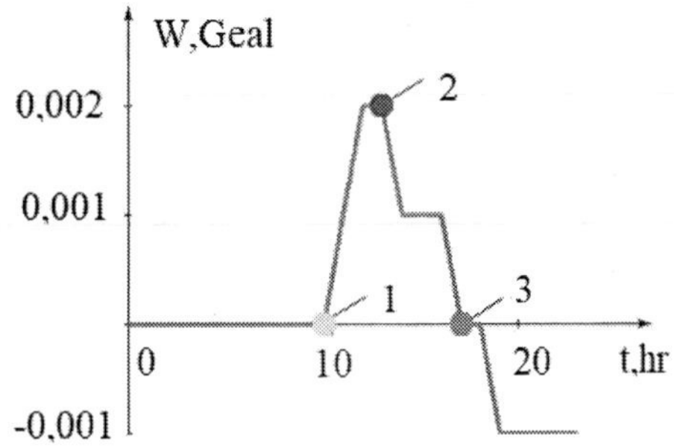

1 - starting point of heat consumption; 2 - maximum heat consumption point;3 -end point of heat consumption.
Figure 2. Result of the heat consumption analysis program.

The application of the above mentioned technologies for hybrid power complexes allows us to significantly increase the energy efficiency of systems. Figure 3 shows the comparison of the amount of thermal energy consumed by the classical control algorithm and the system state prediction algorithm.

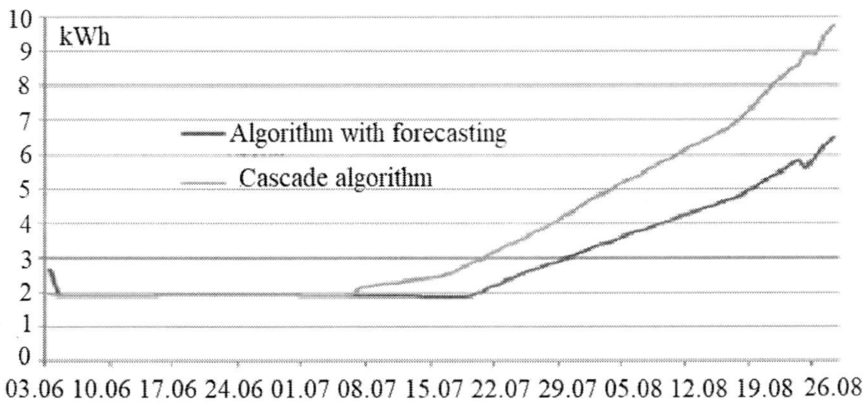

Figure 3. Daily electric energy consumption for the summer 2017.

The authors show that for the Volgograd region the application of digital twin technology and the implementation of the control algorithm with state prediction allows to increase the efficiency of hybrid energy complexes by 16%and for other regions of the Russian Federation up to 20% in case of high non-uniformity of energy consumption during the day.

Application of the detailed description of the equipment allows using digital twins not only for control of the system energy efficiency indicators, but also to carry out actions on virtual commissioning, to switch to the maintenance of the system according to the condition of the equipment, as well as to create simulators for training of the personnel serving the complex [7].

4. Conclusion

Constant feedback from the operating product allows us to evaluate the behavior of the object in real time, predict the behavior of the physical system for earlier detection of faults, prevent sudden failure of the equipment, which generally increases the efficiency of the decisions made. It becomes possible to find the most optimal scenarios of technological processes and to predict deviations in the operation of equipment on the basis of statistical models and analysis. The use of digital twins made it possible to develop a simulator simulating the main processes taking place in the soil heat pump system. The use of a ground heat pump simulator allows computer experiments to be performed to investigate the different modes of operation of the Polygon, as well as to form the qualifications and skills of students in the field of non-traditional and renewable sources of electric and thermal energy. The use of digital twins also allows the implementation of a renewable energy combination management system to improve the energy efficiency of hybrid power complexes. The algorithm developed by the authors for selecting combinations of renewable energy sources takes into account the current values of equipment operation parameters, the forecast of energy consumption, as well as the forecast of power generation by each of the components of the complex.

Calculations show that the application of the proposed algorithms allows us to obtain an increase in the energy efficiency of renewable energy complexes up to 20% for objects having high daily non-uniformity of energy consumption.

References

[1] Boldyrev I A 2013 Study of the efficiency of functioning of combinations of non-traditional andrenewable sources of electric and thermal energy (scientific article) *Alternative energy and ecology* **14** pp49–51
[2] "Industry 4.0 just about complicated" No 154 (September 2018)
[3] "Energovector" Newspaper 2019 No 11
[4] Kokorev D S and Yurin A A 2019 Digital doubles: concept, types and advantages for business *Colloquium-journal* **10(34)**
[5] https://www.smart-energy.com/renewable-energy/new-report-explores-digital-twins-wind-energy-industry
[6] Boldyrev I A and Kuponosova V S 2018 Program for modeling the operation of a ground heat pump *Patent RU 201861953*. **8**
[7] Sultanov M M, Boldyrev I A, Lunenko V S, Smirnov A A and Menshikov P D Program for the organization of verification of knowledge and skills of performing technological operations for the operation of instrumentation and control Systems *Computer program RU 2019661005*
[8] *"Energorynok"Journal* **12**
[9] https://blogs.sap.com/2016/06/07/how-the-internet-of-things-helps-utility-companies
[10] https://www.ge.com/digital/stories/bc-hydro-gains-contextual-insights-two-million-smart-meters
[11] Suganthi L, Iniyan S, Samuel A A 2015 Applications of fuzzy logic in renewable energy systems: A review*Renewable and Sustainable Energy Reviews* **48** pp 585-607
[12] Olatomiwa L, et al. 2016 Energy management strategies in hybrid renewable energy systems: A review *Renewable and Sustainable Energy Reviews* **62** pp 821-835
[13] HiraoN, OhsakiA, Kubota T and Fukumiya K 2018 Demonstrating a predictive control system for a ground source heat pump: A case study from an industry-university-government cooperation project for renewable energy technologies in Japan *IEEE International Conference on Industrial Technology (ICIT)* Lyon Francepp 933-937 doi: 10.1109/ICIT.2018.8352303
[14] Wu D, et al. 2015 A control architecture to coordinate renewable energy sources and energy storage systems in islanded microgrids *IEEE Transactions on Smart Grid* **6(3)** pp 1156-1166
[15] Matos J G, Silva F S F and Ribeiro L A S2015 Power control in isolated microgrids with renewable energy sources and energy storage systems *IEEE Transactions on Industrial Electronics* **62(6)** pp 3490-3498
[16] Bouzid A M, et al. 2015 A survey on control of electric power distributed generation systems for microgrid applications *RenewableandSustainable Energy Reviews* **44** pp 751-766
[17] Kjellsson E, Hellström G and Perers B 2010 Optimization of systems with the combination of ground source heat pump and solar collectors in dwellings *Energy* **35(6)** pp 2667-2673

[18] Wang X, et al. 2010 Experimental study of a solar-assisted ground-coupled heat pump system with solar seasonal thermal storage in severe cold areas *Energy and Buildings* **42(11)** pp 2104-2110
[19] Bi Y, et al. 2004 Solar and ground source heat-pump system *Applied Energy* **78(2)** pp 231-245

Acknowledgments
The research is funded by Russian Federation public contract No. 0720-2020-0025 "Technique development and method analysis for ensuring power system object security and competitiveness based on the digital technologies."

IOP Conf. Series: Materials Science and Engineering 1035 (2021) 012024 doi:10.1088/1757-899X/1035/1/012024

Simulation of the static synchronous compensator for the electrical system with non-linear load

A V Makarov[1,3] , T V Makarova[2], D S Morozov[1]

[1]Department for Electrical Equipment, Institute of Automation and Electronic Instrumentation, Kazan National Research Technical University named after A. N. Tupolev, Chetaev Str., 18a, Kazan, 420126, Russia
[2]Department for Automation and Control, Institute of Automation and Electronic Instrumentation, Kazan National Research Technical University named after A. N. Tupolev, Chetaev Str., 18a, Kazan, 420126, Russia

[3]Corresponding author: AVMakarov@kai.ru

Abstract. This paper presents the STATCOM model in PSIM, which compensate reactive power and current harmonics from the non-linear load. Reactive power consumption and harmonics current of non-linear load given from measurement. The model of STATCOM is a multi-level voltage inverter. Constant system voltage, power factor and harmonics currents are controlled. All modes are realized and plots are plotted.

1. Introduction
STATCOM control strategy is categorized into primary control strategy and secondary control strategy. The primary control strategy offers device reactive power control, constant system reactive power control, power factor control, and constant system voltage control, while the secondary control strategy offers [4].

2. Simulation background
In order to verify the load compensation performance of STATCOM, a professional power electronic simulation software PSIM is used to build a simulation model for an electrical system with a non-linear load. The simulation model is shown below. According to the two different operation modes, two results of simulation schemes were presented to demonstrate the simulation results and STATCOM efficiency.

The rated voltage of the voltage source is 35kV. The non-linear load is connected to the 35kV bus bar. One set of ±130Mvar STATCOM is connected to the same 35kV bus bar. The main technical parameters of 130Mvar STATCOM are shown in table 1. The wiring diagram is shown in figure 2 and the electric circuit of STATCOM power cell are shown in figure 3.

Table 1. STATCOM Comparison total harmonic distortion before and after compensation.

parameter	value	unit
Rated capacity	130	MVar
Rated current	2145	A
Rated voltage	35	kV
Wiring connection	Delta	-

3. Simulation model

Model of STATCOM with non-linear load is shown in figure 1-3.

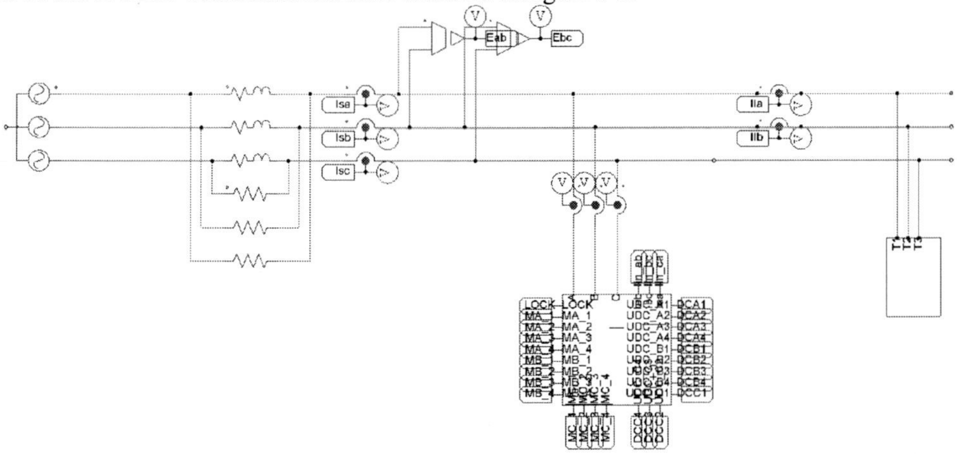

Figure 1. Equivalent electric circuit model with STATCOM and non-linear load.

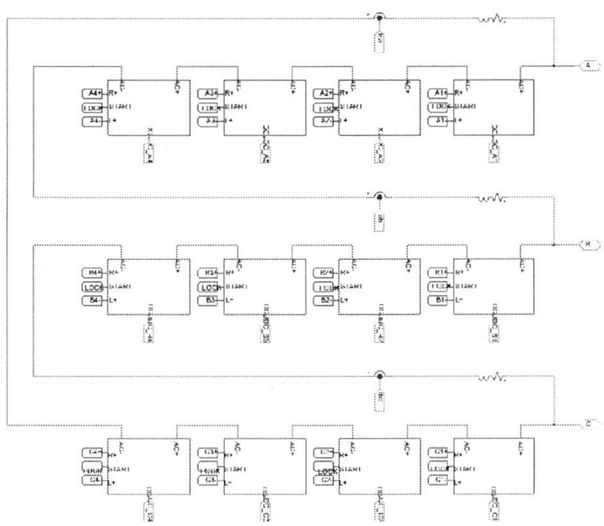

Figure 2. Model of delta connected STATCOM.

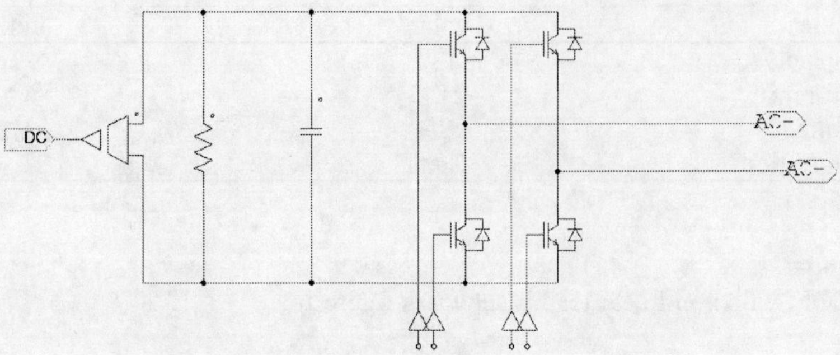

Figure 3. Model of one STATCOM power cell.

4. Non-linear load compensation simulation

4.1. Reactive power compensation

The simulation waveform of the 35kV bus voltage before compensation is shown in figure 4. The load active power *P* and the reactive power *Q* curve are shown in figure 6 (power is calculated as a 20ms average of one power frequency cycle).

Analysis of the above waveforms, the load reactive power fluctuations are large, the instantaneous power factor fluctuates between 0.78 and 0.99, and the average of the load active power is 105 MW.

After compensating with 130 Mvar STATCOM, the 35kV bus voltage simulation waveform is shown in figure 5. The grid active power *P* and reactive power *Q* curve are shown in figure 7 (power is calculated as a 20ms average of one power frequency cycle), and the average active power is 119MW.

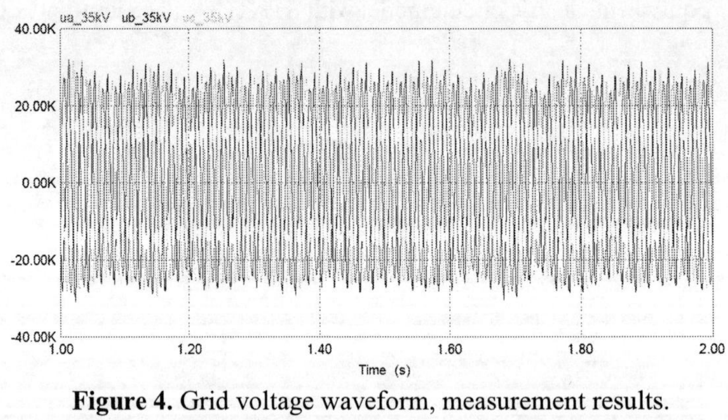

Figure 4. Grid voltage waveform, measurement results.

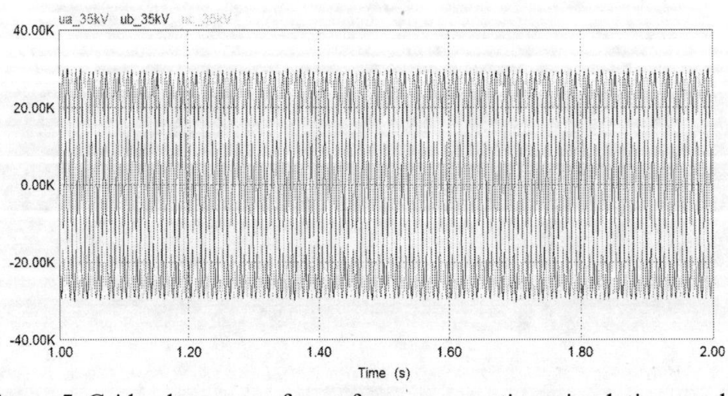

Figure 5. Grid voltage waveform after compensation, simulation results.

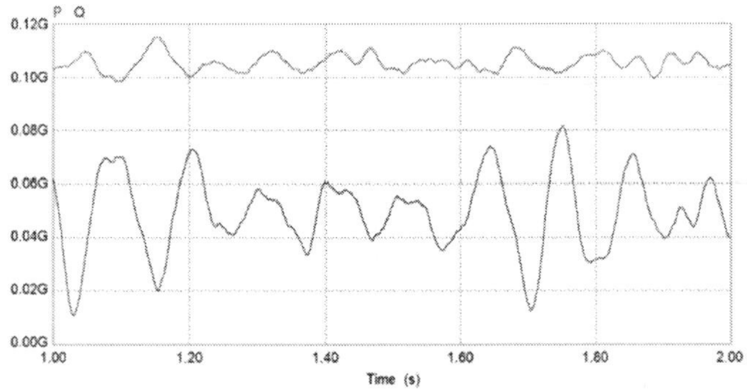

Figure 6. Active and reactive power, measurement results.

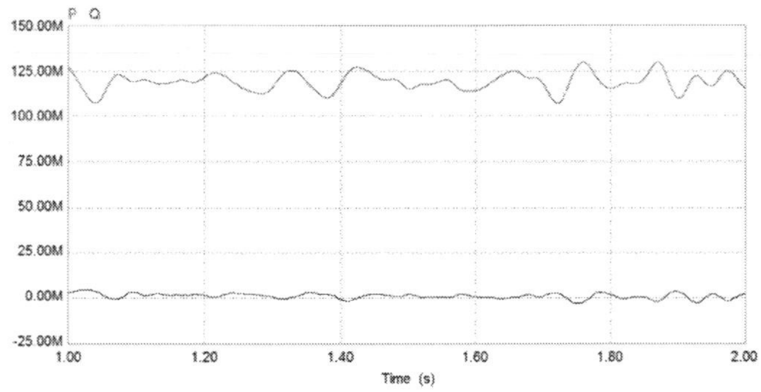

Figure 7. Active and reactive power after compensation, simulation results.

After analyzing the above waveforms, after the STATCOM is put into operation, the reactive power fluctuation becomes smaller, the maximum reactive power is 4.71 Mvar, and the instantaneous power factor is higher than 0.998. The reduction of reactive power results in a significant reduction in grid voltage fluctuations, which in turn leads to an increase in active power and an increase in the active power of 13% for the non-linear load.

4.2. Harmonics compensation

After the STATCOM is started, the harmonic current in the load will be detected, and the currents of the same magnitude and opposite phase are compensated. The uncompensated load current spectrum is shown in figure 8. The compensated grid current spectrum is shown in figure 9. The STATCOM output current waveform is shown in figure 10.

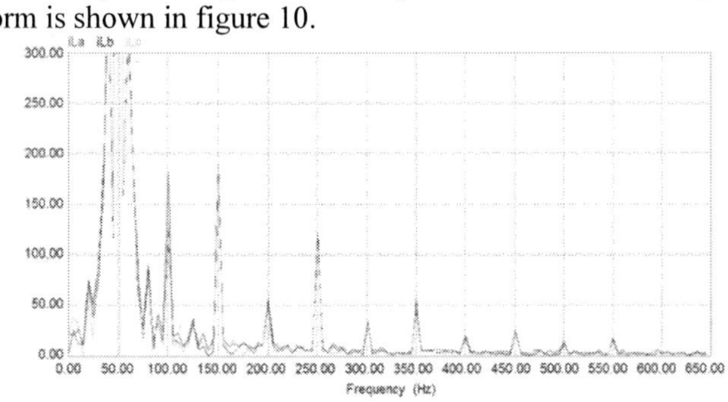

Figure 8. Load current spectrum, measurement results.

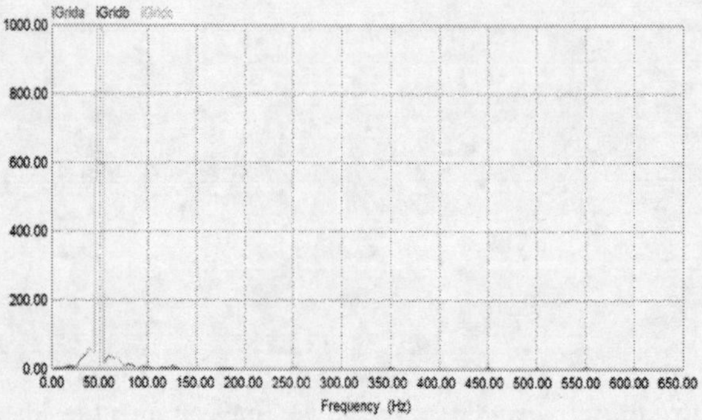

Figure 9. Load current spectrum after compensation, simulation results.

Analysis results of the total harmonic distortion rate before and after the compensation at the STATCOM connected point are shown in table 2.

Table 2. Comparison total harmonic distortion before and after compensation.

	THDu	unit
Before compensation	5.5	%
After compensation	1.8	%

It can be seen from figure 8 that there is a large harmonic component in the STATCOM output current. After the compensation of STATCOM, the harmonic current in the load is basically canceled, and the harmonic distortion rate of the grid current waveform becomes low.

After the STATCOM started the operation. However, considering the filter quality factor and the partial harmonic design, the filtering effect is limited, especially for the second harmonic, it is difficult to have a good filtering effect.

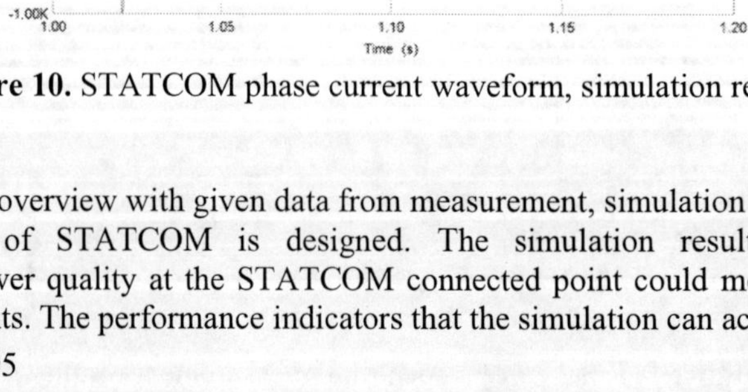

Figure 10. STATCOM phase current waveform, simulation results.

5. Conclusions

For operating projects overview with given data from measurement, simulation verification for the two compensation mode of STATCOM is designed. The simulation results indicate that after compensation, the power quality at the STATCOM connected point could meet and exceed the site assessment requirements. The performance indicators that the simulation can achieve are as follows.

— Power factor ≥ 0.995

— Total Harmonic Distortion $\leq 1.8\%$

The simulation shows that after the compensation device is put into operation, the reactive power is reduced, the bus voltage fluctuation is reduced, and the stability of the bus voltage will cause the active power of the system to be improved.

References

[1] Makarov A, Ferenets A and Metelev I 2019 Simulation 35 kV 35 MVar STATCOM in EMTDC PSCAD *Journal of Physics: Conference Series* **1415** International Conference on Innovation Energy 2–3 October 2019 Perm Russian Federation
[2] Acha E, Fuerte-Esquivel R C, Ambriz-Perez H, Angeles-Camacho C 2004 *FACTS. Modeling and Simulation in Power Networks* A john Wiley & Sons LTD Publ. 403 p
[3] Hingorani, Narain G 1999 *Understanding FACTS: Concepts and Technology of Flexible AC Transmission System* IEEE Press 433 p
[4] Mohan Mathur R., Rajiv K Varma 2002 *Thyristor-based FACTS Controllers for Electrical Transmission systems* A John Wiley & Sons Inc. Publ. 518 p
[5] Pedro Rodrigues, Vítor A Morais, António Martins and Adriano Carvalho 2019 STATCOM Simulation Models for Analysis of Electrified Railways, *IECON 2019 - 45th Annual Conference of the IEEE Industrial Electronics Society* Lisbon Portugal DOI: 10.1109/IECON.2019.8927256
[6] Dekka A, Wu B, Fuentes R L, Perez M and Zargari N R 2017 Evolution of Topologies Modeling Control Schemes and Applications of Modular Multilevel Converters *IEEE Journal of Emerging and Selected Topics in Power Electronics* **5(4)** pp 1631-1656
[7] Chicco G and Mazza A 2019 100 Years of Symmetrical Components Energies **12(3)** pp 1-20.
[8] Shahnia F, Rajakaruna S and Ghosh A 2015 *Static Compensators (STAT-COMs) in Power Systems* Singapore:Springer

Reliability assessment of gas supply to individual consumers using digital modeling

G N Martynenko[1], N A Petrikeeva[1], S A Gorskikh[1,2] and A A Gorskikh[1,2]

[1] Voronezh State Technical University, 20 let Oktyabrya Str., 84, Voronezh, 394006, Russia
[2] OJSC "Gazprom gas distribution Voronezh", Nikitinskaya Str., 50A, Voronezh, 394018, Russia

Abstract. In this work, an algorithm for assessing the reliability of gas supply for individual enterprises was developed and a block diagram of the application program was compiled. The calculation was made in accordance with the principles of the theory of the probability of disconnecting consumers using digital modeling.

1. Introduction

Under the criteria of reliability of gas distribution in science, it is considered to be the transportation to consumers of the specified costs of the target product, in particular natural gas, with uniquely defined parameters in certain operating conditions during the designated time period. A feature of municipal gas distribution systems is that they quickly replace worn-out components and elements with new ones. However, to justify reliability, a finite period of time is necessary so that during it there is no significant reconstruction or restoration of the engineering system [1, 7].

The social aspects of failures in the gas supply system are very important. They are taken into account primarily because modern gas supply systems must ensure the supply of natural gas to household, municipal and industrial consumers in a continuous mode and without significant interruptions.

The theoretical postulates are that systems must operate without failures and fail-safe, which is quite difficult to achieve for gas services in a state of the constant economy. This implies the need to look for simpler ways to increase the reliability indicator of the system through its structural redundancy. Currently, according to the rules of the system, the latter should go into emergency mode. This phenomenon occurs when one or several elements fail. Gas is supplied to more consumers, but less than the calculated values, given the short duration of the emergency. In emergency situations, the limited gas supply is determined by the reserve power of the system, that is, its throughput.

2. Assessment of the reliability of gas supply to individual large consumers

The basic concept for assessing the health of an element or system is the assessment of the probability of failure-free operation during a given time interval t.

Gas supply systems - repaired systems. In this regard, they are characterized by maintainability. The maintainability indicator is that period of time for which a defect (malfunction) of the system can be eliminated. In theory, there is a fundamental characteristic of the reliability of system elements - this is the frequency of its failures (the rate of occurrence of emergency situations). It is considered as the probability of failure of any node in the system at time dt following the time of failure-free operation t.

The failure rate λ is determined experimentally. For most nodes in the system, the failure rate is divided into three periods. First, the first period is observed - the running-in period. During the running-in period, components that have latent defects usually fail. The rate of such failures decreases over time.

At the end of the running-in period, the probability of failure of one or another element is almost constant. The second period follows - the period of normal operation. This is the main time the cell operates when the failure rate is constant. In the third period - the period of operation, the inevitable aging of system nodes occurs. The wear and aging of components and parts begin to manifest itself to a greater extent. The period of operation is more characterized by an increasing failure rate.

All the periods considered above are very relevant for non-repairable systems. The failure rate also applies to non-repairable or rarely repaired systems with long trouble-free operations. As for gas distribution systems, they are more than repairable. However, the above pattern is quite true for new gas supply systems. But it will be relevant in consideration until the first failure of network elements [1]. Functional dependencies associated with λ and λ (t) are given in [1]. The considered characteristics are partially used in the development of assessments of the reliability of repaired areas.

When studying the reliability of gas supply to individual consumers, the results of a computational experiment conducted on a computer for the municipal gas supply system of the district of Lipetsk are taken as initial data. The scheme of the network in question is shown schematically in figure 1 [2, 4, 5]. On sections of the network (seven sections), throttle controls are provided. They are located in areas supplying natural gas to specific consumers (a group of consumers) [3].

If a failure is detected within the system, operating personnel are forced to disconnect certain consumers. We denote such a consumer by the letter *i*. Then, to assess the failure rate of this *i* consumer, this parameter should be determined during the year. If we involve the theory of probability and its postulates, then the disconnection of the *i-th* consumer during the year will be determined by the following relationship

$$z_i^{dis} = 1 - \ell^{-\sum_k w_k} \tag{1}$$

where $\ell^{-\sum_k w_k}$ - the sum of all available parameters of the failure flow of elements by *i* consumer, which will lead to a shutdown.

The standard value of the probability of disconnection is normalized and defined as $z_{dis\ i}^{norm} = 0.005$. The following condition must be met: $z_i^{dis} \leq z_{dis\ i}^{norm}$. If it is not fulfilled, then the probability of failure is significant, therefore, it is necessary to reduce the length of networks in a certain area or the amount of equipment. The total network failure flow is determined by the formula

$$\sum_k w_k = w_1 + \cdots + w_k \tag{2}$$

where w_k- failure flow parameter for each branch.

The probability of failure in the *i-th* section is defined as

$$w_i = w_r \cdot l_r + w_o \tag{3}$$

where w_r – failure rate, l_r – branch length, w_o – the probability of equipment failure on the branch in question.

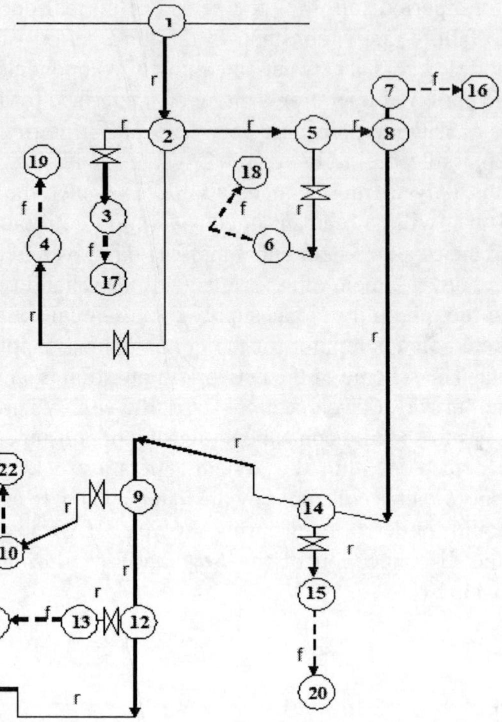

Figure 1. District gas network structural graph.

Figure 1 shows the binary network structure of a gas distribution network with real sections r and a complex of subscriber subsystems *f* [2]. The energy nodes of the settlement zone of the system were determined - 3,4,6,7,10,11,13,15 and the energy nodes of the complex of subscriber subsystems - 16, 17, 18, 19, 20, 21, 22, 23.

To solve the problem, the graphical user interface of the application program [6] is used to assess the reliability of gas supply to individual consumers. The block diagram of the program is shown in figure 2.

In order to calculate the probability of a consumer disconnecting, it is necessary to enter the appropriate values of the failure coefficient, branch length and equipment failure probability for the corresponding sections in the program. Table 1 shows the required values to determine the likelihood of consumers disconnecting through the network of the district of the city of Lipetsk [2].

Table 1. Land Data.

Plot	Failure rate	Branch length	Equipment failure probability
1-2	0,002	0,4	0
2-4	0,002	0,87	0,0003
2-3	0,002	0,08	0,0003
2-5	0,002	0,46	0
5-6	0,002	0,31	0,0003
5-8	0,002	0,49	0
8-7	0,002	0,15	0

8-14	0,002	1,05	0
14-9	0,002	0,92	0
9-10	0,002	0,37	0,0003
9-12	0,002	0,36	0
12-13	0,002	0,05	0,0003
12-11	0,002	0,79	0,0003
14-15	0,002	0,08	0,0003

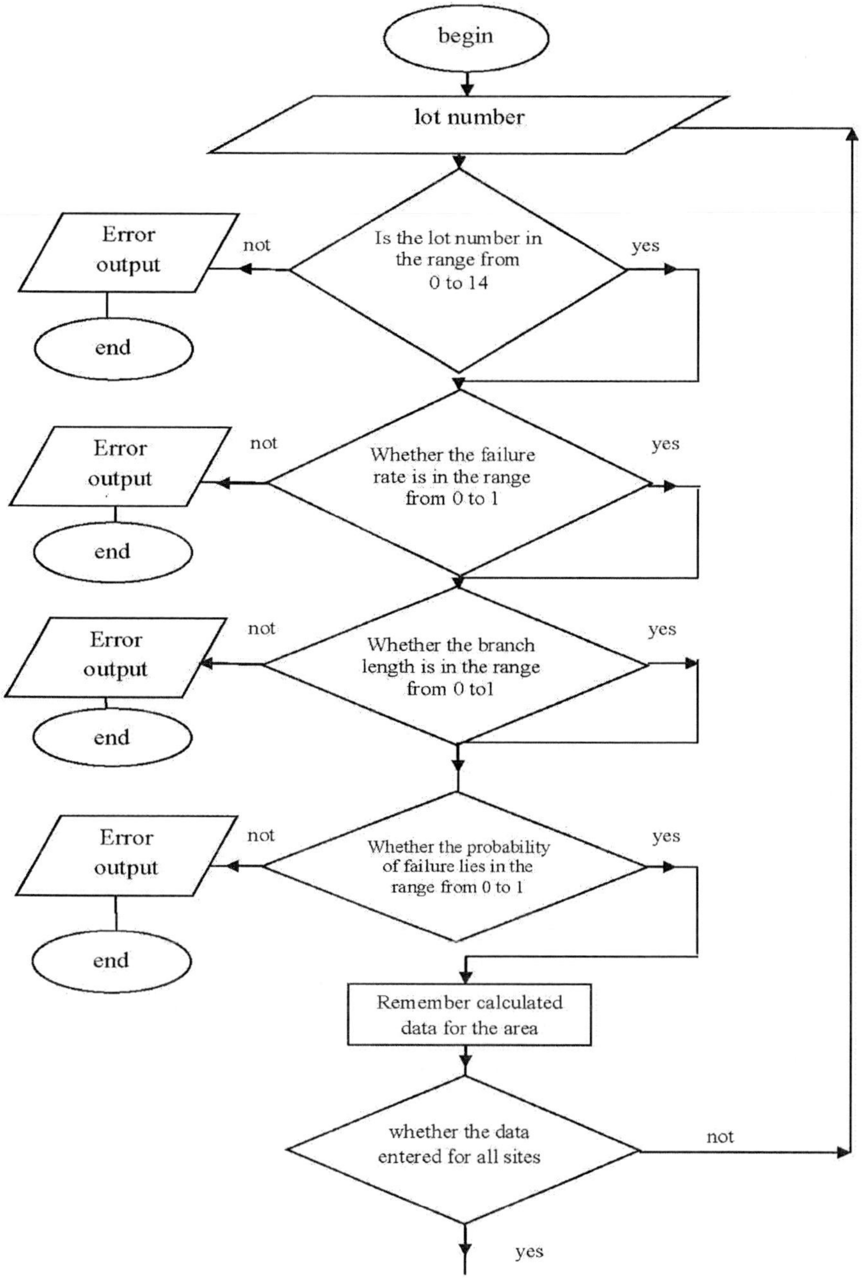

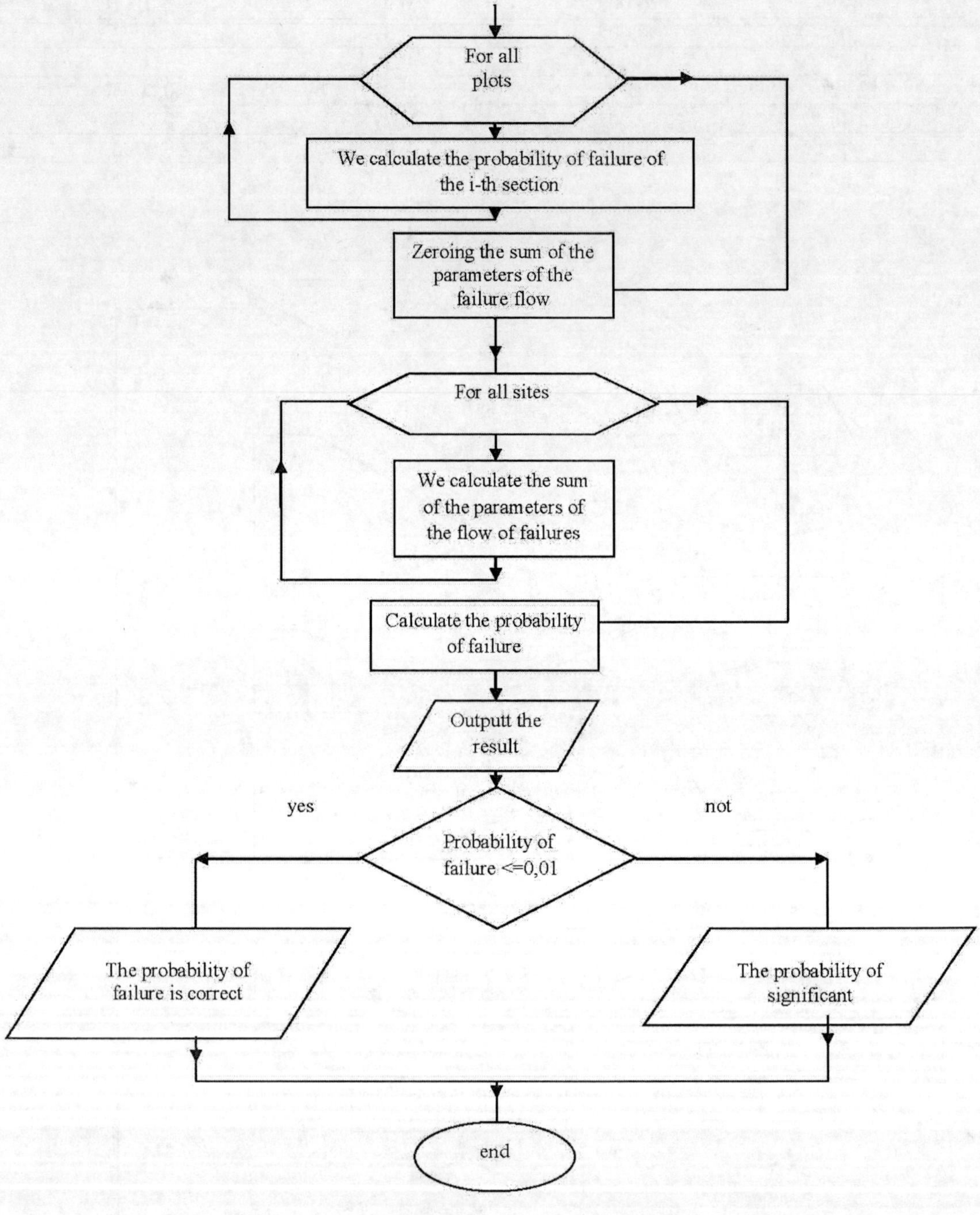

Figure 2. Block diagram of a computing application program.

According to relevant data, using a software product, we calculate the probability of disconnecting consumers for chain branches of the city gas supply system in the district of Lipetsk. Table 2 shows the results of the calculations.

Table 2. Calculation results.

Branch	Probability of disconnection
1-2-3	0,0012
1-2-4	0,0028
1-2-5-8-7	0,0029
1-2-5-6	0,0026
1-2-5-8-14-15	0,0049
1-2-5-8-14-9-10	0,0076
1-2-5-8-14-9-12-13	0,0077
1-2-5-8-14-9-12-11	0,0089

The probability of disconnecting consumers during the year in chain directions 1-2-3; 1-2-4; 1-2-5-8-7; 1-2-5-6; 1-2-5-8-14-15 deadlock network amounted to less than the normalized value, and in the directions 1-2-5-8-14-9-10; 1-2-5-8-14-9-12-13;1-2-5-8-14-9-12-11 exceeded the normalized indicator. Consequently, the probability of failure is not significant in five directions, which means it can be argued that these chain directions correspond to the required reliability standards. The remaining areas will lesser meet the requirements for reliable operation, which will allow you to pay attention to specific areas in advance and take measures to detect failures.

3. Conclusion

As a result of the research, an algorithm for assessing the reliability of gas supply for individual enterprises was developed and a block diagram of the application program was compiled. The program was developed in the environment of Borland Delphi 7 in the programming language Delphi Language. The calculation was made in accordance with the postulates of the theory of probability of disconnection of consumers. The calculation results confirmed the adequacy of the developed algorithm and application program.

References

[1] Ionin A A 1989 *Gas supply* (Moscow: Stroyizdat) P 415

[2] Martynenko G N 2004 Modeling of processes of operational management of urban gas supply systems based on factor analysis: dis. for the degree of Cand. of tech. sciences. (Voronezh) P 182

[3] Martynenko G N, Panov M Y, Shcherbakov V I and Davydova I P 2004 Optimal synthesis of hydraulic pipeline systems in the field of operational management *News of higher educational institutions. Construction.* **2(542)** pp 78-83

[4] Panov M Ya, Martynenko G N and Dmitriev I A 2008 Algorithm for identifying the hydraulic characteristics of controlled chokes on the branches of the structural graph of subscriber subsystems *Scientific Bulletin of the Voronezh State University of Architecture and Civil Engineering. Construction and architecture* **3(11)** pp 106-112

[5] Martynenko G N and Gnatyuk S N 2012 Operational management of a gas distribution system based on a perturbed state model Scientific journal. *Engineering systems and structures* **1(6)** pp 36-42

[6] Martynenko G N, Strizhko S V and Pershin P A 2010 Creation of an optimal regime of gas consumption by means of operational management Scientific journal. *Engineering systems and structures* **1(2)** pp 105-108

[7] Majitov Z Z and Kalandarov N O 2017 Reliability assessment of gas mains *Equipment.Technologies. Engineering* **4(6)** pp 21-27

Assessment of the intensity and flow of straight solar radiation falling onto the horizontal surface of a room

S A Yaremenko, R A Sheps, A A Mershchiev[1] and D V Lobanov
Department of Housing and Communal Services, Voronezh State Technical University, 14 Moskovsky Avenue, Voronezh 394026, Russia

[1]Corresponding author: sasha__1990@mail.ru

Abstract. The paper presents the results of experimental studies of straight solar radiation in the Central Chernozem region of Russia. The intensity and flow of straight solar radiation falling onto a horizontal surface of a room were evaluated. Measurements were made of the flow of straight solar radiation on a horizontal surface through a translucent structure. Experimental data were obtained using a high-class measuring device in the period from July 1 to August 1. The measurement results showed an intense wave-like change in the solar radiation flux over 8 hours. The paper describes a method for estimating the intensity and flow of straight solar radiation falling onto a horizontal surface of a room. This method of evaluating daytime radiation allows one to determine the density of the total and straight radiation flux. Using this estimate to calculate solar installations used in the Central Chernozem region, their efficiency will increase by 10%.

1. Introduction

Renewable energy sources are of interest at all levels of the world community. Renewable energy sources have gained significance in energetics and politics. This is due to the increased interest in the environmentally safe and sustainable energy supply of mankind in the future [1, 2].

Accurate prediction and calculation of solar radiation flows falling onto a horizontal surface at a specific geographical point are not possible in practice, since meteorological information is not always available and reliable [3, 4]. To calculate the characteristics of solar installations, it is necessary to use the data from experimental measurements of solar radiation parameters in the considered location. The most reliable information is the amount of straight solar radiation falling onto a horizontal surface.

The characteristics of solar radiation are often presented in different ways, but the following information is important for understanding and using solar radiation data:

1) the results of a momentary measurement or for a certain period;
2) the measurement time interval;
3) the measured radiation (straight, diffuse, or total);
4) the orientation of the receiving surface to the cardinal directions;
5) the averaging period (month or day).

Most of the measurements of the incident solar radiation flux include direct and scattered ones and relate to the horizontal receiving surface. Such measurements can be made using thermoelectric pyrometers. They are used to measure the most popular characteristics of solar radiation. The sensitivity of the sensors of such devices does not depend on the wavelength in the spectrum of solar radiation, and they are protected from external influences. In addition, these devices must not be sensitive to the angle of incidence of radiation. Most of these devices allow one to get instantaneous data series that are not average or integral values. Therefore, data is stored in the device's memory or on a personal computer and integrated graphically.

2. Problem statement

To estimate the flow of incident solar radiation, you can also use the results of measuring the time during which the solar disk in the sky is visible when the sky is clear. The average sum of total monthly day solar radiation is the most available data on the flows of incident solar radiation falling onto a horizontal surface. Similarly, hourly amount of the total radiation received by the horizontal surface is available. Data from ground-based measurements of solar radiation fluxes made under the auspices of the world meteorological organization are presented in the archives of the World Radiation Data Center. However, it should be noted that the accuracy of the data obtained differs from the corresponding modern requirements, the error of which is from 5% to 20%.

Measurement of solar radiation flows falling onto inclined or horizontal surfaces is important in determining the energy input of solar collectors. For a long time, the traditional unit for measuring the flux density of incident solar radiation was the calorie/cm^2. Currently, the more familiar unit of measurement is MJ/m^2 or W/m^2.

3. Conducting experiments

Detailed data on solar radiation is necessary for calculating the dynamic behavior of solar installations. They are also used to model the long-term operation of installations. In this regard, we measured the flow of direct solar radiation onto a horizontal surface through a translucent structure. The measurements were made using a high-end device from July 1 to August 1, 2019.

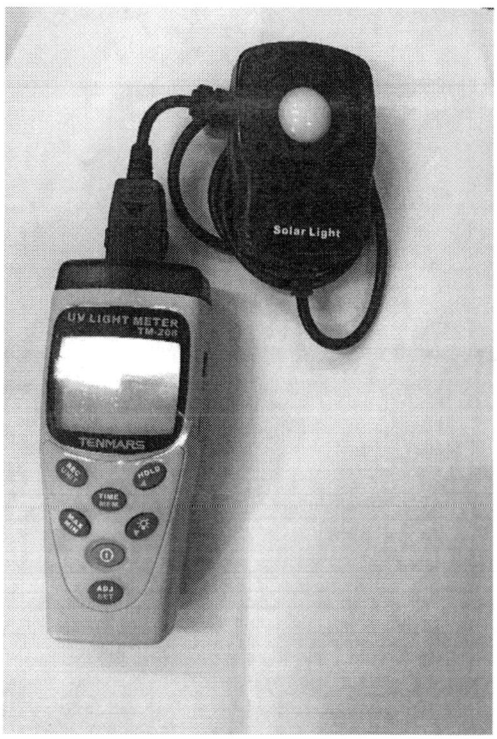

Figure 1. Solar radiation measurement device.

Measurements were made on the horizontal surface of the experimental model of the room with a size of 1×1×1 m. The model of the room was made of light structures with one of the vertical walls completely made of glass. The model fully meets the requirements for thermal protection for the climatic parameters of the Voronezh region and is shown in figure 2.

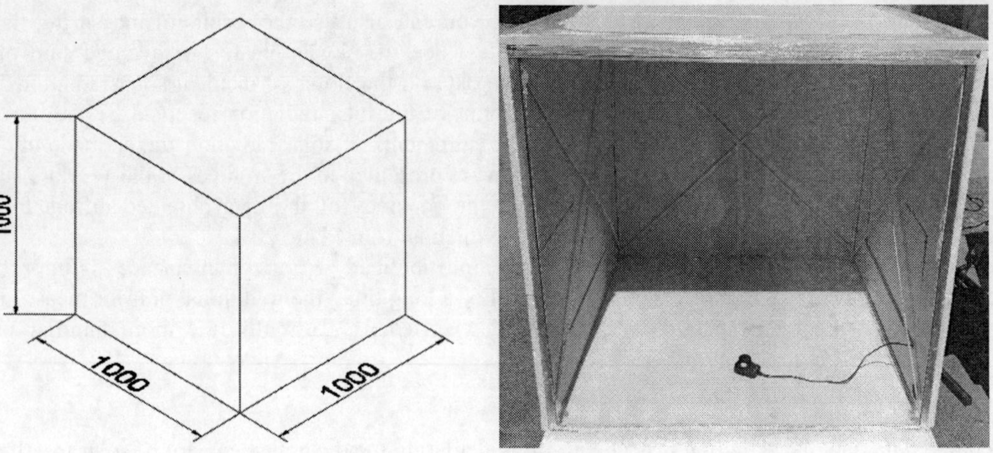

Figure 2. Experimental model.

The measurement results showed an intense wave-like change in the solar radiation flux over 10 hours. Figure 3 shows the measurement graphs.

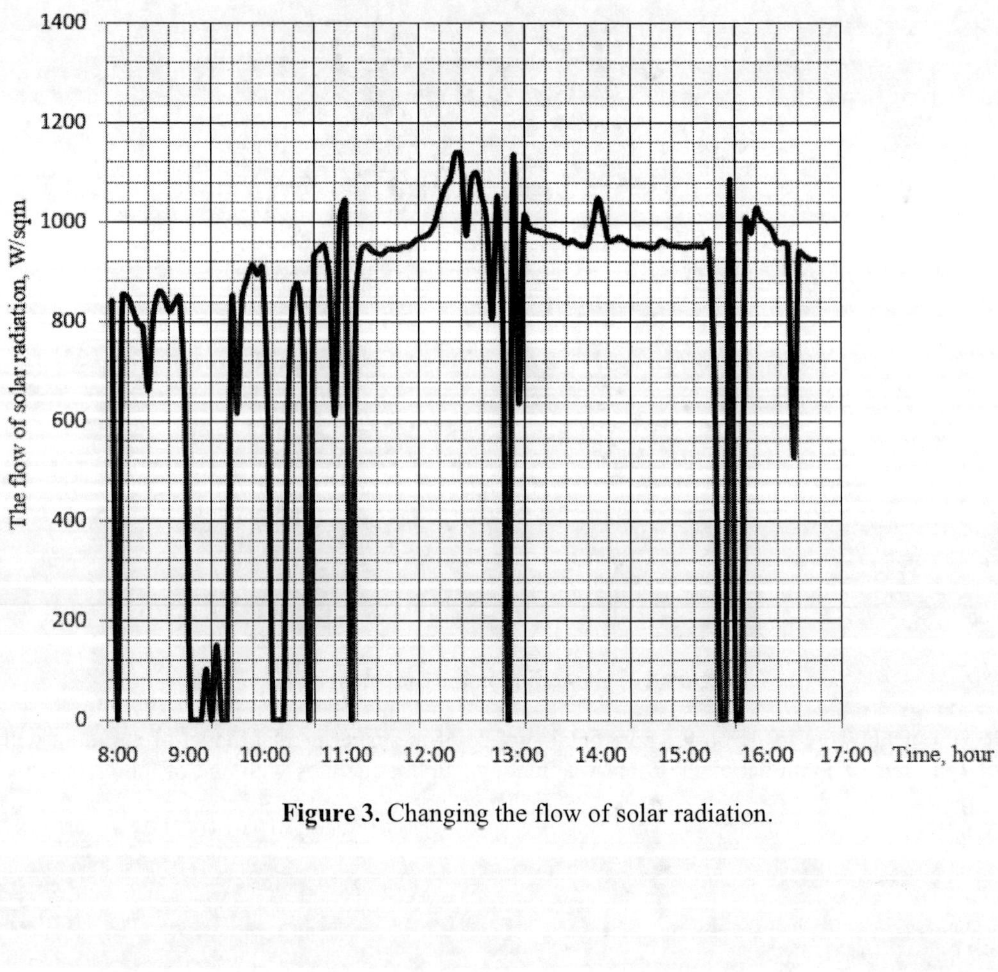

Figure 3. Changing the flow of solar radiation.

These measurements are the best source of information for estimating the average values of incident solar radiation. Using the obtained values helps avoid large errors in the engineering method of calculating the performance of solar installations [5, 6].

4. Calculations

The state of the atmosphere and its mass changes during the day. The scattering and absorption of incident solar radiation also depend on time. When calculating the flows of solar radiation falling onto a horizontal surface, it is necessary to use the concept of a standard (clean) sky. In [7], Professor of the Massachusetts Institute of technology Hoyt Clark Hottel proposed a method for assessing the flow of direct solar radiation propagating through a transparent atmosphere, which considers the Zenith angle and height of the measurement site. However, the method allows calculations for standard atmospheric conditions and four types of climate. In this case, the atmospheric transparency coefficient for direct solar radiation τ_b is equal to the ratio of the scattered radiation flux to the flow of the transatmospheric radiation falling onto a horizontal surface and is determined by the formula:

$$\tau_b \equiv a_0 + a_1 \exp\left(\frac{-k}{\cos\theta_z}\right), \quad (1)$$

where $\cos\theta_z$ - the cosine of the sun's Zenith angle [8, 9], and is determined by the following formula:

$$\cos\theta_z = \cos\phi\cos\delta\cos\omega + \sin\phi\sin\delta\ , \quad (2)$$

a_0, a_1, k - the constant coefficients for a standard atmosphere with a visibility of 23 km are calculated from the coefficient values a_0^*, a_1^*, k^*, which are defined for altitudes in the atmosphere less than 2.5 km by the following expressions:

$$a_0^* \equiv 0.4273 - 0.0821(6 - A)^2, \quad (3)$$

$$a_1^* \equiv 0.5055 + 0.0595(6.5 - A)^2, \quad (4)$$

$$k^* \equiv 0.2711 + 0.01858(2.5 - A)^2, \quad (5)$$

where A – the height of the observation in kilometers.

The transparency of the standard atmosphere for direct radiation can be determined for any Zenith angle and any height. As a result, the flow of direct radiation to the normal surface of the beam in a clear sky is determined as follows:

$$G_{cnb} \equiv G_{on}\tau_b,\ \mathrm{W/m^2}, \quad (6)$$

where, G_{on} – the solar radiation flux density, $\mathrm{W/m^2}$.

But to get a total radiation flux falling onto a horizontal surface, it is necessary to estimate the intensity and flow of direct solar radiation in clear sky conditions. In [10], the following empirical relationship was proposed between the transparency coefficient for direct and scattered radiation in a clear sky:

$$\tau_d \equiv \frac{G_d}{G_0} = 0.271 - 0.294\tau_b, \quad (7)$$

where $\tau_d = G_d / G_0$ - the ratio of the scattered radiation flux to the transatmospheric flux falling onto a horizontal surface. The above equation can be used to estimate the flow of direct solar radiation and its intensity.

Thus, the flow of radiation falling onto a horizontal surface can be found as follows:

$$G_{cd} \equiv G_{on}\cos\theta_z\tau_d,\ \mathrm{W/m^2}, \quad (8)$$

Based on the measurements, we estimate the transparency of the atmosphere for the direct flow of solar radiation for each hour (the middle of an hour) and calculate the flow of radiation falling onto the horizontal surface. We calculated τ_d for the time interval from 10 to 11 hours (i.e. for 10:30). Using the height of the standard atmosphere equal to 0.27 km we get the following coefficient values:

$$a_0^* \equiv 0.4273 - 0.0821(6 - 0.27)^2 = 0.154,$$

$$a_1^* \equiv 0.5055 + 0.0595(6.5 - 0.27)^2 = 0.736,$$

$$k^* \equiv 0.2711 + 0.01858(2.5 - 0.27)^2 = 0.363.$$

The cosine of the Zenith angle of the sun, for a given time, is $\cos\theta_z = 0.507$.
Then we find the atmospheric transmittance by formula (1):

$$\tau_b = 0.154 + 0.736\exp\left(\frac{-0.363}{0.507}\right) = 0.513.$$

From equation (6), the flow of direct radiation to the surface normal to the beam in a clear sky, according to the results of measurements, will be equal to:

$$G_{cnb} \equiv 980 \cdot 0.513 = 502.74 \text{ W/m}^2.$$

In this case, using formula (7), we find the ratio of the scattered radiation flow to the flow of the transatmospheric falling onto the horizontal surface.

$$\tau_d \equiv 0.271 - 0.294 \cdot 0.513 = 0.12.$$

Substituting the obtained value in equation (8) we get:

$$G_{cd} \equiv 980 \cdot 0.507 \cdot 0.12 = 59.62 \text{ W/m}^2.$$

5. Conclusion

When using this method of estimating and calculating the radiation flux falling onto a horizontal surface for calculating solar installations used in the Central Chernozem region, the efficiency and accuracy of selecting such installations on a real area will increase, which will give a subsequent increase in productivity of up to 10%.

References

[1] Sheps R A, Yaremenko S A and Pereslavtseva I I 2017 Influence of solar energy on thermal protection of constructions *Proc. of 1st Russian Scientific and Technical Conf. Devoted tothe Memory of the Dr. Sci. Tech., Professor A A Sander: Energy Saving and EnergyEfficiency at the Industrial Enterprises Housing and Communal Services* pp 199-209

[2] Sheps R A, Portnova N V, Shchukina T V and Pereslavtseva I I 2018 The use of solar radiation in the application of energy-efficient walling *Bulletin of Voronezh State Technical University* **14(3)** 46-51

[3] Sheps R A, Kushchev L A, Shashin A V and Lobanov D V 2017 Influence of dustiness of enclosing structures on the ability to absorb solar energy *Privolzhsky Scientific J.* **4(44)** 51-59

[4] Fathi N Y and Samer A 2016 View factors of flat solar collectors array in flat, inclined, and step-like solar fields *Trans. ASME. J. Sol. Energy Eng.* **138 6** 061005/1-8

[5] Vorotyntsev A V, Bolgov V A, Sheps R A and Shchukina T V 2018 Economic efficiency of the walling with the heat storage capacity *Economics and Entrepreneurship* **7-9** 55-62

[6] Beccali M, Cellura M, Longo S and Guarino F 2016 Solar heating and cooling systems versus conventional systems assisted by photovoltaic: Application of a simplified LCA tool *Sol. Energy Mater. and Sol. Cells.* **156** 92-100
[7] Hottel H C 1976 A simple model for estimating the transmittance of straight solar radiation through clear atmospheres *Solar Energy* **18** 129
[8] Iqbal M 1983 *An Introduction to Solar Radiation* (Toronto: Academic)
[9] Spencer J W 1971 Fourier series representation of the position of the sun *Search* **2(5)** 172
[10] Liu B Y H and Jordan R C 1960 The interrelationship and characteristic distribution of straight, diffuse and total solar radiation *Solar Energy* **4(3)** 1

IOP Conf. Series: Materials Science and Engineering 1035 (2021) 012027 doi:10.1088/1757-899X/1035/1/012027

Hardware solutions for connecting static capacitor banks in high voltage electrical networks

V V Babenko, I A Khaichenko, Yu V Nefedov[1] and V N Krysanov

Department of Electric Drive, Automation and Control in Technical Systems, Voronezh State Technical University, 14 Moskovskiy Avenue, Voronezh 394026, Russia

[1]Corresponding author: nefedowjurij@yandex.ru

Abstract. This work is dedicated to finding effective hardware solutions for implementing the power part of reactive power compensation units for high voltage electric networks. It shows the possibility of cooperative usage of thyristor and electromechanical key elements for switching static capacitor batteries in high-voltage networks. Based on the suggested hardware solutions main technical and cost characteristics of thyristor capacitor units are determined. The rationality of their use in industrial high-voltage electrical networks of 6-10 kV is concluded. It is shown that the usage of hybrid thyristor capacitor units allows one to reduce the expenses, to provide the required range of accuracy of reactive power regulation, to cut power losses in the network, to guarantee normalized voltage level in the load nodes and to increase the reliability and duration of power equipment life cycle.

1. Introduction

Nowadays it becomes highly important for a country's successful development to decrease the usage of the resources and to derive benefit from rational use of all types of means to the uttermost. In the face of this challenge, power generating industry is to complete various tasks which includes the following:

- to increase significantly the efficiency in the electric power industry and to reduce the losses in networks;
- to improve the quality and reliability of the electric industry processing by the advanced management of electric networks.

It makes sense to take full advantage of using the Smart Grid concept (creating an "intelligent network" of active and adaptive transferring and dispatching control systems) to address these tasks. Smart Grid includes flexibility, accessibility, reliability, and cost-effectiveness of the electric network management [1-5]. It is well-known that the role of intellectualization of the high-voltage distribution electric networks (6-35 kV) is downplayed at present.

Currently the electric network element (distributing regional power supply networks) is the foundation of the economic progress and welfare of our country. The growth of large cities, the expansion of energy consuming industrial enterprises, an increase in the density of electrical loads pose a new challenge for electric grid companies in reliable and efficient transportation and distribution of electricity.

The electrical power systems of these facilities make up the lion's share of all distributing networks (with the exception of the ones of 0.4 kV) and stay the closest to direct consumers. However, the process of intellectualization in distributing electric networks of medium voltage (6-35 kV) is left unaddressed. Even though the percent of technological losses of electric energy in distributing electric networks with a voltage of 6-10 kV averages 8-12% of the amount of electricity supplied to the network. The amount of electricity loss is determined by the parameters of the electrical circuit, network design and load conditions.

The calculations for 10 kV networks have shown that energy losses depend significantly on the amount of reactive power transmitted to consumers through the network elements. For example, when the power factor (tan φ) changes from 0.5 to 0.8, the energy loss increases by about 20%. This task can be easily resolved with the help of intellectualization of distributing electric networks by applying the usage of modern automatic reactive power compensation systems. These devices are highly demanded, and they can significantly increase the level of energy efficiency of distributing electric networks [6].

2. Compensation of reactive power in high-voltage power networks

The above stated was acknowledged by scientific research, including the analysis of large statistical data of the city electric network. We analyzed the performance of "Voronezh Gorelektroset" in this paper. It is the largest territorial grid organization in Voronezh, occupying about 70% of the retail market for electric energy and having a complex technical infrastructure and serving more than 9,500 electricity consumers. The analysis covers two-time ranges of the "Voronezh Gorelectroset": April - June 2014-2018 and December 2014-2018. For each feeder of eighty-three main power distribution stations, we carried out measurements during 3 months in the spring-summer time period and one month in the winter period.

As a result of studying the obtained measurements database, we established a significant correlation between the system load and the load of the main electricity consumers by power distribution points of the hydroelectric power station. We used the data to calculate diurnal variation of reactive power factor in power distribution stations. Figure 1 shows the values of two of the 83 analyzed power distribution stations. The diagrams show diurnal variation of active power (yellow curve), reactive power (blue curve), present power factor (tgφ grey curve) and controlled power factor (tgφ = 0.35 orange line).

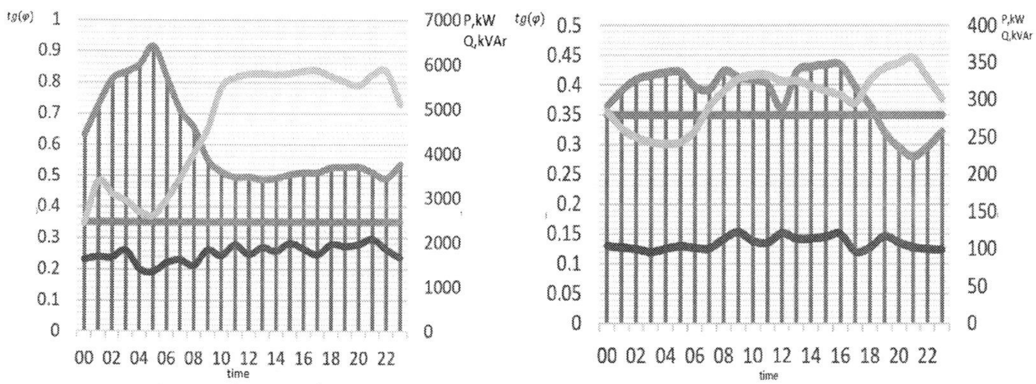

a) December b) June

Figure 1. Diurnal variation of active and reactive power in power distribution stations.

Analysis of the diurnal variation of reactive power factor displays that in the course of Smart Grid concept there are good reasons to implement the controlled reactive power compensation for maintaining a high level of energy efficiency. Modern technologies offer several types of hardware solutions for reactive power compensation:

- synchronized electric machine compensators (SC);
- controlled shunt reactors CSR (including those with unregulated sections of capacitors);
- static reactive power compensators (STATCOM);
- controlled thyristor capacitor units (TCU).

All these hardware solutions for the controlled reactive power compensation were carefully analyzed from the theoretical point of view and put into practice [7-10].

Synchronized electric machine compensators (SC) do not always successfully combine their main functions of electric drive and reactive power generation and are not completely reliable and do not have large specific mass-dimensional indicators. Controlled shunt reactors and thyristor groups of power capacitor banks connected by thyristor groups have various modulations (also called SVC - Static VAR Compensator, figure 2), good adjusting characteristics, but have an increased cost due to the use of semiconductor switches [11].

It is worth mentioning that the phase-pulse method of regulating the current in the reactor generates harmonics into the network. This makes it necessary to implement a filter system, providing some

IOP Conf. Series: Materials Science and Engineering 1035 (2021) 012027 doi:10.1088/1757-899X/1035/1/012027

additional obstacles associated with the appearance of higher harmonic and possible resonance, due to the presence of capacities in the SVC installation.

Today's level of development of power conversion technology allows one to use the powerful static devices for the best possible control of electric power facilities at the level of the FACTS technology platform. To perform such a function hardware solution with the ability to consume or return reactive power are required. The devices such as static reactive power compensator (STATCOM) are employed, which are operating on the principles of pulse-width modulation (PWM), with the mandatory installation of a filter between the converter and the network (figure 3). STATCOM devices can also work in the active filtering mode.

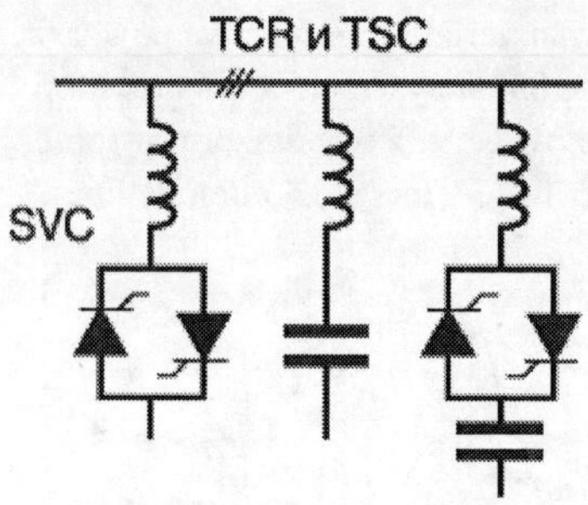

Figure 2. SVC Compensator in RP structure.

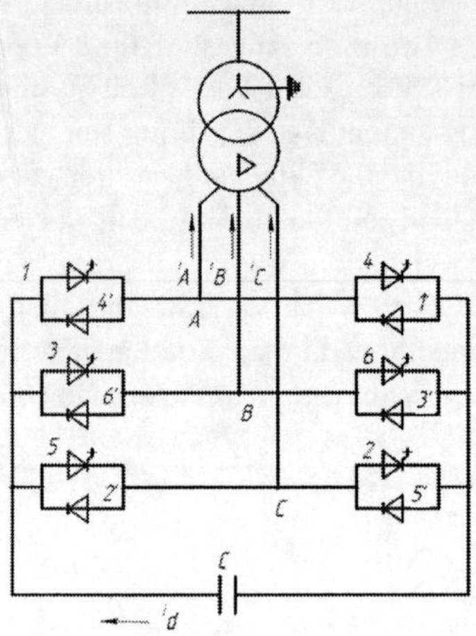

Figure 3. STATCOM structure.

With the multifunctionality and adaptability of such devices, their widespread use is restricted by their high cost (mainly because of the need to use either fully controlled thyristors 1-6 in conjunction with 1'-6 'diodes, or complex artificial switching nodes) [12-14].

Controlled thyristor capacitor units (TCU) use thyristors groups to connect the capacitor banks. This allows one to avoid a number of issues that can arise when connecting static capacitors to the network with electromechanical keys. In accordance with the requirements of reactive power regulation during the operation of the power capacitor banks, when regulating the stages, the capacitor banks undergo frequent switching. In this case, large inrush currents arise, significantly (up to 250 times) exceeding the nominal value and significant power surges. This results in the need of using the specially designed high-speed electromagnetic contactors or high-cost vacuum circuit breakers. However, with an average daily connection frequency of more than 10-50, it is advisable to use power capacitor banks hardware solutions. Many circuit design solutions have been developed by now for connecting power capacitors to the network, a great amount of companies have launched the production of TCU.

It should be noticed that, along with sufficient functionality, the disadvantage of such solution is the high cost of high-voltage semiconductor equipment, which increases significantly with an increase in the number of power capacitor banks stages (which is directly related to the quality of reactive power control).

3. Hybrid switching method of power capacitor banks

For high-voltage networks (6, 10, 35 kV), the most reasonable are reactive power compensation installations using a hybrid method of switching power capacitor banks with a network [15, 16]. This method resides in the usage of semiconductor power keys (thyristor, thyristor-diode groups, bidirectional thyristor), shunted by the contactor (figure 4). Power keys 3, 4 perform the function of connecting the power capacitor bank to the network at the most favorable moment (the passing of the phase voltage line through zero). After that, the corresponding signal from the control system is supplied to contactor 2 and it shunts keys 3, 4, and control signals are taken from them. This long-term mode of the TCU work is maintained until the power capacitor bank is turned off. Then a corresponding signal is supplied to turn on power keys 3, 4, and with a slight delay - a signal to turn off the drive of contactor 2. Its shutdown occurs, in fact, in a de-energized (off-the-line) mode (the load current continues to flow through included power keys 3, 4). Next, a signal from the control system comes for a guaranteed shutdown of the battery during a fraction of a half period of the supply voltage. Thus, mechanical keys 2 operate in an extremely light, low- loaded mode. Therefore, as contactor 2, high-voltage switches (vacuum) and even disconnectors with a high-speed electromagnetic drive can be used.

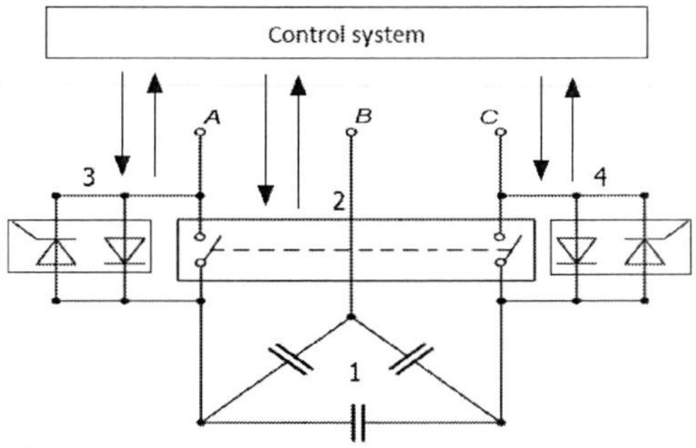

Figure 4. Structure of reactive power compensation unit based on a hybrid connection of power capacitor banks.

Practising a hybrid method of switching capacitor bank 1, the requirements for power keys 3, 4 can be significantly lowered. Due to the fact that the semiconductor device (thyristor, bidirectional thyristor, diode) can function in overload mode for a short time, it is possible to reduce the value of the installed power of the power keys by 2-10 times [17]. The volume of reducing of the installed power (and, consequently, the cost) of power keys 3, 4 directly depends on the speed of contactor 2. The faster the bridging of power contacts of all three phases occurs in it, the more significant the reduction in the installed power of keys 3, 4 can be. Moreover, the cost of equipment will be reduced due to the lack of a conventional cooling system for semiconductor devices, which does not required in TCU switching modes.

The proposed hardware solutions for using hybrid TCU have a wide range of variations in the structure and specific design of the power unit of reactive power compensation installation. The best possible implementation scenario of hybrid TCU is selected depending on the specific requirements for regulating the mode of the electric network (range and accuracy of regulation of the RP), its main parameters (voltage level, value of the required RP value, compliance with standard quality parameters of voltage quality standardized by GOST (All-Union State Standard)), limitations on the cost of the main power equipment, and its resource of work.

It can be noted that the implementation of any option will reduce the cost of the installation (by 1 kvar of compensated RP) in comparison with typical units of reactive power compensation using high-voltage contactors and vacuum circuit breakers [18]. Moreover, the more sections the power capacitor bank has, the lower the unit cost of 1 kvar of compensated RP in comparison with similar contactor units of reactive power compensation [19]. This effect is achieved by usage of one common thyristor switching unit for the disconnector system of individual sections of the QSG3, QSG4 battery. Figure 5 shows a single-line connection diagram for a 2-section power capacitor bank based on a hybrid thyristor switching.

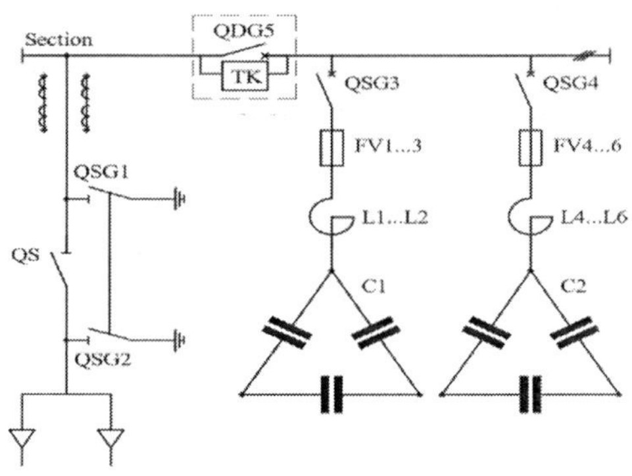

Figure 5. Single-line connection diagram for a 2-section power capacitor banks based on hybrid TCU.

As a rule, contactor units of reactive power compensation have similar structure and parameters. For example, two sectional power capacitor banks with 4 power levels (100 and 200 kvar) was installed at one of the 10 kV power distribution station of Voronezh Hydroelectric Station. Based on the minimum requirements of GOST 13109 to maintain the normalized value of the reactive power coefficient tgφ at a level of no higher than 0.35 and the actual load schedule on the connections of this power distribution station, the approximate number of switching sections of the units of reactive power compensation should be about 50 per day. With this switching mode, after 2-3 years of service there is a need to replace high-voltage contactors based on the resource run out factor. Therefore, the control system of the units of reactive power compensation based on the digital microprocessor controller of the DCRG "LOVATO Electric" series is configured for the number of switching no more than 5-10 times a day. In this mode the vacuum contactors used in the installation are guaranteed to ensure the required quality of RP compensation during the entire period of operation (at least 10 years) [20].

Along with two vacuum contactors of the KVT-10-6.3 / 630D-3 U3 type, the installation contains: 2 KEP1-6.3-ZUZ capacitors, 6 fuses PKT 1CO-VK-6 / 7.2-31.5-50 UZ, 6 reactors RMV-360-60-250, 2 current transformers TOL-10-1-2-0.5 / 10R-50/5 U2, 4 disconnector RVZ-10/400, and a control system.

To implement the hybrid TCU principle based on such an installation, minimal changes in the power part are required (to remove two vacuum contactors and add one QSG5 disconnector and a three-phase power key of thyristor switching, figure 5) and it is necessary to reprogram the microprocessor controller of the control system accordingly (its functionality allows providing the necessary operation algorithm of the power part of a two-stage power capacitor bank).

Given the cost of two high-voltage contactors (87,000 rubles), disconnectors (14,000 rubles) and a three-phase key of thyristor switching (based on either domestic thyristors T471-200 and diodes D371-250, or ABB thyristors of the 5STP series, and diodes 5SDA) approximately 32,000 rubles, we can estimate the cost reduction during the modernization of units of reactive power compensation at 476,000 rubles during its lifetime. At the same time, it is possible to ensure the required range and accuracy of regulation of the RP, reduce the loss of energy efficiency in the network, provide a normalized voltage level in the load node and increase the reliability.

4. Conclusion

1. Nowadays in the light of the increasing losses of electrical power in the networks of 6-35 kV it is necessary to expand the usage of up-to-date energy-saving technologies in the terms of Smart Grid concept.
2. An analysis of the load graphs of a typical city power grid shows the possibility of a significant reduction in electricity losses due to automatic compensation of reactive power on the 10 kV buses of the distribution points of the HPP. After analyzing the diagrams of the load of the city's electric grid it becomes evident the possibility to cut the electricity losses due to the automatic compensation of reactive power in 10 kV connection of distribution stations of hydroelectric power stations.
3. Existing means of reactive power compensation have insufficient operating life cycle of switching equipment, which does not allow one to provide the required network power factor, according to the criterion of minimum power loss.
4. There are effective hardware solutions of connecting static capacitor batteries to a high voltage network based on thyristor keys.
5. The suggested usage of hybrid thyristor capacitor units in electric networks of 6-10 kV will allow one, at lower costs, to ensure the required range and accuracy of regulation of the RP, reduce the loss of energy efficiency in the network, provide a normalized voltage level in the load node and increase the reliability.

References

[1] 2009 Decree of the Government of the Russian Federation dated November 13, 2009 no. 1715-R "Energy strategy of Russia for the period up to 2030"

[2] 2010 *European Technology Platform Smart Grids*. Strategic Deployment Document for Europe's Electricity Networks of the Future

[3] 2003 *«Grids 2030». A National Vision for Electricity's Second 100 years* (Office of Electric Transmission and Distribution of USA Department of Energy)

[4] 2009 *Smart Grid System Report* (U. S. Department of Energy)

[5] Gusev S, Shakarian Yu and Novikov N 2013 Development of FACTS devices *Electric Power Industry: Today and Tomorrow* **1** 7-20

[6] Galyaev A N and Shevchenko I V 2010 Problems of improving energy efficiency in the electric power industry *Finance and Credit* **11** 8-13
[7] Timofeev A S 2010 *Compensation of Reactive Power* (Novokuznetsk: Siberian State Industry University) p 67
[8] Krysanov V N and Zaitsev A I 2016 *Energy-Saving Technologies in Distributed Electric Power Networks* (Voronezh: Voronezh State Technical University) p 223
[9] Debs A S and Larson R E 1970 Dynamic estimator for tracking the state of a power system *IEEE Transactions on Power Apparatus and Systems* **4**
[10] Glanzmann G 2005 *Flexible Alternating Current Transmission Systems* (Zurich: Power Systems Laboratory ETH)
[11] Songhai Y and Shaoting W 1987 Application of SVC to improve transmission capacity *IEEE Joint Conference on High Voltage Transmission System in China*
[12] Milyutin A Yu, Temnikov E A and Gottfried P A 2017 *Compensation of Reactive Power using STATCOM* (Moscow: MTSNO) p 122-129
[13] 1999 *Static Synchronous Compensator (STATCOM)* (ElectraN185 08.1999)
[14] Trokawa S and Horiuchi T 1987 Transmission line loss minimization *IEEE Joint Conference on High Voltage Transmission System in China*
[15] Krysanov V N and Demikhov A V 2009 Optimization of ways to connect a capacitor bank *Electrotechnical Complexes and Control Systems* **4(16)** 53-58
[16] Alieva L F, Krysanov V N, Rubin Ya S and Bykov V E 1990 *Device for Connecting a Capacitor Bank in Electrical Networks* AC #1576983 A1 BI # 25
[17] http://www.vishay.com/docs/94630/vs-vsk4156.pdf
[18] Vagin G Ya, Sevostyanov A A and Yurtaev S N 2012 On the choice of sources of reactive power at industrial enterprises *Industrial Energy* **4** 35-37
[19] Krysanov V N and Ivanov K V 2018 Analysis of the effectiveness of hardware solutions based on the FACTS technology platform in distribution electrical networks *Proc. of the Int. Scientific and Technical Conf. "Technogenic Energy Security and Energy Resource Saving"* (Moscow: MIEE) pp 40-41
[20] GOST 27389-87 Condenser installations for reactive power compensation. terms and definitions. general technical requirements

IOP Conf. Series: Materials Science and Engineering 1035 (2021) 012028 doi:10.1088/1757-899X/1035/1/012028

Improvements of choice of the water-chemical regime of reserve osmosis installation

M V Odoevtseva[1] and D A Spiridonov

Department of Thermal Power Engineering, Moscow Power Engineering Institute, 69 Lenina Avenue, Volzhsky 404110, Russia

[1]Corresponding author: odoevcevamv@mail.ru

Abstract. The effectiveness of monitoring the composition and intensity of deposits on reverse osmosis membranes is shown. The use of test installations of the NGO "NPK Mediana-Filter" speeds up and significantly reduces the labor intensity of monitoring deposits in the conditions of industrial operation of reverse osmosis installations. The effectiveness of determining the composition and intensity of deposits on the membrane surface using the energy dispersive analysis method was revealed. The effectiveness of commercial formulations for cleaning intensely contaminated membranes is not always sufficient. The use of ready-made solutions from various manufacturers excludes the possibility of a targeted selection of the concentration of a solvent to remove deposits from the membrane surface, provided that these solutions are safe for the operating parameters of the membranes. The monitoring results made it possible to implement a targeted choice of compositions of detergent mixtures, their combinations and the regime of their effect on deposits on membranes, provided that the pH of the solutions is stable throughout the entire cleaning time. Controlling the intensity of the formation of deposits according to the data of test installations will reduce the laboriousness of choosing the optimal water-chemical regime of an industrial reverse osmosis plant, which will reduce the rate of formation of deposits, increase the service life of membrane elements of an industrial water treatment plant with a consistently high quality of demineralized water.

1. Introduction

The interest in using reverse osmosis in desalting water is caused by the low quantity of consumed reagents (acids, alkalis, sodium chloride) and the consistently high quality of purified water [1]. Over time, during operation, reverse osmosis (RO) membrane elements are exposed to contamination by suspended or slightly soluble substances present in the source water [2]. The nature and speed of deposits on the membrane surface depends on the ionic composition of the source water and the stability of the pre-treatment system.

2. Research methodology

To determine the composition and quantity of pollutants the following was carried out:

- an analysis of the source water and prediction of possible pollution;
- a visual inspection of the ends of the membrane elements;
- an analysis of the composition of acidic and alkaline solutions after chemical washing of the test setup;
- dismantling (opening) of the membrane element and direct analysis of deposits from the

membrane surface.

To comply with the operating conditions, the production and test plants were connected in parallel, to the concentrate of the first stage of the reverse osmosis installation (ROI) (figure 1) [3].

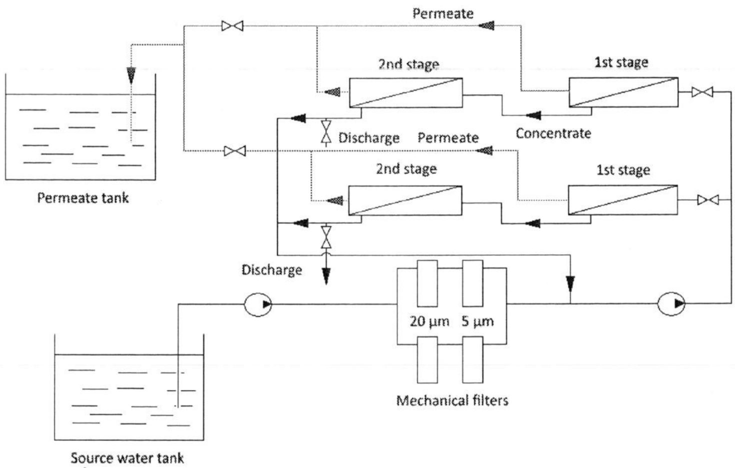

Figure 1. Hydraulic diagram of the production ROI.

3. Key results

In the course of the work, the following parameters of the test setup were monitored: flow rates, pressure on mechanical filters, conversion value, and selectivity of membrane elements. After 800 hours of operation of the test setup, the membrane selectivity decreased by 19%, the salt content of permeate increased by 15 times (table 1).

Table 1. Test results when connecting a test setup.

Indicators	Source water	Concentrate	Permeate	Source water	Concentrate	Permeate
	Beginning of work (7.05.18)			Ending of work (4.06.18)		
SDI	8.81	-	-	9.65	-	-
Total hardness, meq/dm³	3.8	6.8	0.04	3.9	5.2	0.58
Calcium hardness, meq/dm³	2.8	4.9	0.03	3.0	3.8	0.36
Magnesium hardness, meq/dm³	1.0	1.9	0.01	0.9	1.4	0.22
Total alkalinity, meq/dm³	2.9	4.8	0.2	2.9	3.8	0.6
Chlorides, mg/dm³	26.6	50.2	1.06	38.4	-	8.90
Salinity, mg/dm³	690	1421.8	15.30	745.9	1025.7	41.90
pH	6.7	7.7	5.7	7.0	7.6	6.3
Flow, dm³/h	39.2	28.6	10.5	38.5	32.7	5.8
Recovery, %		27.0			15.0	
Selectivity, %		96			77	

IOP Conf. Series: Materials Science and Engineering 1035 (2021) 012028 doi:10.1088/1757-899X/1035/1/012028

An increase in the salt content of permeate was a criterion for the chemical washing of membranes. To implement chemical cleaning, the membrane element was connected to a chemical washing station [3].

Chemical washing of membrane elements was carried out in several stages:

- the circulation of an alkaline solution;
- soaking the membrane element, followed by the circulation of the alkaline solution;
- the acid solution circulation;
- soaking the membrane element with a subsequent circulation of the acid solution;
- washing the membranes with distilled water to pH = 6-7 after the circulation of alkaline and acid solutions.

The duration of chemical washing of the membrane element of the test setup was from 6 to 19 hours.

The composition of the deposits was determined by the pollution of the washing solutions (table 2) and by the analysis of precipitation from the surface of the membranes (table 3).

Table 2. The composition of the contaminants of washing solutions.

Indicators	Alkaline solution	Acid solution
Suspended solids, mg/dm^3	28	35
Total iron, mg/dm^3	1.0	1.4
SiO_2, mg/dm^3	16.3	-

Deposits from the membrane surface were determined using the energy dispersive method [6] on a Versa 3D instrument at Volgograd State Technical University. The results of the analysis of deposits on the surface of the membranes of the test setup are shown in table 3. Studies of deposits allowed us to clarify the composition of the deposits on the surface of the membranes of the industrial ROI. It was revealed that organic substances, compounds of silicon, iron, and calcium accumulate on the membranes. The data obtained are consistent with the results of the analysis of models of washing solutions of the test setup (table 3).

Table 3. Results of the energy dispersive sediment analysis method.

Element	Before chemical washing	
	Percentage, %	Measurement error, %
Carbon	13.32	11.23
Oxygen	46.25	8.7
Sodium	0.73	14.88
Magnesium	1.55	9.2
Silicon	14.47	4.75
Potassium	1.98	4.96
Calcium	3.1	4
Manganese	0.49	19.11
Iron	9.17	3.07

Chemical cleaning of the test ROI membrane elements with solutions of Nalco Company, NPK Mediana-Filter (model 1, 2) do not provide significant removal of deposits (table 4). Ineffective removal of deposits determines the low quality of demineralized water, high energy consumption of the ROI, expenses for chemical reagents and own needs of the industrial plant.

The research results (model 3) showed that the effectiveness of chemical cleaning of membrane elements depends on the contact time of the washing solutions with the membrane surface and the composition of the solution, as determined by the manufacturer. The advantages of these solutions are determined not only by a change in the composition of the solution but also by the stability of the pH value throughout the entire cleaning time of the membrane models.

Photos of membrane surfaces after the chemical cleaning of the membranes are shown in figure 2.

A B

Figure 2. Photos of membrane surfaces after chemical cleaning (A – model 1, B – model 2).

The research results showed the high efficiency of the washing solutions of NPK Mediana-Filter (model 3). The mass of removed contaminants is 63% greater than when using another washing solution (model 2). The pH value was kept stable throughout the entire cleaning time of the membrane models. With an increase in the duration of chemical cleaning of the membranes, the mass of washed deposits increases (model 3).

The effectiveness of commercial compositions for cleaning intensely contaminated membranes is not always sufficient. The use of finished compositions of various manufacturers eliminates the possibility of directional selection of the concentration and solvent to remove deposits from the membrane surface.

The mass of washed deposits and the pH of the solutions after the chemical cleaning of the membranes are shown in table 4.

Table 4. Change in weight of membrane models using a washing solution.

Model no.	Cleaning solution manufacturer	Trademarks for washing solutions	Total exposure time of detergent solutions, h	The mass of washed sediment per 1 m^2, g	pH of solutions before/after chemical cleaning
1	"Nalko Company"	PC-33, PC-67 PC-77	6	16	11.8/9.9 2.1/6.75
2	"NPK Median-Filter"	MF – A – T10 MF – B – L20	6	20	12.1/10.35 2.0/2.0
3	"NPK Median-Filter"	MF – CRO – 220 MF – CRO – 218	12	26	11.8/11.4 2.1/1.6
4	"NPK Median-Filter"	NaOH MF – CRO – 218	19	42	11.0/10.4 2.1/1.7

The maximum number of deposits is removed from the membrane surface during alkaline cleaning with a NaOH solution (pH = 11.0) for 11 hours and with an acid solution of NPK Mediana-Filter - MF - CRO - 218 (model 4). In this mode, the mass of washed deposits increases by 1.5 times.

4. Conclusions

The use of test facilities of NPK Mediana-Filter is an effective way to determine the activity of formation and the composition of the contamination of the membrane elements of an industrial ROI. Determination of the composition of deposits on the membrane elements ensures the selection of the optimal chemical treatment regime for the reverse osmosis production plant. High-quality cleaning of the membranes will reduce the rate of formation of deposits, increase the life of the membrane elements of an industrial water treatment plant with a consistently high quality of demineralized water.

References

[1] Balandina A G, Khangildin R I, Ibragimov I G and Martyasheva V A 2015 The development of membrane technologies and their application wastewater treatment chemical and petrochemical enterprises *Oil and Gas Business* **5** 336-375

[2] Pervov A G Andrianov A P and Telitchenko E A 2004 Influence of biological contamination on the operation of reverse osmotic and ultrafiltration membrane elements *Critical Technologies. Membranes* **1** 3-18

[3] Panteleev A A, Bobinkin V V, Larionov S Yu, Ryabchikov B E 2017 Methodological approaches to conducting pilot and proof tests on reverse-osmosis systems: Results of comparative studies *Heat Power Engineering* **10** 92-98

IOP Conf. Series: Materials Science and Engineering 1035 (2021) 012029 doi:10.1088/1757-899X/1035/1/012029

Experimental obtaining characteristics of current transformers for developing their models and determining the time of saturation of the magnetic circuit

M R Batmanov, A E Evdakov, G A Filatova, A E Petrov[1] and A A Yablokov

Department of Electric Power Systems Automatic Control, Ivanovo State Power Engineering University, 34 Rabfakovskaya, Ivanovo 153003, Russia

[1]Corresponding author: usetheforcealex@yandex.ru

Abstract. Electromagnetic current transformers (CT) are subject to saturation and residual magnetization of a magnetic circuit. This phenomenon has an impact on the operation of devices of commercial accounting of electric energy and relay protection and automatics (RPA). Saturation of the current transformer's magnetic core can lead to emergency situations due to incorrect operation of the protections and excessive disconnection of the protected connections. At the moment, Russian norms do not regulate the work in transient modes, including short circuits with an aperiodic current component. The Ivanovo State Power Engineering University (ISPEU) conducts research on the topic of determining the time to saturation of CT at short circuit. The purpose of the work was to develop an algorithm for monitoring saturation and determining the time to saturation of the CT magnetic core, in order to ensure the correct operation of the RPA in all modes. The article is devoted to the method of construction and research of mathematical models of CT for calculating the time to saturation. The general structure of the algorithm for determining the time to saturation is presented.

1. Introduction

During operation, the conversion error of current transformers (CT) may not meet the requirements of the declared accuracy class due to distortion of the shape of the current curve due to the influence of the residual magnetization of the magnetic circuit, which may appear after disconnecting the short circuit. This phenomenon has a significant impact both on the operation of the commercial electricity metering system and on the operation of relay protection and automation devices (RPA). Accordingly, incorrect accounting of electricity supplied or consumed may be maintained for a long time. However, more significant commercial and operational implications happen during the wrong action of relay protection and automation.

The issue of CT saturation and modelling of transients in secondary CT chains has been given close attention in the USSR and the world since 1960-80s. A large number of works by such specialists as I. M. Sirota, A. D. Drozdov, V. E. Kazansky, L. V. Baginsky, V. I. Novash and their students were devoted to the problems of CT operation in transient short-circuit modes [1]. Measures were developed to eliminate saturation of the magnetic circuit, including the use of CT with a voltage of 750 kV and 1150 kV with a non-magnetic gap, and methods for calculating loads on the CT were developed [2]. To eliminate the influence of CT saturation, CT with new accuracy classes were used in foreign practice. RPA algorithms were developed and applied everywhere, implementing the correct operation of protection with a speed of about half a period of industrial frequency in conditions of deep CT saturation.

However, the results of the theory of CT operation in transient modes are not taken into account in existing domestic standards on CT. Several major accidents, including those in 2018 in the Crimea and 2014 at the Rostov nuclear power plant, caused the release of a new series of standards for measuring CT and a number of works on the topic of CT saturation, designed to ensure the correct operation of the RPA connected to electromagnetic CT, at short circuit, including when there is an aperiodic component of the current [3, 4].

For a number of large power facilities in the Russian Federation, it has become mandatory to monitor the state of the CT and estimate the time to saturation of the CT at short circuit. In accordance with the requirement of JSC "System Operator of the Unified Electric Power System" ("SO UEPS"), it is necessary to calculate the time to saturation of the CT at approximately 100 power facilities by 31.12.2020.

Therefore, it is important to develop and study an algorithm for determining the time to saturation of the current transformer magnetic core to assess the correct operation of RPA devices and current transformers in transient modes. This paper presents the results of the development and verification of CT simulation models designed to work out algorithms and principles for determining the time to saturation, as well as the calculation of the time to saturation of CT type TZLM, available in the laboratory of the Ivanovo State Power University.

2. Mathematical models of CTs

For the development and verification of models, a CT of the TZLM type was selected (figure 1). The TZLM Passport data required for modelling are summarized in table 1.

Figure 1. CT of type TZLM.

Table 1. TZLM parameters.

Type of magnetic core steel	Length of the average power line of the magnetic circuit L, cm	The cross-sectional area of the core S, cm^2	The number of turns of the secondary winding
3411	33.9	7.2	25

CT models were developed in the RSCAD program environment of the RTDS hardware and software complex, which has earned its trust and has been repeatedly tested for the purpose of modelling objects and modes of electric power systems and RPA devices.

In the course of the work, three CT models were developed and studied (figures 2 – 4):

1. CT simulation model based on Faraday's electromagnetic induction equations. The main ratios for calculating the output value (EMF of the secondary CT winding) are shown below (1) [5]:

$$F_{sum} = I_1 \cdot w_1 - I_2 \cdot w, \quad H = \frac{F_{sum}}{L}, \quad H \to B,$$
$$\Phi = B \cdot S, \quad \Psi_2 = \Phi \cdot w_2, \quad E_2 = -\frac{d\Psi_2}{dt}. \tag{1}$$

where B – the induction in the core (T); H(t) –the magnetic field strength (A/m); F_{sum} – the total magnetization (A); L_m –the average length of the magnetic lines (m); I_1 – the primary current (A); w_1 – the number of turns in the primary winding; I2 – the secondary current (A); w_2 – the number of turns in the secondary winding; Φ – the magnetic flux in the core (Wb), S – the area of the core (m^2); Ψ_2 – the flux linkage of the secondary winding; E_2 – the EMF induced in the secondary winding (V).

In this model, the conversion of magnetic induction B from the strength H is performed by piecewise linear approximation of the main magnetization curve (MMC).

Calculated by (1) EMF of the secondary CT winding (at a given primary circuit current) is the control value in the model of the secondary CT circuit, consisting of the resistances of the secondary CT winding and the specified secondary load (figure 2b).

2. Simulation model of the CT based on the equations of dynamics of the Kadochnikov magnetic core remagnetization (2) [6, 7, 8]:

$$\begin{cases} \dfrac{dB}{dt} = r \cdot (1 - \dfrac{B^2}{B_s^2}) \cdot [H(t) - H_{st}(B)] \\ H(t) = \dfrac{F_{sum}(t)}{L}; \\ H_{cm} = f(B); \\ F_{sum}(t) = I_1(t) \cdot w_1 - I_2(t) \cdot w_2; \\ \Phi = B(t) \cdot S; \\ \Psi_2 = \Phi(t) \cdot w_2; \\ E_2 = -\dfrac{d\Psi_2}{dt} \end{cases} \tag{2}$$

where r – the losses on the eddy currents and magnetic viscosity; Bs – the magnetic induction at the core saturation; H_{st}=f(B) – the static hysteresis loop.

In the TZLM model, based on equations (2), losses in the core steel for hysteresis and eddy currents were approximated by a constant active resistance connected in parallel to the magnetization branch. The resistance can be obtained in various ways, for example, calculated based on the geometry of the core, or experimental based on hysteresis loops.

The conversion of magnetic induction B from the strength H is performed by piecewise linear approximation of the specified magnetization characteristic (for example, the main magnetization curve of the MMC). This representation of the H_{st} function (MMC and accounting for active losses in the core) gives a fairly accurate representation of the magnetization of the magnetic core steel [9].

3. Simulation model based on the Gills-Atherton equations (built-in model of the RSCAD software environment, Current Transformer block).

The non-linearity of the magnetization branch in the Current Transformer block is set either by the MMC (for known geometric parameters of the core) or by the Volt-Ampere characteristic (VAC).

The input values of simulation models are the primary current (if necessary, one can also model the electrical part of the primary circuit), and the output values are the EMF of the secondary circuit. The physical conditions for unambiguity are the number of turns of the primary and secondary circuits, the length of the average magnetic line, and the cross-sectional area.

IOP Conf. Series: Materials Science and Engineering 1035 (2021) 012029 doi:10.1088/1757-899X/1035/1/012029

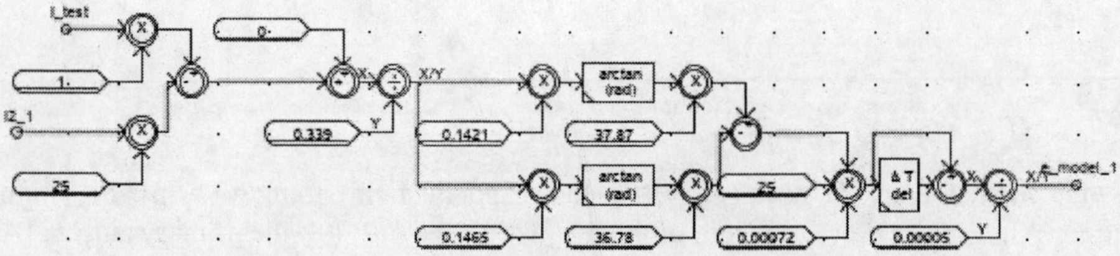

a)

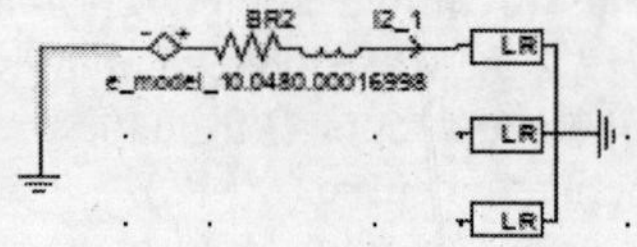

b)

Figure 2. Faraday's model.

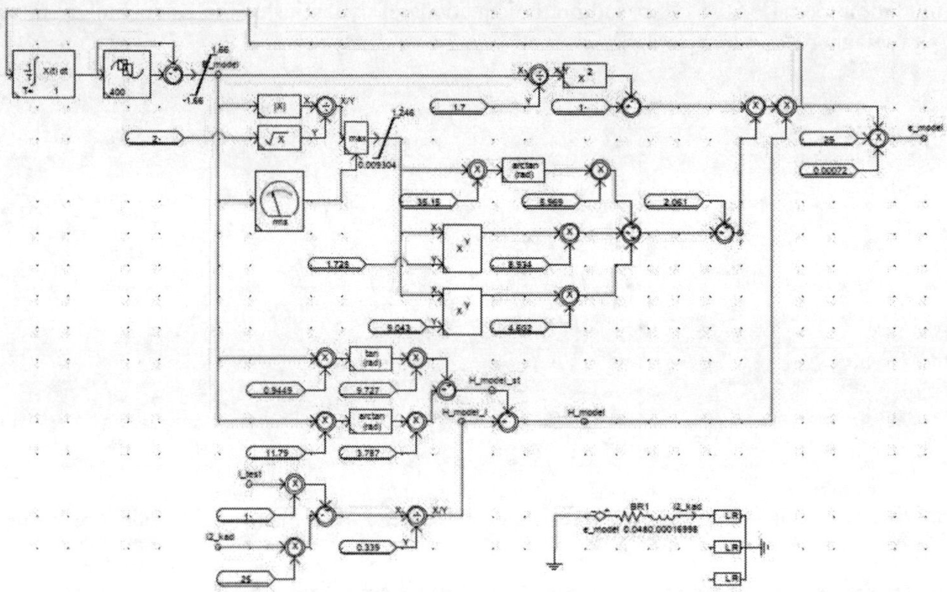

Figure 3. Model of Kadochnikov.

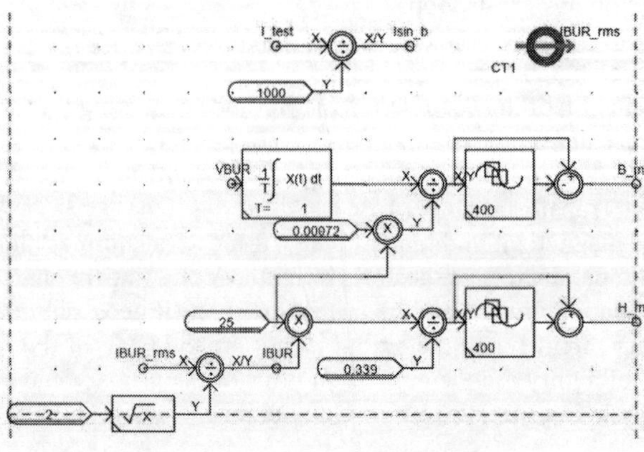

Figure 4. Gills-Atherton model (CT1 block, below-mathematical blocks for calculating H and B of the magnetic core of the secondary current of the CT and the secondary voltage with the model CT – B_mm and H_mm).

Thus, all three models contain a function describing the non-linearity of the magnetization branch (for example, the MMC), which can be taken from reference data or obtained experimentally. Due to the fact that the characteristics of CT samples of the same type made with cores made of electrical steel may differ significantly, CT models with experimentally obtained characteristics will have greater accuracy.

3. Determining parameters for CT models

To create models, the parameters of current transformers (steel magnetization curve, CT hysteresis loop parameters, secondary winding resistance, etc.) are required. These parameters were obtained using two methods.

3.1. Removing CT characteristics using Omicron CT Analyzer

CT Analyzer is a test system for protective and measuring CT that meets the IEEE and IEC standards.

Using the indirect measurement method, the voltage signals are fed to the secondary winding, and the CT parameters are measured (figures 5-6). Based on the measurement results, the characteristics are calculated (for example, the volt-ampere characteristics VAC).

Figure 5. Test facility of a CT TZLM type.

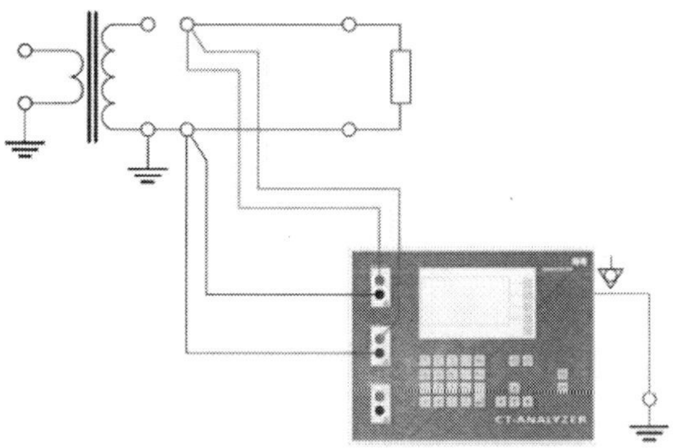

Figure 6. Scheme of the experiment to obtain the VAC.

Based on the passport data obtained in the CT-Analyzer VAC report (according to IEC 60044-1) and calculated expressions (3 – 4), the main magnetization curve of the window of the transformer under study was constructed (figure 7)

$$H = \frac{I_{CT\,RMS} \cdot w_2 \cdot 1.414}{L};$$

$$B = \frac{U_{CT\,RMS}}{4.44 \cdot 50 \cdot w_2 \cdot S}.$$

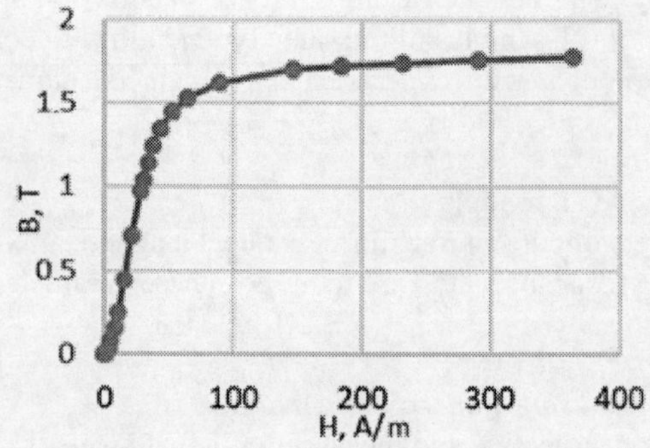

Figure 7. Estimated MMC for CT TZLM type (obtained with CT Analyzer).

The standard test program with CT Analyzer allows you to get the basic parameters for building CT models, but it is not intended for obtaining partial hysteresis loops and calculating the resistance of the magnetization branch.

3.2. Obtaining TZLM characteristics using a real time simulation system RTDS

The experiments were conducted as follows:

1. Assembling the circuit in accordance with figure 7 or figure 8 and supplying the primary current (from Omicron 356 CMC or PONOVO amplifiers) to the primary winding of the transformer under study: the number of turns of the primary winding N=20, the measuring noninductive shunt 0.1 Ohms, the resistance in the arms of the voltage divider 1.8 kOhm and 25.7 kOhm;
2. The primary current source control from Test Universe (for Omicron CMC 356) or RSCAD (for PONOVO PAC60Ci);
3. The removal of magnetization characteristics, primary current and secondary voltage waveforms (using RTDS GTAI analogue cards) when increasing the primary current from 6 A to 100 A;
4. Processing and comparing the experimental results obtained using Omicron CMC 356 and PONOVO PAC60Ci as the sources of the primary current with the MMC obtained using Omicron CT Analyzer.

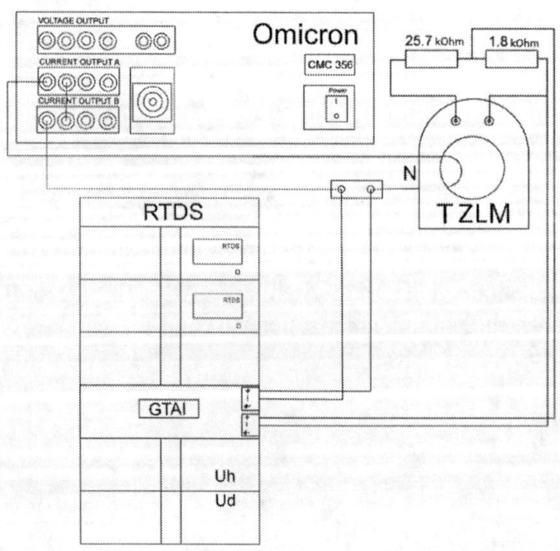

Figure 8. Scheme for conducting experiments with Omicron CMC 356.

IOP Conf. Series: Materials Science and Engineering 1035 (2021) 012029 doi:10.1088/1757-899X/1035/1/012029

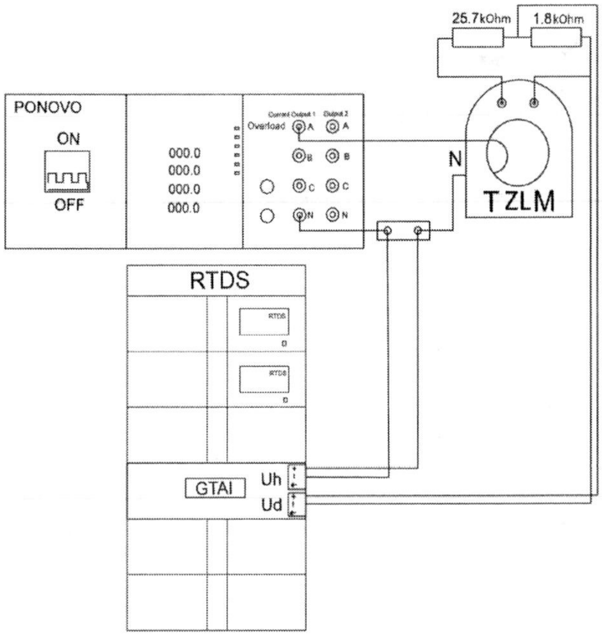

Figure 9. Scheme for conducting experiments with PONOVO PAC60Ci.

The appearance of the control program and out-put results is shown in figure 10.

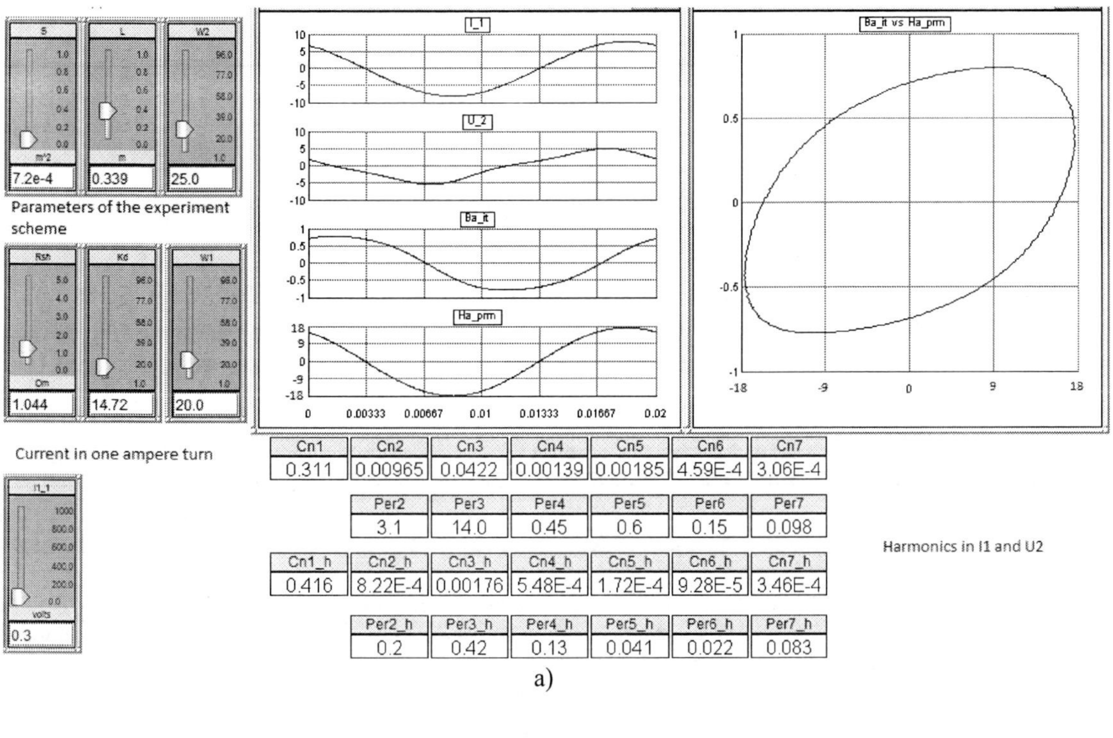

a)

IOP Conf. Series: Materials Science and Engineering 1035 (2021) 012029 doi:10.1088/1757-899X/1035/1/012029

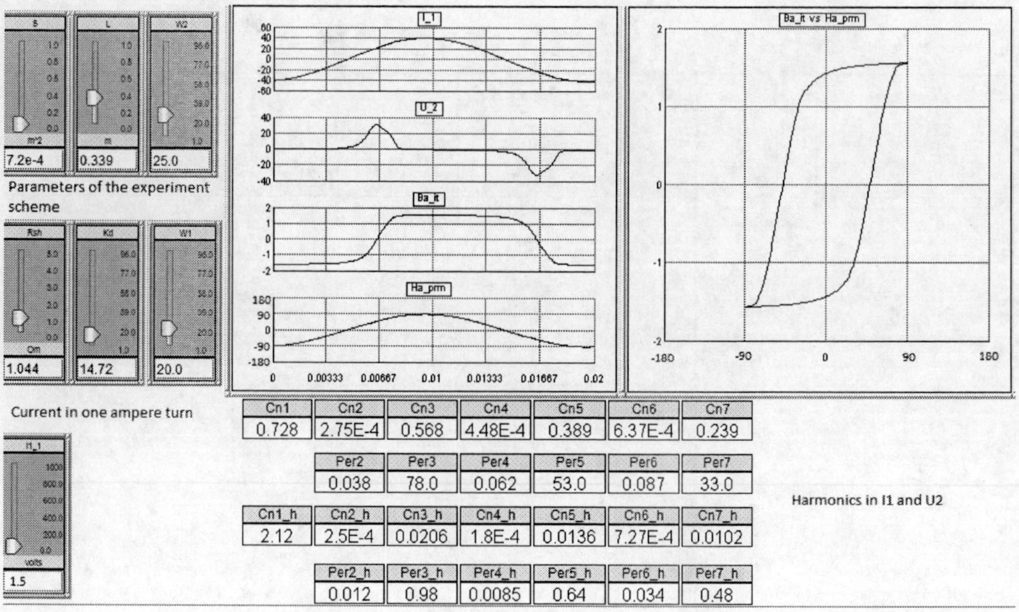

b)

Figure 10. Appearance of the program in RSCAD: a – the data for small primary currents of TZLM; b – the data for primary currents of TZLM more than 20 A.

The following experimental data were obtained:

- the time dependencies of the primary currents I_1(t), secondary voltages U_2 (t);
- the time dependences of Ba_it(t) induction and Ha_prm(t) strength of the TZLM magnetic core;
- Ba_it(Ha_prm) private hysteresis loops.

The dependences are obtained when the core is magnetized in the XX mode and the primary current changes from 6 A to 100 A with a frequency of 50 Hz.

Additionally, the spectrum of the signals of the primary current and secondary voltage CT was calculated.

The results of the experiments with the primary current supply with Omicron CMC 356 and PONOVO-PAC60Ci generally coincided, however, when using Omicron CMC 356, the noise level in the primary current and secondary voltage signal was significantly lower, and it was possible to produce more primary current while maintaining its sinusoidality.

Based on the obtained Ba_it(Ha_prm) dependencies, the TZLM MMC is constructed (figure 11). The points on the MMC corresponded to the vertices (maxima of induction or magnetization) of the partial hysteresis loops. The vertices of the partial loops were different when constructing the maximum of induction or magnetization for loops that have the shape of an "ellipse" (for small values of primary currents, figure 10a).

Next, we compared the MMC obtained using CT Analyzer (orange curve in figure 12) and RTDS (blue curve in figure 12).

Comparison of MMC obtained on Omicron CT Analyzer complexes (recalculation of VAC in MMC) and RTDS (MMC obtained as the geometric location of the vertices of partial hysteresis loops when the core is magnetized) showed their discrepancy, which can be caused by the following:

- errors introduced by voltage dividers and current shunts when conducting RTDs circuits;
- errors introduced by the analog-to-digital Converter (ADC) RTDS;
- errors induced by the interference in the measuring circuit in the experiments with the RTDS;
- fewer points for building a MMC when conducting experiments with RTDS, and less accuracy of removing the" knee" of the MMC;

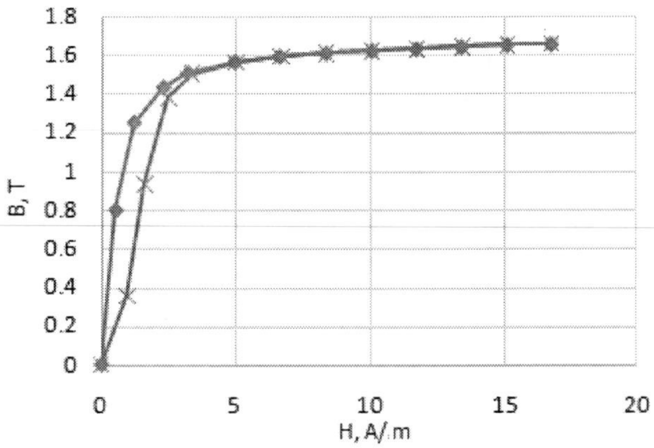

Figure 11. MMC TSLM (obtained with RTDS), the blue curve – the maximum for induction B, the purple one – the maximum for magnetization H.

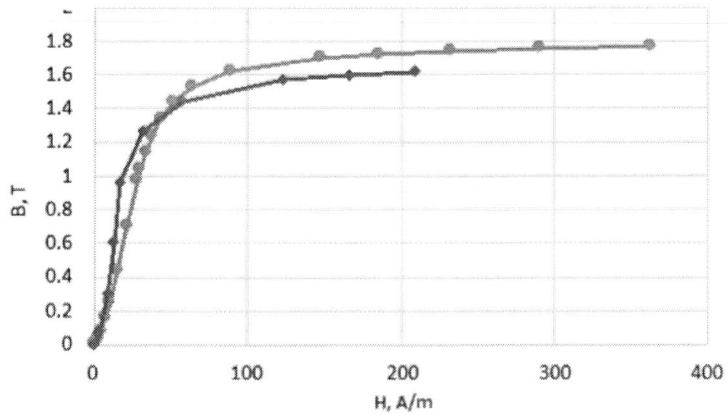

Figure 12. Comparison of TZLM MMC obtained using CT Analyzer (orange curve) and RTDS (blue curve).

- the general method of constructing a MMC using partial hysteresis loops – in experiments with RTDS, the MMC was constructed in 2 ways, with a maximum of B and a maximum of H (not shown in the figures). The method of construction affects the type of characteristic on the linear section, but does not affect the estimate of the calculated saturation induction. Omicron CT Analyzer builds VAC based on integral values of currents and voltages.

Further, there are the models used the characteristics (VAC and MMC) obtained by Omicron CT Analyzer.

3.3. Determination of saturation induction by the received MMC

Based on the obtained experimental data (VAC and MMC), the saturation induction (inflection point) for TZLM was determined: 1.29 T (Omicron CT Analyzer data), 1.1 T (RTDS data).

4. Verification of CT models

For verification of CT models:

1. Using the scheme of the experiment (figures 8-9); the current is supplied using RTDS and amplifiers (Omicron CMC 356 and PONOVO PAC60Ci).
2. The output signals of the studied CT (physical signal) and simulation models of CT (according to claim 2) are compared.

All simulation models were tested at the same values of the primary values. An example of the experiment results is shown in figure 13, where the upper waveform is the primary current TT, oscillogram 2 – the secondary current with model 1, oscillogram 3 – the secondary current with model 2, waveform 4 – the secondary current with model 3.

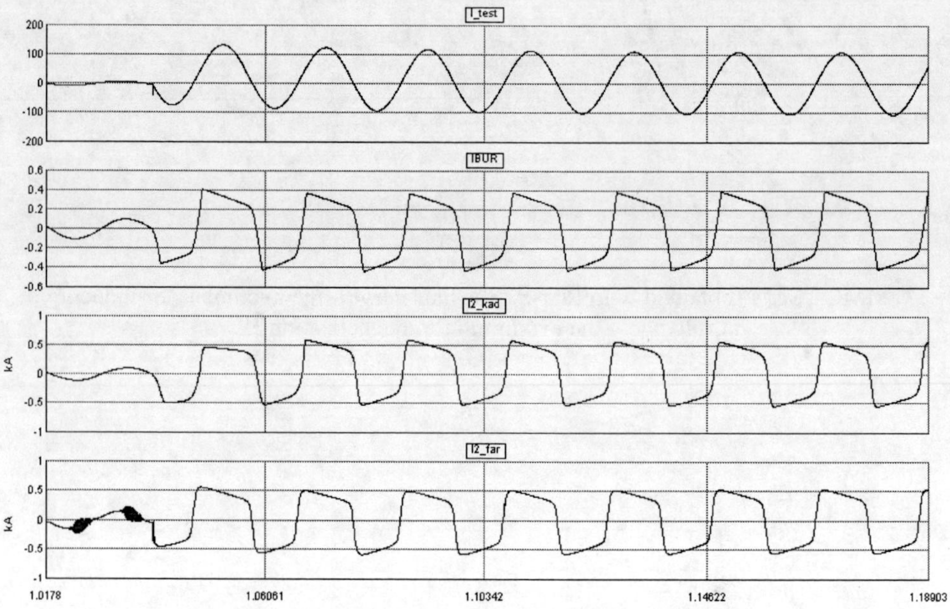

Figure 13. Verification of CT TZLM type models.

In all three models of TT, the experimental characteristics obtained according to clause 3 were used.

In the studied modes (no-load and secondary load CT 1 Ohm), the secondary currents of all models differ by no more than 10 %. The models were accepted as reliable. They can be used, for example, to obtain the MMC and then calculate the saturation induction when conducting a physical experiment is not possible.

5. General algorithm for determining the time to saturation

The block diagram of the algorithm is shown in figure 14.

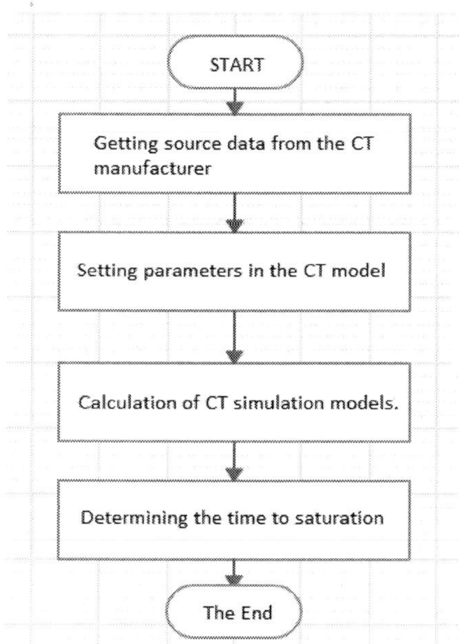

Figure 14. Block diagram of the algorithm.

General method for calculating the time to saturation of CT:

1. Getting the original CT data from the manufacturer. If a CT sample is present, the experimental characteristics (magnetization curves or VAC, windings resistances) are taken using Omicron CT Analyzer and RTDS.
2. Getting initial data on the CT operating mode short-circuit currents (maximum values of the short-circuit current amplitude and the constant attenuation of the aperiodic component), the secondary load of the CT.
3. Input of the received initial data into the CT simulation models (section 2 of the article).
4. Calculating CT models based on the received initial data and characteristics. Conducting computational experiments to obtain the CT MMC or VAC, CT secondary currents at specified modes and secondary loads.
5. Based on the calculation of the model CT, determining the time to saturation of the CT, in order to ensure the correct operation of the RPA.

6. Conclusion

Simulation models of CT in the RSCAD environment were developed and verified, which allow simulating both steady-state and transient modes with sufficient accuracy, taking into account the influence of saturation and residual magnetization of the magnetic core. Based on the research, an algorithm for determining the time to saturation was developed. The models can be used to determine the saturation time taking into account the real characteristics of electromagnetic current transformers, the axial magnetization, the multiplicity of the short-circuit current, etc. Using these models, it is also possible to check the stability of various RPA algorithms during saturation and residual magnetization of the CT.

References

[1] Stogniy B S, Rogoza V V and Chernenko V A 1970 Similarity conditions and physical modeling of current transformers in steady-state and transient modes *Electricity* **10** 68-72

[2] Korolev E P and Liberzon E M 1980 *Calculations of Permissible Loads in Current Relay Protection Circuits* (Moscow: Energiya)

IOP Conf. Series: Materials Science and Engineering 1035 (2021) 012029 doi:10.1088/1757-899X/1035/1/012029

[3] Vorobyov V S 2015 On the incorrect operation of RPA devices in transient modes when current transformers are saturated *Meeting of the Non-Profit Partnership "Scientific and Technical Council of the Unified Energy System"* (Moscow) pp 12-43
[4] Throughout the Crimea, electricity was cut off: the site is a Digital substation, http://digitalsubstation.com/blog/2018/06/13/vo-vsem-krymu-otklyuchilos-elektrichestvo/ (accessed: 06.07.2019)
[5] Yablokov A A, Lebedev V D and Evdakov A E 2019 Simulation of saturation and residual magnetization of a magnetic core in current transformers *RELAVEXPO-2019* (Cheboksary) pp 194-197
[6] Evdakov A E, Yablokov A A and Filatova G A 2019 Development and research of a simulation model of an electromagnetic current transformer taking into account the effects of saturation and residual magnetization of the magnetic core *Relay Protection and Automation* **3** 24-29
[7] Yablokov A A, Evdakov A E, Filatova G A and Petrov A E 2019 Development, implementation in the information system and testing of the algorithm for determining saturation and residual magnetization of current transformers *Electric Power through the Eyes of Youth 2019* (Irkutsk) pp 96-99
[8] Kadochnikov A I 2007 *Dynamic Loops of Magnetic Hysteresis* (Ekaterinburg)
[9] Filatova G A 2013 Application of the equation of dynamics of remagnetization for modeling zero-sequence current transformers *Materials of the XVII Int. Scientific and Technical Conf. "Benardos Readings"* (Ivanovo: ISPU)
[10] http://dynamics.com.ru/pages/proverka-transformatorov-toka-s-ispolyzovaniem-kompleksa-retom-21

IOP Conf. Series: Materials Science and Engineering 1035 (2021) 012030 doi:10.1088/1757-899X/1035/1/012030

Numerical study of the effects of pressure gradient on the heat transfer in the KMS heat exchanger channel

N Petrova[1] **and A Tsynaeva**

Department of Heat Supply and Ventilation, Samara State Technical University, 244 Molodogvardeyskaya str., Samara 443100, Russia

[1]Corresponding author: nadej.orlowa2013@mail.ru

Abstract. The work is devoted to studying the influence of the pressure gradient on the heat transfer intensity in the channel of the KMS air heater. The aim of the work is to develop methods for improving the heat engineering characteristics of a heat exchanger using a pressure gradient. Numerical simulation is implemented based on the RANS approach using the Code Saturne software package using the k-w-sst turbulence model. The simulation conditions were set in accordance with the manufacturer's instructions for the KMS-2 type air heat exchanger. In the work, the heat transfer coefficient was studied in the presence of a longitudinal pressure gradient in the heat exchanger channel from the gas path. As a result of the simulation, data were obtained on the effect of the pressure gradient on the heat transfer intensity for different working fluids in the channel of an advanced heat exchanger.

1. Introduction

Heat exchangers are widely applied in microclimate maintenance systems and production plants. Some examples include ventilation plants employed for warm-air heating in public and manufacturing buildings. Moreover, heat exchangers can be used for heating (cooling) technical gases (carbon dioxide, helium, nitrogen, and others). They are divided into recuperative (direct-flow or counter-flow) and regenerative. However, these units have performance limitations that concern gas flow as the thermal physical properties of the gas (air in microclimate maintenance systems) are significantly worse than the characteristics of the heating or cooling heat carrier in these units. Furthermore, the increase of gas flow (air) velocity in the heater aimed to enhance the heat transfer intensity is limited by the noise-level control requirements for operation of the units of microclimate support systems [1, 2]. The existing velocity limits reduce the possibility of increasing the heat transfer coefficient from the air flow. This in turn leads to an unchanged heat transfer coefficient that is indicative of the efficiency of the heat exchanger.

The enhancement of heat transfer intensity is possible due to the application of active and passive methods of enhancement. The active method implies that the heat transfer is influenced by external factors, such as electric field, pulsation, blowing, sound field, etc. The passive method includes the use of intensifiers that act by changing the topography of the heat transfer surface (fins, dimples, pins, and pressure gradient). It is known that the superposing of flow pulsations [3] in the separation area or other influences contribute to the enhancement of heat transfer intensity. As the main method of heat transfer enhancement in the designed heater, it is suggested to use a variable section of channels in which a longitudinal pressure gradient is realized. The interest in this method of increasing heat transfer intensity can be explained by the fact that the use of this method of enhancement does not complicate the development of new heater constructions.

It should be noted that the pressure gradient sign has a significant impact on heat transfer. Therefore, when designing heat exchangers in ventilation systems, it is necessary to choose the method of performance estimation that will give the most correct results. It is important to note rather high errors of calculation of heat transfer characteristics determined on the basis of dimensionless equations (about 30%). Thus, while developing new samples of heat transfer equipment, using only engineering calculations is not enough to assess the characteristics of heat transfer processes. Therefore, when developing new models of heat exchangers-heaters it is recommended to use the numerical method of research. This research method does not require significant material costs and makes it possible to get the real picture of the object under study. The influence of pressure gradient on heat transfer intensity was analysed by numerical simulation of the heater channel from the air flow side using the RANS approach based on free software [4, 5] provided that the channels of this heat exchanger (KMS-2) are divergent or convergent.

In spite of the fact that at present there are some works [6-9] devoted to the research of influence of pressure gradient on heat transfer, some lack of data for operating conditions of the KMS-2 heat exchanger was revealed. Thus, in work [6] the Reynolds number range is Re=5000-14000, the acceleration factor is $K=1.4\times10^{-6}$–2×10^{-5}, i.e. the characteristics of KMS-2 heat exchanger operation will correspond to a very narrow range of data, especially since the acceleration factor may not correspond to the parameters of the proposed KMS-2 heat exchanger modernisation.

In work [7] a numerical study of heat transfer in a diffuser was conducted with an opening angle $\alpha=3°$, relative channel length Ld=10–100 calibres, Pr=0.026-80 (working masses: air Pr=0.71, helium-xenon mixture Pr=0.24, water Pr=5.40, transformer oil Pr=55.7, mercury Pr=0.024). Nevertheless, in work [7], the Reynolds number range 10^4–10^5 does not correspond to the operating conditions of the KMS-2 heat exchanger (Re=30000-6000). In addition, although the numerical study in work [7] was conducted using the RANS approach, another model of turbulence was used which may give insufficiently accurate results in the selected range Re=3000-6000.

Work [8] is devoted to the numerical study of heat transfer in a flat channel with a converging channel and relative length of the converging section $Lk/h1$=13.33, $Lk/h2$=18.18–100, acceleration parameter $K=(3–13)\times10^{-6}$ and Re=10500. In the proposed study, the geometrical characteristics of the examined converging and diverging channels have other parameters.

2. Numerical study

The design scheme of a heat exchanger with gradient channels that we suggested for microclimate maintenance systems for air heating (cooling) is shown in figure 1 [9].

In order to check the adequacy of the obtained results of the numerical study, they were compared with the results of experimental studies of the influence of turbulence and pressure gradient on the characteristics of heat transfer in the boundary layer given in [10]. The analysis carried out in comparison with the results of numerical modelling [11] and experimental data [10] showed that deviations in the flow rate and pressure parameter do not exceed 10%. It should be noted that there is also a qualitative coincidence of results on the distribution of the heat exchange along the length of the plate [11].

IOP Conf. Series: Materials Science and Engineering 1035 (2021) 012030 doi:10.1088/1757-899X/1035/1/012030

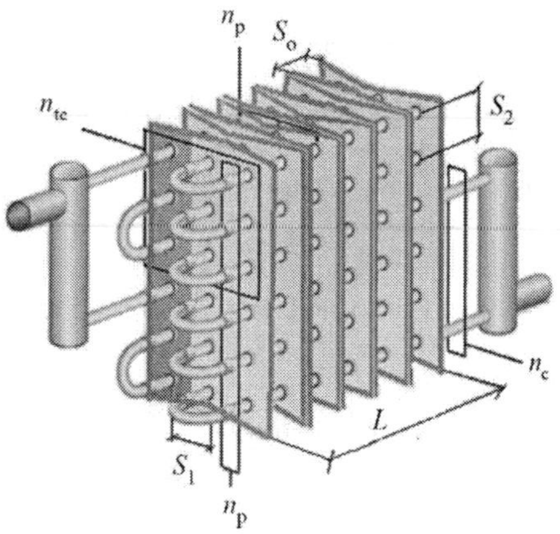

Figure 1. Design scheme of the suggested heat exchanger with a longitudinal pressure gradient: L– the length; S_o – the fin spacing; S_1 – the longitudinal tube spacing; S_2 – the transverse tube spacing; n_p – the number of tubes in a row; n_c – the number of circulation loops; n_{tc} – the number of tubes in the circulation loop.

During numerical research, a heat exchanger of type "KMS" of the manufacturer JSC "TST" was chosen [12] as a basic model of a heat exchanger. In this unit, the heating medium moves through metal finned tubes while the heated air is directed between the plates with dimensions of 136×117 mm located in increments of 5 mm [12].

A modified heater channel formed between two adjacent plates with a longitudinal pressure gradient through which the heated air passes was chosen as the object of the research. It is assumed that the heat exchanger channel from the air flow side is diverging or converging. The geometry of the calculation area is shown in figure 2. To create a pressure gradient, each of the plates deviates from the longitudinal axis by 3°. The geometry of the calculation area was developed on the basis of cloud software Onshape [13] used for free subscription.

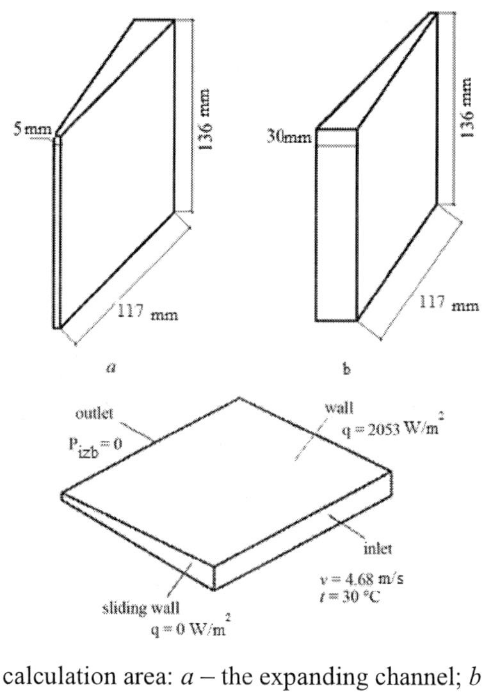

Figure 2. Geometry of the calculation area: a – the expanding channel; b – the converging channel.

The working grid was created with the help of Tetrahedron Negten (3D) algorithm, Negten 1D-2D in Salome software package [5] distributed under a free licence and providing access to open source code. The grid is constructed with near-wall viscosity layers and consists of 795911 elements: three-dimensional (3D) elements - 732709 (tetrahedrons - 403645, prisms - 329064); two-dimensional (2D) objects - 62322; nodes - 880.

The carried out analysis of the grid convergence has shown that the reduction of the number of computational cells to fewer than 700 thousand in this problem leads to the appearance of unreasonable fluctuations of flow characteristics. Cell sizes are specified in the range from 0.0001 to 0.01 m. The characteristics of the near-wall viscosity layers in the grid are determined on the basis of the Courant number by the value of y+=1, the value of the initial velocity and equivalent channel diameter; the total thickness of the δ layers is equal to 2.7×10^{-4} m, the number of layers *nsl*=6, the layer expansion coefficient is given equal to 1.5.

The following modelling conditions were set on the calculation area (figure 2): the working mass is air whose density depends on temperature. The air temperature at the inlet is assumed to be equal to the temperature of five cold days for Samara (Russia) (-30 °C), the air velocity at the inlet of the heater model range "KMS-2" takes values of v_1=3.36 m/s, v_2=4.68 m/s, v_3=6.7 m/s, the Reynolds numbers on the minimal section are equal to 3000, 4177, 6000, respectively. The turbulence level is set to Tu=8.4%. On the channel side walls which are heated plates, a heating rate is set. It is determined according to the operating conditions of the KMS heater at a water velocity of v_w=0.25 m/s. The temperature difference in the heat exchanger Δt_{cp} is measured according to the formula:

$$\Delta t_{cp} = \frac{\Delta t_g - \Delta t_m}{\ln(\frac{\Delta t_g}{\Delta t_m})} = 86.8\ °C\ . \quad (1)$$

The temperature of the heating agent at the inlet and outlet of the unit is t_{g1}=95°C, t_{g2}=70°C, respectively; the temperature of the air at the inlet t_{m1}=–30°C and t_{m2}=20°C at the outlet of the heater. It follows from the performed calculations that the specific heat flow is set equal to q_1=1832 W/m², q_2=2053 W/m², q_3=370 W/m² at the corresponding air inlet velocities. At the outlet of the channel, a free-flow discharge condition is set (see figure 2).

For the realisation of numerical modelling on the basis of the RANS approach with the application of the *k-w-sst* turbulence model, the software package Code Saturne [4] was used. The solution was obtained on the basis of Navier-Stokes equations, energy and state presented in [14, 15]. Works [9, 11, 14-16] have shown the rightfulness of using this approach to solving the problem studied in the work.

3. Simulation results

As a result of the numerical study, the primary characteristics of the flow are determined: temperature, pressure, velocity, energy of flow pulsation. On the basis of the data obtained, local values of the heat transfer coefficient along the channel length for the KMS-2 heater are calculated:

$$\alpha_{xi} = q_j / (T_{wi} - T_{fi}), \quad (2)$$

where α_{xi} – the local heat transfer coefficient; T_{wi} – the wall temperature; T_{fi} – the flow temperature.

In order to further verify the adequacy of the results, figure 3 shows the results of comparison of the data of the numerical study and the calculation of the local heat transfer coefficient according to Zukauskas formula [17] taking into account the high turbulence and acceleration of the flow:

$$Nu_x = Nu_p \cdot (v_x / v_0)^{\ b}, \quad (3)$$

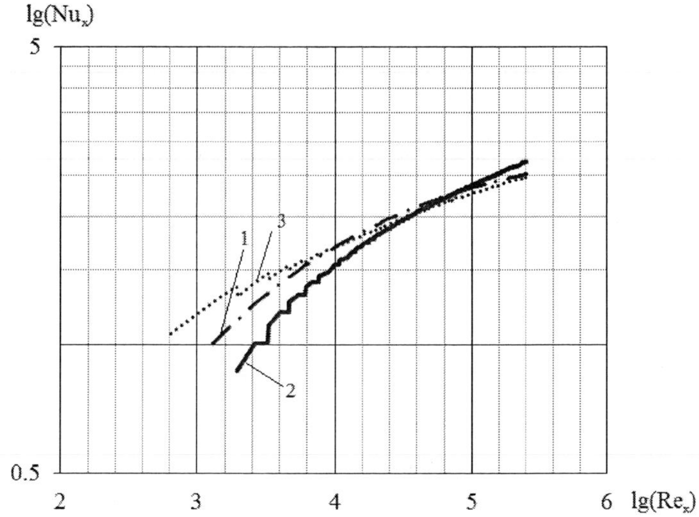

Figure 3. Local values of heat transfer coefficient along the length of the converging channel at v=6.7 m/s, $dP/dx < 0$: *1* – the calculation by Zukauskas formula [18]; *2* – the calculation by formula (4) from work [19]; 3 – our numerical modelling.

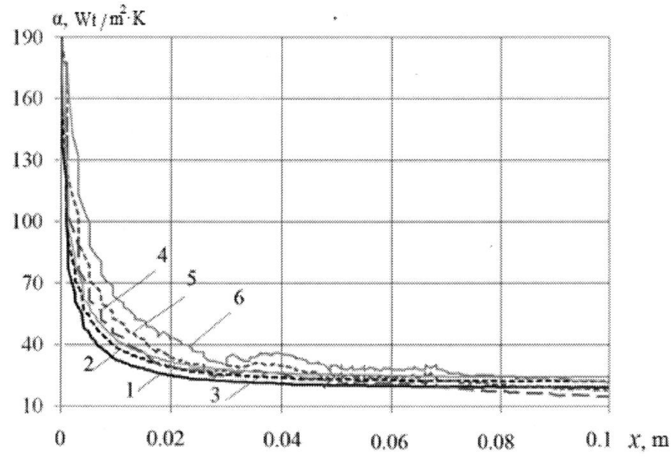

Figure 4. Distribution of local values of the heat transfer coefficient along the length of the channel: *1* – Re_d=3000, $dP/dx < 0$; 2 – Re_d=4177, $dP/dx < 0$; *3* – Re_d=6000, $dP/dx < 0$; *4* – Re_d=3000, $dP/dx > 0$; 5 – Re_d=4177, $dP/dx > 0$; *6* – Re_d=6000, $dP/dx > 0$.

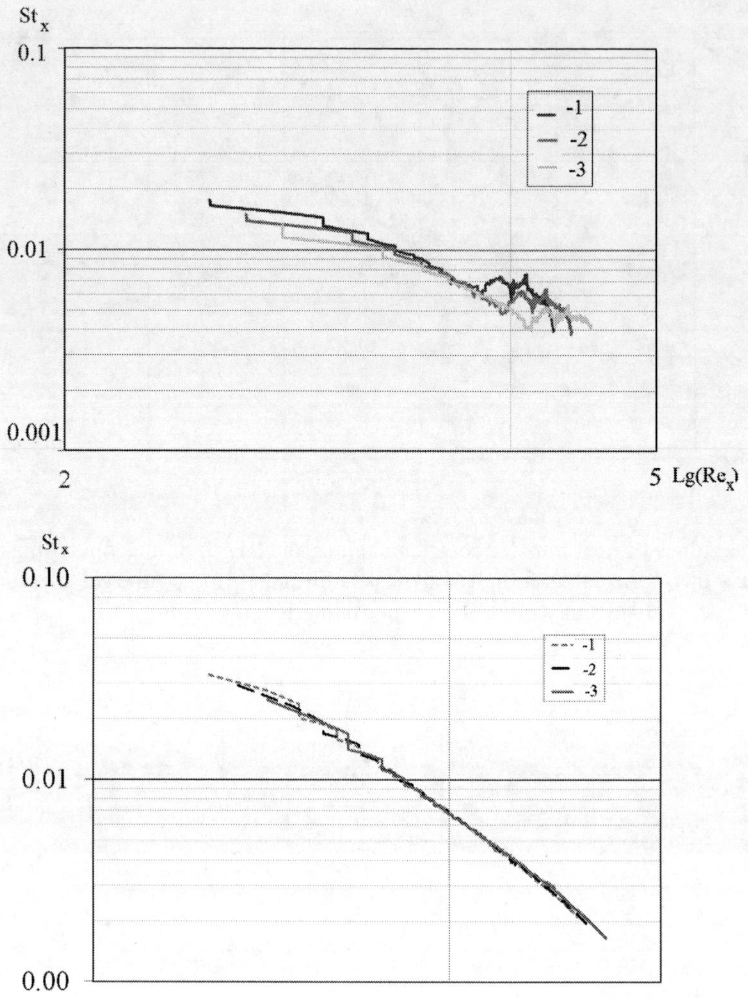

Figure 5. Distribution of the local heat transfer coefficient along the length of the channel: *a* – at $dP/dx > 0$ and the velocity at the channel inlet *1* – v=3.36 m/s, *2* – v=4.68 m/s, *3* – v=6.7 m/s; *b* – at $dP/dx < 0$ and the velocity at the channel inlet *1* – v=3.36 m/s; *2* – v=4.68 m/s; *3* – v=6.7 m/s.

In addition, figure 5 shows the results of the calculation for the converging channel with an opening angle from 2° to 17° by formula (4) presented in [19]:

$$\mathrm{Nu}_x = 0.024 \cdot Re_x^{0,8}, \tag{4}$$

where Re_x – the local value of Reynolds number; x – the distance from the channel inlet. The analysis of figure 3 shows that the deviation of the calculation results by the dimensionless equations and the numerical modelling method is within the margin of error of dimensionless equations. It should be noted that expression (4) will give better results at higher values of Re_d.

Figure 4 shows the results of numerical simulation of local values of heat transfer coefficient along the length of the channel at different Reynolds numbers in the presence of a negative and positive longitudinal pressure gradient.

It was found that the maximum value of the heat transfer coefficient was achieved at higher velocities. The growth of heat transfer intensity is mostly due to the value of air flow velocity.

Since the influence of velocity significantly affects the intensity of heat transfer, the results are summarised according to Stanton's criterion for positive and negative pressure gradients. Figure 5 shows the dependence of the St number on the Reynolds number along the length of the expanding and narrowing channel.

The results are generalized not only by local values but also by the average values of the heat transfer coefficient. Figures 6, 7 present the average values of the heat transfer coefficient and the Nusselt number for the expanding and narrowing channel (the Reynolds number is determined for the minimal section of the channels).

As a comparison, figure 7 shows the results of the heat transfer estimation by the known dimensionless equations (Mikheeva, Petukhova, Gnielinski) for channel flow [19]. It should be noted that the results of calculation by Petukhov formula [19] whose range of applicability at Re_d=10000-5×10^6 are given considering that it is Petukhov formula's modification that represents Gnielinski expression for Re_d=2300–5×10^6 [19].

According to the results of the numerical study presented in figure 6, the average values of the heat transfer coefficient at $dP/dx < 0$ are 12-16% lower than at $dP/dx > 0$ depending on the Reynolds number. Figure 7 shows that the heat transfer intensity in a channel with a gradient with $dP/dx < 0$ will be lower than for a flow with $dP/dx=0$ (heat transfer in a channel without pressure gradient is determined by the known dimensionless equations of Mikheev and Gnielinski). For a flow with $dP/dx > 0$, there is an area (Re=3000-5000) in which the heat transfer intensity is higher than for a flow with $dP/dx = 0$ (see figure 9). The Petukhov formula gives somewhat overestimated results as its area of application is Re_d=10^4–5×10^6 which does not correspond to the operating conditions of the KMS-2 heat exchanger.

Since it is known that the cost of compensation for pressure losses significantly depends on the pressure gradient sign, while developing a modified design of heat exchanger should be performed not just converging or diverging channels, but combined channels. In combined channels, the length and angle of opening (or narrowing) of individual sections will determine the efficiency of heat exchange in the unit. Therefore, this raises the importance of elaborating a simple engineering design methodology of such heat exchangers on the basis of diagrams or the dimensionless equations.

It is known that, in the presence of a pressure gradient, heat transfer is significantly impacted by the nature of the working masses. One more cycle of numerical experiments has been carried out whose results are shown in figure 8. This figure presents data on the distribution of local heat transfer coefficient in the channel (at the inlet v=10 m/s) in the presence of positive and negative pressure gradient for gases of different physical nature (working masses: air, helium, carbon dioxide). The research results (see figure 8) are presented in relative coordinates: St_x/St_{00}, where $St_{00}=0.03 \cdot Re^{-0.2} \cdot Pr^{-0.4}$ – Stanton number for undisturbed flow and $x=x_i/L$, where x_i – the distance from the channel inlet, L – the channel length. Line $St_x/St_{00}=1$ in figure 8 shows relative values of heat transfer for a gradientless flow since in a gradientless flow $St_x=St_{00}$.

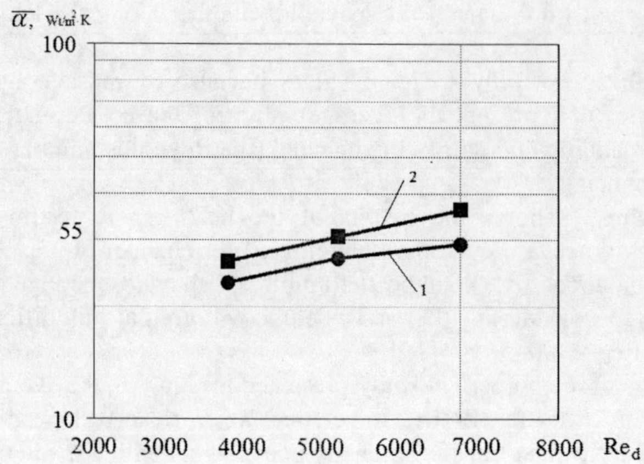

Figure 6. Channel heat transfer coefficient: *1* – at $dP/dx < 0$; *2* – at $dP/dx > 0$.

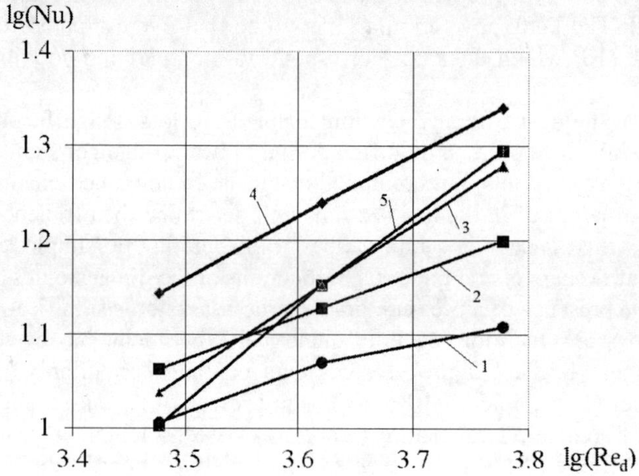

Figure 7. Heat transfer in the channel: *1* – the numerical modelling at $dP/dx < 0$; *2* – the same at $dP/dx > 0$; *3* – the calculation by Petukhov formula [19]; *4* – the calculation by formula (5); *5* – the calculation by Mikheev dimensionless equation for air.

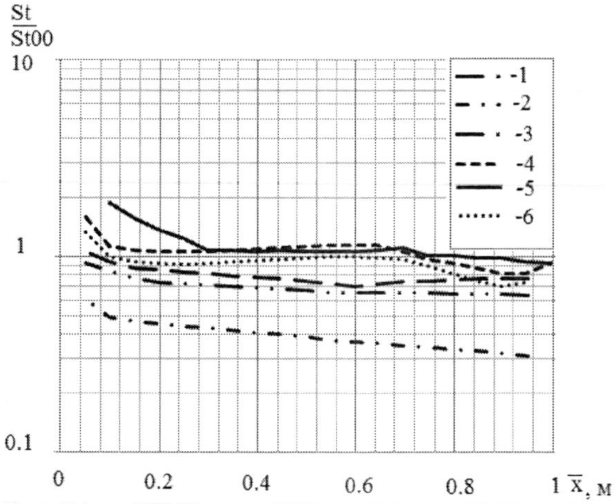

Figure 8. Distribution of the local heat transfer coefficient for different working masses at inlet velocity $v = 10$ m/s with positive pressure gradient: *1*, *2*, *3* – air, helium, carbon dioxide, respectively, at $dP/dx < 0$; *4*, *5*, *6* – air, helium, carbon dioxide, respectively, at $dP/dx > 0$.

According to the results presented in figure 8, the heat transfer intensity at $dP/dx > 0$ will be higher than at $dP/dx < 0$. Moreover, the data in figure 8 show that the presence of a pressure gradient has a significant effect on the heat transfer compared to a gradientless flow. A negative pressure gradient leads to some reduction in heat transfer compared to a gradientless flow; the minimum values of heat transfer coefficient are observed at $dP/dx < 0$ for helium used as a working mass. With a positive longitudinal pressure gradient, the heat transfer coefficient can be higher than in a gradientless flow. The heat transfer intensity for air is higher than for carbon dioxide but lower than for helium at $dP/dx > 0$.

4. Conclusion

As a result of numerical modelling of the KMS-2 heat exchanger-heater channel (assuming that the channel is made in the form of a converging or diverging channel) with a pressure gradient, it was found that the heat transfer intensity significantly depends on the gradient sign.

Thus, the average values of heat transfer coefficient at $dP/dx < 0$ are 12 – 16% lower than at $dP/dx > 0$ in the range of Reynolds numbers Re=30000–6000 and the acceleration coefficient $K=8.04\times10^{-6}$–1.56×10^{-5}. The use of various working masses shows that a negative pressure gradient leads to some reduction in heat transfer compared to a gradientless flow; the minimum values of heat transfer coefficient are observed at $dP/dx < 0$ for helium used as a working mass. With a positive longitudinal pressure gradient at $dP/dx > 0$, the heat transfer coefficient can be higher than in a gradientless flow and the heat transfer intensity for air is higher than for carbon dioxide, but lower than for helium.

When developing the design of a heat exchanger with modified channels, it is assumed to perform not just converging or diverging channels, but combined channels, in which the length and angle of opening (or narrowing) of individual sections will determine the efficiency of heat exchange in the unit. Therefore, this raises the importance of elaborating a simple engineering design methodology of such heat exchangers on the basis of diagrams or the dimensionless equations.

References

[1] *Set of Rules 60.13330.2012 2012 Heating, Ventilation and Conditioning* (Moscow: Minregion Rossii) p 62
[2] *Set of Rules SP 51.13330 2011 Sound Protection* (Moscow: Minregion Rossii) p 46
[3] Davletshin I A, Mikheev N I and Molochnikov V M 2008 Heat transfer in a turbulent separation

region with superimposed stream pulsations *Thermophysics and Aeromechanics* **15(2)** 215–222
[4] *Code_Saturne* https://www.code-saturne.org/cms/ (accessed 02.02.2018)
[5] *SALOME Platform* http://www.salome-plat- form.org/ (accessed 02.02.2018)
[6] Tanaka H, Kawamura H, Tateno A and Hatamiya S 1982 Effect of laminarization and retransmission on heat transfer for low Reynolds number flow through a converging to constant area duct *Journal of Heat Transfer* **104** 363–371
[7] Luschik V G and Reshmin A 2018 Heat transfer enhancement in a plane separation-free diffuser *High Temperature* **56(4)** 569–575
[8] Luschik V G, Makarova M S and Reshmin A I 2019 Laminarization of flow with heat transfer in a plane channel with a confuser *Fluid Dynamics* **54(1)** 67–76
[9] Petrova N P and Tsynaeva A A 2019 Numerical study of heat transfer in a heat exchanger channel with a longitudinal pressure gradient *Heat Processes in Technology* **11(12)** 532–540
[10] Epik E Ya 2006 Effect of turbulence and longitudinal pressure gradient on heat transfer in a turbulent boundary layer *Proc. of the 4th Russian National Conference on Heat Transfer* vol 2 (Moscow: MPEI) pp 270–273
[11] Petrova N P and Tsynaeva A A 2018 Design and study of exchangers with pressure gradient heat intensifiers *Urban Planning and Architecture* **8(3)** 137–144
[12] *KMS and KMB heaters* http: //zao-tst.ru/kalorifery-kms-kmb.html (accessed 21.11.2019)
[13] *OnShape* https://www.onshape.com/ (accessed 01.02.2018)
[14] Tsynaeva A A and Nikitin M N 2017 Fluid dynamics and heat transfer in a curved channel with imposed unsteadiness of the flow *Paperss of Academenergo* **1** 42–49
[15] Tsynaeva A A and Nikitin M N 2016 Numerical modeling of rectangular channel with shallow dumbbell dimples based on Code Saturne *Proc. of ISP RAS* **28(1)** 185–196
[16] Tsynaeva A A, Razorenov S E and Belaya V V 2017 Numerical modeling of heat transfer of channel with shallow curly dimples *Proc. of ISP RAS* **29(5)** 329–344
[17] Zhukauskas A A 1982 *Convective Transfer in Heat Exchangers* (Moscow: Nauka) p 472
[18] Vasilev F V 1983 *Calculation and Experimental Study of Local and Averaged Heat Transfer Characteristics during Turbulent Flow of Coolant in Direct, Diffuser and Confuser Channels* Ph. D. Diss. (Bryansk) p 223
[19] Lienhard J H and Lienhard J H 2011 *A Heat Transfer Textbook* (Cambridge, Massachusetts: Phlogiston press) p 755

IOP Conf. Series: Materials Science and Engineering 1035 (2021) 012031 doi:10.1088/1757-899X/1035/1/012031

The main problems and solutions to the management of the electric power system by the example of metropolis power supply

N Sh Chemborisova and M Yu Pogorelets[1]

Department of Electric Power Systems, National Research University "Moscow Power Engineering Institute", 14 Krasnokazarmennaya str., Moscow 111250, Russia

[1]Corresponding author: mpogorelets@gmail.com

Abstract. Developments of electric power systems of Russia as well as foreign electric power systems are on the way of the annual complications. This trend can be well observed in the example of the development of electric power systems of large agglomerations - metropolises. We should also mention that along with this complication new labor-intensive tasks arise - increasing the reliability, improving the management of modes and ensuring the survivability of electric power systems, in this regard, we can distinguish the following factors:

- increasing the number of interconnected power facilities;
- growth of the maximum installed capacity of the power system;
- complication of the electrical network;
- complication of ensuring the required level of controllability of objects;
- increasing of short circuit currents.

Thus, designers of electric power systems often use the principle of multi-factority in their design, which ultimately just affects the complexity of electric power systems, leading to a large number of configurations of network circuits that have different capabilities, properties, technical and economic indicators.

1. Introduction

Since energy supply is one of the primary tasks of life support and life-sustaining activity, it must comply with the highest requirements for its quality and reliability. The most obvious example of an electric power system (EPS) that takes into account all current requirements can be a complex EPS of a metropolis (MP).

Not only at present but also in the near future, high-quality and trouble-free power supply to consumers, will be largely determined by the trends, features and principles of building metropolises' power systems. MP power systems not only cover and coordinate the power supply of the entire city but also effectively synchronize the parallel operation of all power sources, both internal and external. The number of inhabitants of modern metropolises exceeds one million, and the number of cities with a population of more than fifteen million as of 2018 exceeds 15 cities in the world. Thus, metropolises are characterized by the accumulation of a large number of people in a relatively limited territory, very diverse utilities, technologically large and complex volumes and types of production, etc. In this regard, it can be argued that most MPs are deployed in areas with a relatively warm and mild climate, but in general, they are characterized by a variety of environmental conditions.

Consequently, in the development of the electric grid economy and electrical power generating system (EPGS) of MP in present circumstances, the following requirements must be observed - the organization of reliable, sustainable and economic energy supply of MP and the entire infrastructure of giant urban entities.

The most important conditions for achieving this goal are the diversity of energy resources, methods for their receipt, delivery and use in the MP, non-alternative and rigid restriction of violations of ecological interaction with the environment, ensuring stringent requirements for energy conservation and energy efficiency.

The management of the regimes of electric power systems of large agglomerations - metropolises (EPSMP) is becoming ever more complicated every year. The following problems can be identified: territorial extent, a constant increase in the load of consumers, a variety of time ranges of control cycles, restrictions on independent and dependent variables, non-linearity of state equations, increased requirements for accounting settings and the formation of automatic control systems, relative certainty of the initial calculation information. For these reasons, different requirements should be applied to the energy system of MP than to any other energy systems.

The main difference of the metropolis energy system is the presence of a large number of various energy objects: generation, cable and overhead networks, consumers of various reliability categories, which reveals a tendency to increase short-circuit currents and makes it quite a labor-consuming process to calculate PSP settings; significant interconnection of electric, heat and gas supply modes affects a high share of combined production of electric and heat energy; a large number of consumers of the first and second reliability rating, including life-support facilities that do not allow interruptions in power supply, and other equally. Their density in the metropolis is significant.

In a large urban agglomeration, there is a great interdependence between electricity, heat and gas supply. The power system is characterized by a complex topology of the electric network, a high current load of its elements, and a high level of s/c current. Given the above, the time to restore such a power system after an accident should be much less than in a conventional power system. It is necessary that the loss of at least one element of the system leads to the smallest consequences, since in a metropolis completely different requirements for the viability of the energy system. There is also limitedness of the territory. It is especially worth noting that often in modern metropolises there is simply no place for the construction of energy facilities. If generating facilities, apart from modern facilities of renewable energy sources, can still be placed outside the city, then substations supplying such a metropolis cannot. And this is the basis for clearly verified technical, economic and regime calculations, to assess what we are going to build here: a generation, a power load-center substation, or an ordinary, cable or overhead networks, what the capacity of this object or the line capacity should be. These criteria form the main requirements for EPSMP.

The property of metropolis power systems compared to a traditional power system is that with a significant density of the grid and generating complex, this power system can have a wide range of bottlenecks, as well as insufficient equipment with emergency automation systems, which in fact leads to unforeseen emergency situations. For example, an accident in the Moscow EPS in 2005.

Optimal control of the EPSMP mode is the most important condition for its high-quality operation, first of all, reliability and efficiency. In this connection, the requirements for the energy systems of metropolises include many conditions that cannot be applied to traditional energy systems. So, it is worth to increase the period of development of the scheme and development program from five years to seven, the development of additional sections in schemes and programs with power supply schemes of the power network of 110 kilovolt and above, fuel supply, heat and power supply to distribution networks with a voltage class of 6–35 kV.

The increased requirements are supposed to be applied to the internal high-voltage network of metropolises and its electrical connections with the UES of Russia. For example, the output of electric networks providing communication between the objects of the electric grid economy of the metropolis and the IPS should be at least 20-25% of the total load of consumers of the metropolis

The energy system of a metropolis must have at least three basic electrical substations with a voltage of 220 kV, connected to other power systems, and also connected for mutual redundancy in the “ring”, which must be equipped with uninterruptible power supplies that provide the substation's own needs during the day or more. At the same time, the total installed capacity of autotransformers of the highest voltage class at one reference substation cannot exceed 20% of the maximum load capacity of a metropolis. In contrast to existing conventional power systems, in metropolises, when designing new and reconstructing existing power facilities, it is advisable to include the most severe design disturbances. For example, simultaneous shutdown of all generators at the power plant, a shutdown of the switchgear of any voltage class at a substation or power plant, simultaneous shutdown of cables of a power network passing through one collector. Thus, in the design of EPSMP it is advisable to use inserts and direct current transmissions.

2. Problems and principles of building power supply in metropolises using new technologies and equipment

2.1 Problems of energy supply of metropolises

Currently, there are a number of cardinal problems in the energy supply of the MP:
a) lack of generating capacity;
b) lack of distribution networks at 110 ÷ 220 kV, which bring the power to the concrete MP areas;
c) unreliability of power supply to consumers of distribution networks of 6 (10) and 0.4 kV;
d) instability of all three directions - generation, transmission, distribution;
e) lack of effective consolidation between them and with energy consumption;
f) unsatisfactory pricing, etc.

From this point of view, the analysis of the listed directions is carried out separately and together. Despite the fact that they are based on competing requirements and conditions, they are generally organically combined and consolidated by the unity of solving the common problem - ensuring reliable power supply to the MP.

Requisite conditions for the modern energy of metropolises are constancy and continuity:

maintaining all facilities and structures in a healthy state, at the level necessary to meet consumer demands;

modernization of existing energy facilities;

development, implementation and effective use of innovative technologies in all aspects of production, transmission, transformation, distribution and consumption of electricity, without exception.

These technologies should not only take into account the current objective need of the MP electric energy complex, but also comply with the advanced evolutionary directions of the domestic and industrial sectors, guarantee indisputable reliability of power supply, comply with environmental safety requirements, be cost-effective, and at the same time fully comply with the requirements and demands of consumers.

2.2 Relocation of power plants in the foreseeable future beyond the lines of the metropolis

The construction of large thermal and nuclear power plants outside the metropolis area agglomeration solves the problem. This is one of the possible ways of developing MP energy. Undoubtedly, this will entail a number of cardinal problems. First of all, this is the creation of "deep entry" systems and the restructuring of the heat supply system. According to many experts, urban thermal power plants need to be radically modernized. Now the problem is being solved by installing and commissioning new effective equipment, CCGT and other technologies.

At the same time, there are problems of building and providing power supply to the border areas of the agglomeration. A striking example of this is the Moscow region, where large sources and consumers of electricity are located in the satellite suburbs of Moscow. Only environmentally friendly sources of energy based on renewable energy resources, including non-traditional individual ones, should work mainly within the megalopolis, especially to solve the problems of providing consumers with heat.

2.3 Expanding the range of distributed generation sources

This direction can be considered as an alternative to the above. These are cogeneration and trigeneration power plants of small and medium capacity that use natural gas as fuel and provide consumers with electricity, heat and cold. They can work independently or together with the main network. In the latter case, an important condition is the optimal combination of joint operation of a particular power plant and network, which allows obtaining the maximum coefficient of primary fuel use and, accordingly, the minimum harmful impact on the environment.

A significant hindrance in the development of this direction in Russia is the lack of legislative instruments on joint work and mutual settlements of traditional electricity suppliers, on the one hand, and, on the other, producers (consumers) using distributed generation sources. Moreover, the former are currently actively rejecting and inhibiting the adoption of objective decisions in this direction, in most cases, without taking into account the arguments, assessments and references to the positive experience of the European Union, Asian and American countries. Nevertheless, it is clear that the need for the development of distributed generation is obvious, and the near future time will present itself.

2.4 Concentration of energy production directly inside the territory in existing metropolises

This direction is typical for the organization of the power supply of existing MPs. To a greater extent, in domestic large cities, this is due to the need to organize district heating.

In the examples of building electric networks and power supply systems for the largest Russian cities of Moscow, St. Petersburg, Samara, Yekaterinburg [1], and this regularity can be traced distinctively. In foreign MPs, this trend is manifested to a lesser extent due to the geographical location and, consequently, a milder, warm and hot climate than in Russia.

2.5 Complete solution to the power load-center substation

This is relevant both now and in the future. In other words, in the MP, it is necessary to transport large flows of electricity to power MP consumers from sources located in the agglomeration or at a distance and outside. One of the most well-known solutions for modern EPSMPs, which impose severe restrictions on the dimensions of the switchgear, is power load-center substation at voltages of 110 ÷ 330 kV into the territory of the MP in power consumption centers [2, 3]. At the same time, the possibility of using SWYD is almost completely excluded in new construction.

They not only violate the environment but also increase electromagnetic pollution, both at the industrial frequency and in high-frequency bands, disrupting TV and radio reception, mobile communications, etc. In residential areas, noise effects, such as shots fired when air switches are operating, are also completely unacceptable.

2.6 Creation of modern distribution circuits of a metropolis

Electric networks of the metropolis are created on the basis of innovative principles of their construction: accessibility, outstripping demand for consumption, reliability, environmental friendliness, aesthetics of the urban territory, etc. This problem is considered in the work for the ENEPS of Russian MPs, namely Moscow, St. Petersburg, Yekaterinburg, Samara, as well as characteristic fragments of the power supply of their agglomerations, and, in addition, existing and promising fundamental solutions and schemes that ensure a high level of reliability due to reservation [4].

2.7 Exchange of urban overhead line (OHL) on cable lines (CL)

This major global problem of EPSMP development is solved on the basis of modern types of OHL with polymer insulation (CLP - cross-linked polyethylene, EPR - ethylene-propylene rubber et al.) and CL, which use the effect of high-temperature superconductivity. The use of CL instead of OHL is important due to strict prohibitions and restrictions on OHL in networks of all voltages in the megalopolis and above.

The smaller area of the CL, its greater reliability in comparison to OHLs (if all the necessary conditions for selecting the necessary cables for the transmission of the required power at the design stage and compliance with the laying technology at the installation stage are met), as well as the absence of supports and hanging wires determine the widespread use of CLs high voltage and ultra-high voltage in the energy systems of large cities, which is observed in the energy systems of the MPs.

At the same time, the cost of constructing a CL of the corresponding voltage class is approximately 15–20 times higher than that equal to it in terms of transmitted power. However, even in spite of this circumstance, CLs based on cables with insulation from CLP are increasingly used [5].

2.8 Short-circuit current limitation in the metropolis ENEPS

In the networks of all metropolises, high levels of SC reach alarming values, being a well-known problem. The situation is made worse by the fact that there are certain difficulties with the serial industrial production of switches, in particular, HV and UHV with tripping currents above 63 kA. The cost of such equipment is more than significant, and their development is unique and still economically unprofitable.

This task is effectively solved when using a complex of relatively new types of electricity-generating equipment and measures, their application is considered in detail [6]. Probably, without their application in the schemes of the long-term development of energy systems of the MP at the level of 2016 ÷ 2020 and further dispensation is almost impossible.

2.9 Implementation of technologies and materials using the HTS effect

These innovations are increasingly being used for cables, in particular, for a power load-center substation and distribution networks, as well as power lines, power transformers, current limiters and other electrical installations and complexes [7].

At voltages of 10–35 kV and an output at the level of traditional of 110–330 kV OHLs in real cases, it is almost irrelevant to them the mutual location of consumers in the metropolis and sources outside of it.

2.10 The development of transmission and distribution networks of the metropolis

It is very difficult to develop electric networks on the territory of existing MPs [8]. First of all, in conditions of high-density development and penetration of the city by various types of communications in some cases, it is practically impossible to place and build substations and power lines, since this requires withholding of significant territories from circulation. High-cost, but effective from all points of view, high-temperature superconductivity (HTS) and polymer CLs can be called as possible ways [9, 10]. As an alternative, we can offer sources of distributed generation using gas-piston and gas-turbine installations, although they, like the well-known diesel generators, are suitable as backup and are preferred in cogeneration and trigeneration complexes. Moreover, if strategic investors can provide financing for these works and purchase equipment, then to overcome other restrictions (environmental, historical, organizational, etc.), even using the latest equipment and technologies state guarantees and effective administrative support of city structures are needed.

2.11 Introduction and development of inserts and direct current transmissions

EPSMP, which are becoming more complex every day, require high-quality supervisory monitoring. Therefore, along with the above mentioned, to enhance supervisory monitoring of AC systems, power management using direct current transmission can be the introduction of a ring or power lines creating a "torn" ring of direct current transmissions with a hierarchical submission system, which will serve a regional power grid or a local center of electricity consumption, for example, a metropolis, using controlled AC to DC converters, in the form of a system of a variety of points of supply/consumption of electricity. Such a system would allow one to make a very stable operation of the grid according to the scheme: a network of alternating current - direct current - alternating current, which at the same time would provide full control over the supply of electricity to all supplied loads, while at the same time providing power control in all incoming and outgoing power lines of alternating current and/or direct current.

IOP Conf. Series: Materials Science and Engineering 1035 (2021) 012031 doi:10.1088/1757-899X/1035/1/012031

According to the considered options for the use of direct current transmissions and DC converting substations, the implementation of measures to improve supervisory monitoring can be represented by the following options:

- a direct current transmission system will allow to "decouple" the local alternating current network from the surrounding alternating current system, where there is an alternating current electric power consumption center with many local alternating current electric loads and one or more distribution network supply lines serving a lot of local alternating current electric loads. Also, there may be a detachable AC load. At least one remote generation producing electricity AC power to a local AC consumption center and/or to a remote AC electrical load via an AC power transmission line, and between the local AC power consumption center and the remote power generating station and the remote electric load, at least partially there is a ring or an incomplete ring of direct current power lines having a plurality of direct current electric loads connected to it.

The DC power line ring provides isolation of AC electricity received from a remote power station from a local AC power consumption center. The first set of AC/DC converters that are electrically connected to a ring of DC power lines that is external to the local AC power consumption center is designed to convert AC power from a remote power plant to DC power that flows through a ring of DC power lines and to supply this DC power to at least, on some other loads from the set of DC electrical loads on the ring of DC power lines. The second set of AC/DC converters, which are electrically connected to a ring of DC power lines, is designed to convert DC power from a ring of DC power lines to AC power, which is directed to both local AC electrical loads and remote electrical loads.

One or more runs of the distribution network in the local AC power consumption center are electrically connected to the ring of DC power lines through the second plurality of AC/DC converters to supply AC power to the local AC power consumption center while decoupling all local electrical AC loads from a remote power station. A power line that is external to the local AC power center is electrically connected to the ring of DC power lines through the second plurality of AC/DC converters for supplying AC power to remote electrical loads, all local AC electrical loads being unleashed from this power transmission operation

- many DC-to-DC converters have been introduced in pre-defined locations on the DC power line ring to provide decoupling, which are designed to decouple from faults on the DC power line ring, preventing interference to the system operation or complete disruption of the system due to these faults.

- at least one local AC power generator is electrically connected to the AC power consumption center through a ring of DC power lines.

- AC electric power produced by a remote power plant is transferred to the local electric network via a power line selected from the group of power lines consisting of high voltage (HV) or ultra-high voltage (UHV) lines, and it can be overhead or cable.

- controlled AC / DC converters direct electricity that is planned to be transferred from a remote power station instead of a distribution located remotely from a local AC power center through a ring of DC power lines.

- at least the system includes one DC power source located at a local AC power consumption center that is electrically connected to the ring of DC power lines. The ring of DC power lines provides isolation of DC power sources from AC electrical loads and provides supervisory control of DC sources.

- in the local AC power consumption center, there are many different distributed sources of DC power generation that are electrically connected to the ring of DC power lines. The ring of DC power lines provides isolation of DC power sources from AC electrical loads and provides supervisory control of DC sources.

- power source converters are AC to DC converters.

- in the local center of AC electricity consumption, there are many different distributed sources of DC electricity generation, which are electrically connected to the ring of DC power lines and are selected from the group consisting of fuel cells, micro turbines, solar photovoltaic batteries, batteries and DC electric micro grids current. The ring of DC power lines provides isolation of DC power sources from AC electrical loads and provides supervisory control of DC sources.

- the second ring of DC power lines, designed to be used as a backup, which is connected to the system and passes along a similar, but on an adjacent route.

- DC power transmission system, providing isolation of the local AC power transmission network from the surrounding AC power system, and it comprises a local AC power consumption center in which there is a plurality of AC electrical loads, an AC distribution line supply line serving a plurality of AC electrical loads, and a local DC power consumption center in which there is a plurality of DC electric loads and DC distribution network supply line serving a variety of DC electrical loads. One remote power station provides at least a supply of AC power to a local AC power consumption center. A ring of direct current power lines having a plurality of direct current electrical loads connected thereto is located at least partially between the local AC and DC power consumption centers and the remote power plant. A ring of DC power transmission lines provides isolation of AC power received from a remote power plant from local AC and DC power consumption centers. The first plurality of AC/DC converters that are electrically coupled to a ring of DC power lines that is external to the local AC power center is for converting AC power received from a power plant to DC power and for providing DC power transformed in one of the AC/DC converters for at least some other loads of the plurality of DC electrical loads on the ring of the DC power lines. The second plurality of AC/DC, which are electrically connected with the ring to a local center electricity consumption of DC power lines AC, designed to convert DC power from the ring transmission lines DC to AC power for its supply to the electrical loads AC. The run line of the AC distribution network in the local AC power consumption center is electrically connected to a ring of DC power lines through a second set of AC/DC converters to supply AC power to the local AC power consumption center while providing isolation of all local electrical loads from the remote power plant. The run line of a DC distribution network in a local DC power consumption center is electrically connected to a ring of DC power lines through a second set of AC/DC converters to supply DC power to the ring of DC power lines.

- the local AC power transmission network contains a variety of distributed DC power generation sources that are electrically connected to a ring of DC power lines. The ring of DC power transmission lines provides isolation of DC power sources from AC electrical loads and provides dispatching control of DC power sources.

A method for providing isolation of a local AC power transmission network from the surrounding AC power system according to one of the proposed options includes a system that has an AC power consumption center with multiple AC electrical loads, one or more supply lines of the distribution network that serve multiple local AC electrical loads, and at least, one remote power plant for supplying AC power to the local AC power consumption center. The method contains the following operations: between the local AC power consumption center and the remote power plant, at least a partial ring of DC power lines is placed, which has a set of DC electrical loads connected to it, and by means of a ring of DC power lines, the AC power received from the remote power plant is separated from the local AC power consumption center. Convert AC power received from a power plant to DC power and provide it for at least some other loads from the set of DC electrical loads on a ring of DC transmission lines.

Converting DC electric power from the ring of DC power lines to AC electric power is carried out, and AC electric power is supplied to a local center of AC electric power consumption, while isolation of all local AC electric loads from a remote power station is ensured.

- in pre-defined locations on the ring of DC power lines, many DC-to-DC converters are installed to provide isolation, which are designed to isolate faults on the ring of DC power lines, preventing system failure due to these faults.

The processes described above are shown in figure 1 and figure 2

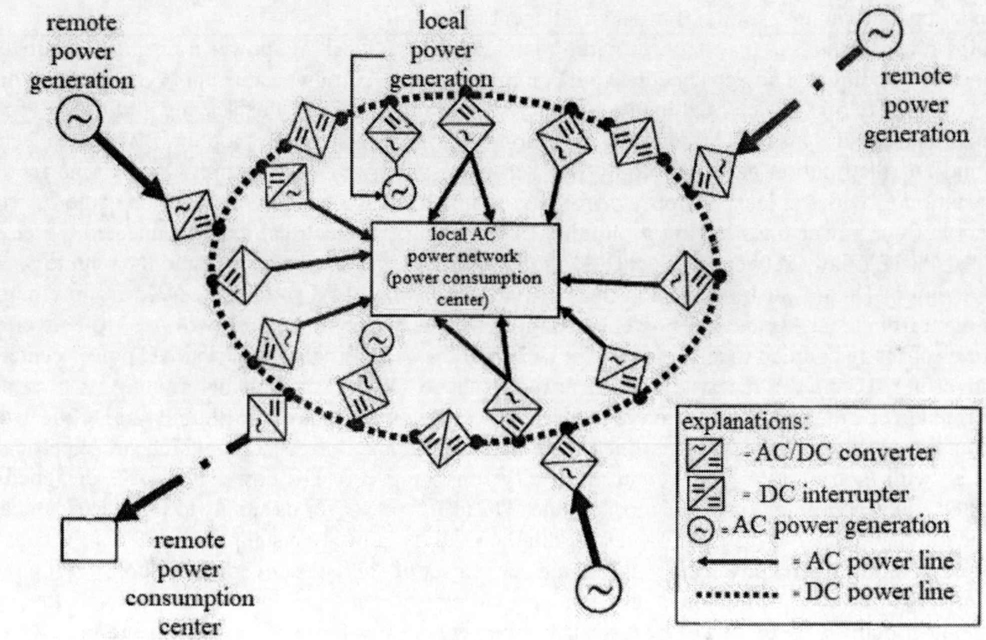

Figure 1. Distribution of electricity based on the use of remote production and consumption centers.

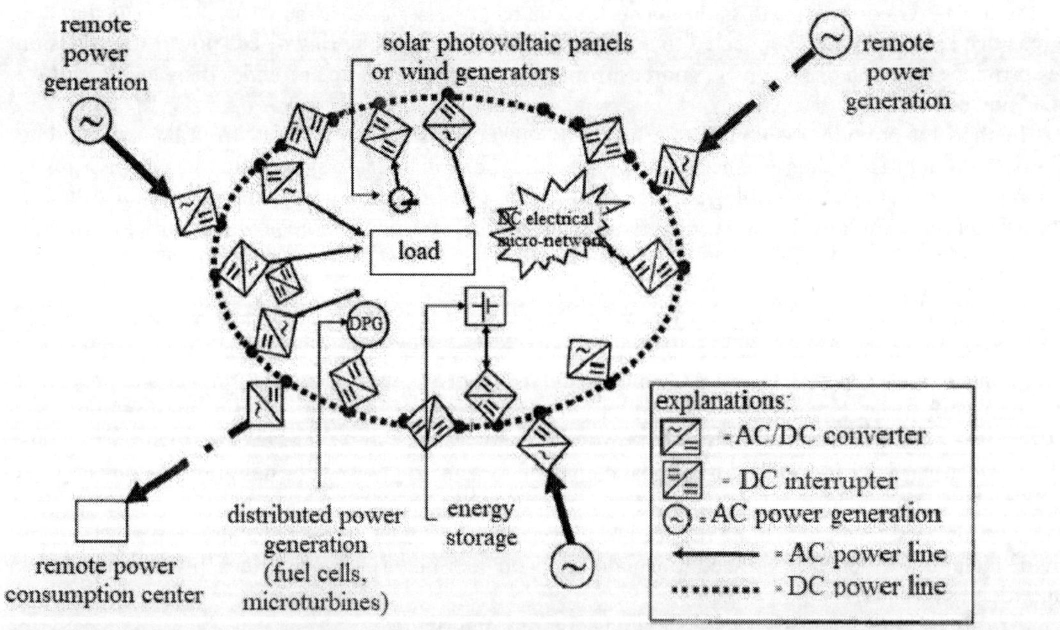

Figure 2. Distribution of electricity, taking into account the use of remote production/consumption centers and local renewable energy sources.

To analyze the proposed options, you can use the 14-node IEEE test circuit (figure 3). The scheme, due to the irregularity of network indicators in different areas, whether it is the district's energy system or metropolis, will not vary.

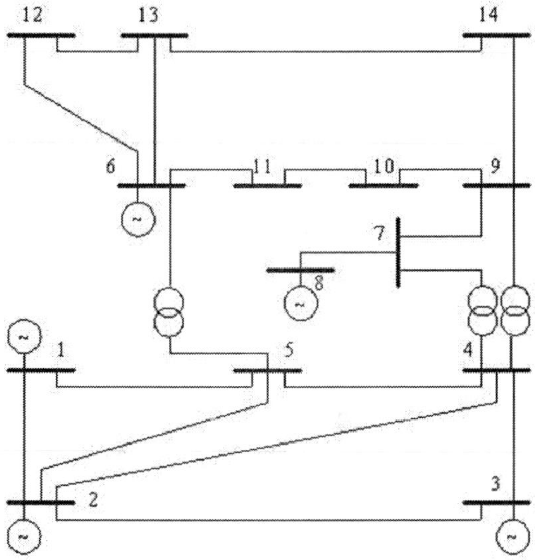

Figure 3. 14-node IEEE test circuit.

The proposed test scheme consists of two areas: the first - have a deficit of generated capacities (nodes 3, 4, 5) and the second - have an excess of generation (nodes 1 and 2).

The main purpose of mode calculations is to determine their parameters that characterize the conditions under which network equipment and its consumers operate. The results of calculations of network modes are the basis for evaluating the quality of electricity issued to consumers, the validity of the considered modes in terms of operation of network equipment, as well as identifying optimal conditions for power supply to consumers.

The initial data in the calculation of the modes of electric networks are the known capacities of consumer substations, the voltage values of power supplies or substations of systems that receive energy through electric networks from power plants, as well as the parameters and interconnection of network elements, on the basis of which the calculated equivalent circuit is compiled. In this case, the characteristic features of the network and the purpose of the calculations, which may be design or operational, are taken into account. It should be noted that in most cases the load in such calculations seems to be constant active and reactive powers. Nevertheless, there are examples when the calculation of the regime is carried out when the loads are taken into account by constant resistances and static voltage characteristics.

Losses as well as reactive power generation take place in the electric network line. The difference between reactive powers at the beginning and end of the line depends on the ratio of losses and reactive power generation.

The result of the computer calculation is the following results (tables 1 - 3):

Table 1. Node data.

No	Type	Unom	cap+	Voltage		Load power		Load generation		Limits the generation	
		kV	ind-	phase (deg.)	module (kV)	P MWt	Q MVar	P MWt	Q MVar	Qmax. MVar	Qmin MVar
1	3	230	-	0.00	243.80	-	-	232.39	-16.55	-	-
2	2	230	-	-4.98	240.35	21.70	12.70	40.00	43.56	50.00	-40.00

IOP Conf. Series: Materials Science and Engineering 1035 (2021) 012031 doi:10.1088/1757-899X/1035/1/012031

3	2	230	-	-12.73	232.30	94.20	19.00	0.00	25.08	40.00	0.00
4	1	230	-	-10.31	234.06	47.80	-3.90	-	-	-	-
5	1	115	-	-8.77	234.49	7.60	1.60	-	-	-	-
6	2	115	-	-14.22	123.05	11.20	7.50	0.00	12.73	24.00	-6.00
7	1	115	-	-13.36	122.08	-	-	-	-	-	-
8	2	115	-	-13.36	125.35	-	-	0.00	17.62	24.00	-6.00
9	1	115	1.44	-14.94	121.43	29.50	16.60	-	-	-	-
10	1	115	-	-15.10	120.86	9.00	5.80	-	-	-	-
11	1	115	-	-14.79	121.54	3.50	1.80	-	-	-	-
12	1	115	-	-15.08	121.35	6.10	1.60	-	-	-	-
13	1	115	-	-15.16	120.79	13.50	5.80	-	-	-	-
14	1	115	-	-16.03	119.09	14.90	5.00	-	-	-	-

Table 2. Legs data.

In	Final	R, Om	X, Om	(cap + ind -)	Ktr
1	2	10.2500	31.3009	0.0998	-
1	5	28.5819	117.9882	0.0930	-
2	3	24.8577	104.7261	0.0828	-
2	4	30.7402	93.2733	0.0643	-
2	5	30.1266	91.9825	0.0654	-
3	4	35.4483	90.4749	0.0242	-
4	5	7.0622	22.2762	0.0000	-
4	7	0.0000	105.8105	0.0000	0.511247
4	9	0.0000	276.2604	0.0000	0.515996
5	6	0.0000	115.8037	0.0000	0.536481
6	11	12.5611	26.3045	0.0000	-
6	12	16.2548	33.8309	0.0000	-
6	13	8.7483	17.2282	0.0000	-
7	8	0.0000	23.2958	0.0000	-
7	9	0.0000	14.5488	0.0000	-
9	10	4.2069	11.1751	0.0000	-
9	14	16.8103	35.7578	0.0000	-
10	11	10.8511	25.4013	0.0000	-
12	13	29.2167	26.4341	0.0000	-
13	14	22.6055	46.0256	0.0000	-

Table 3. Network mode parameters.

In	Final	Qin,MVar	Pin,MVar	Iin,A	Pfinal,MWt	Qfinal,MVar	Ifinal,A
1	2	156.88	-20.4	374.65	-152.59	27.68	372.51
1	5	75.51	3.85	179.05	-72.75	2.23	179.20
2	3	73.24	3.56	176.13	-70.91	1.60	176.29
2	4	56.13	-1.55	134.89	-54.45	3.02	134.53
2	5	41.52	1.17	99.77	-40.61	-2.10	100.13
3	4	-23.29	4.47	58.93	23.66	-4.84	59.56
4	5	-61.16	15.82	155.82	61.67	-14.20	155.82

4	7	28.07	-9.68	73.25	-28.07	11.38	143.28
4	9	16.08	-0.43	39.68	-16.08	1.73	76.89
5	6	44.09	12.47	112.81	-44.09	-8.05	210.28
6	11	7.35	3.56	38.33	-7.30	-3.44	38.33
6	12	7.79	2.5	38.37	-7.71	-2.35	38.37
6	13	17.75	7.22	89.89	-17.54	-6.80	89.89
7	8	0.00	-17.16	81.17	0.00	17.62	81.17
7	9	28.07	5.78	135.56	-28.07	-4.98	135.56
9	10	5.23	4.22	31.94	-5.21	-4.18	31.94
9	14	9.43	3.61	47.99	-9.31	-3.36	47.99
10	11	-3.79	-1.62	19.66	3.80	1.64	19.66
12	13	1.61	0.75	8.48	-1.61	-0.75	8.48
13	14	5.64	1.75	28.24	-5.59	-1.64	28.24

3. Result and discussion

Considering the analysis, we can recommend possible solutions to problems related to the supply of electric and heat energy to the metropolis:

1. The growth of investments in the development of energy systems of the MP in general, including the financing of generation development programs.
2. Optimization of the control system due to the construction of inserts and direct current transmissions.
3. Construction of new and modernization of existing power facilities, including the use of 20 – 35 kV networks.
4. Energy management as a variant of the rigid demand management model.
5. Further development of distributed generation based on renewable sources.
6. Switching in cities to selling electricity at market prices, rather than through subsidy programs.

At the same time, environmental factors related to environmental pollution and stimulating global warming have already acquired special importance and will become more and more significant. For this reason, in almost all the world's MPs, when developing a strategy and tactics for the development of energy, along with the creation of new large sources, the growth trend of renewable energy sources and tight control of electricity demand are clearly visible.

Thus, in modern conditions, organizational, technical, economic and financial aspects of designing and large-scale implementation of promising advanced, innovative environmentally friendly energy technologies, processes and electrical equipment are of significant importance in matters of the power industry of metropolises.

The electric power industry as the basic branch of the economic mechanism of Russia must meet the following requirements and ensure the sustainable development of all areas of the economy: industry, agriculture, science, education, as well as other aspects of the population's life.

The modern electric power industry was created thanks to the tremendous efforts of many generations. To date, the electric power industry should provide:

- standard quality and reliability of electricity consumers;
- prompt connection of new consumers to the electric grids and providing them with the required capacity, which ensures generally reliable development of the Russian economy;
- reducing the risk of emergency disturbances, including the development of major system accidents;
- reduction of direct and indirect losses from lack of electricity supply.

The target of the development of the modern electric power industry in Russia is the implementation of innovative types of electrical equipment and methods for controlling the regimes of large power systems based on the new capabilities of power regulating devices, measurement and information processing systems. This circumstance is especially relevant for power systems of large MPs. Under

IOP Conf. Series: Materials Science and Engineering 1035 (2021) 012031 doi:10.1088/1757-899X/1035/1/012031

current conditions and midterm the population, as well as the main production, cultural, scientific, technical and other resources of the country are concentrated in and around MPs.

To ensure the normal functioning and further successful development of the infrastructure of an MP and centered around resources, it is necessary to be able to provide energy in the required volumes to all existing consumers and to connect and provide electricity to new consumers.

In such conditions, the following requirements are imposed on MP power systems, which will only become tougher in the future:

- the readiness of the electric grid infrastructure to ensure the operation of the wholesale and retail electricity markets, accession of new generating capacities and consumers;
- the reliable operation of power facilities and, as a result, reliable power supply to consumers;
- the profitability of functioning and development;
- the personnel safety and reduction of negative impact on the environment.

The energy systems of MPs, formed, built and implemented today according to traditional principles, are highly likely to not be able to fully meet the requirements that they will have to meet in the short term, which determines the need for a transition to the innovative directions of their development that are formulated above.

All these circumstances complicate the provision and maintenance of the required level of reliability of MP energy systems. It becomes obvious that the construction of traditional outdoor SWYD and OHLs in such conditions is in most cases impossible.

4. Conclusions

1. The problems of the development of EPSMP are directly related to the vitals of MP, the most important of which are the population and communications.
2. Modern metropolises are characterized by a complex of problems of the energy trilemma, which includes: energy, environmental safety and free access to energy, and energy resources in the ENEPS.
3. The electric power industry of a metropolis must be considered taking into account the specifics of regional entities - medium and small villages located in close proximity to the metropolis and forming its agglomeration.
4. We can state a steady trend of decreasing energy, environmental and economic efficiency in the conditions of the predominant use of traditional organic fuels for the energy supply of the metropolis
5. Among a large number of basic principles and principles of the formation of the ESSE of a metropolis at present and in the near future, we can distinguish: the construction of large thermal and nuclear power plants outside the metropolis area; the complete solution to the power load-center substation problem; the creation of modern distribution networks of MP; the expansion of the complex of distributed generation sources; an exchange of urban OHLs on CLs; short-circuit current limitation in the metropolis ENEPS; the implementation of technologies and materials using the HTS effect and others.
6. The effective development of the electric power industry of the MP is determined by the development and adoption of decisions based on a set of the latest technologies and the use of innovative devices both in the present and in the future.

References

[1] Goldstein V G, Khalilov F H, Kokin S E, Kuznetsov D V and Smolovik S V 2015 *Modern Problems of Power Supply of Megalopolises* (Moscow: Energoatomizdat) p 300
[2] Alekseev B A 2009 *Energoexpert* **1** 92–97
[3] Varivodov V N 2008 *Energy Saving* **4** 2-8
[4] Kadomskay K P, Lavrov Y A and Laptev O I 2008 *High Voltage Electrical Equipment of the New Generation. Main Characteristics and Electromagnetic Processes* (Novosibirsk: NSTU Publishing house) p 343
[5] Kovalev G F 2011 *Electric Power. Transmission and Distribution* **1** 64-70
[6] Semenov A S and Kuznetsov N M 2014 *Measurement Techniques* **7** 417-420
[7] Dolgopolov A G and Sokolov S E 2012 *News of Electrical Engineering* **3** 18-22

[8] Dolgopolov A G 2014 *Controlled Shunt Reactors. The Principle of Operation, Construction, Relay Protection and Automation* (Moscow: Energiya) p 120
[9] Todirca S N 2010 *Energoexpert* **5** 56-58
[10] Cherepanov V V and Suvorova I A 2012 *Energy Security and Energy Saving* **5** 12-14

IOP Conf. Series: Materials Science and Engineering 1035 (2021) 012032 doi:10.1088/1757-899X/1035/1/012032

Study of the effect of the thickness of the photosensitive layer of perovskite on its efficiency using SCAPS-1D software

A A Rozhko[1,3], V V Petrov[1] and A V Sayenko[2]

[1]Department of Technosphere Safety and Chemistry, Institute of Nanotechnology, Electronics and Instrumentation, Southern Federal University, 2 Shevchenko str., Taganrog, Russia

[2]Department of Electronic Equipment Design, Institute of Nanotechnology, Electronics and Instrumentation, Southern Federal University, 2 Shevchenko str., Taganrog, Russia

[3]Corresponding author: drerozhko@mail.ru

Abstract. Photovoltaic cells are the best way to convert solar energy into electrical energy by absorbing photons emitted by the sun. This paper presents the results of studies of the effect of the thickness of the absorbing layer of perovskite on the quantum efficiency of the solar cell, as well as on its efficiency. These results were obtained by modelling in the software product SCAPS-1D. Perovskite photovoltaic is getting to be a distinctly predominant option for the conventional solar cells achieving maximum efficiency of 22.1% and more. This work is concerned about the design and analyses of lead-based perovskite solar cell model with the architecture of TiO2/CH3NH3PbI3/spiro-OMeTAD. The analysis of solar cell architecture is done using the Solar Cell Capacitance Simulator (SCAPS). It is a computer-based software tool and is well adapted for the analyses of homo and heterojunctions, multi- junctions and Schottky barrier photovoltaic devices. This software tool runs and simulates based on the Poisson's and continuity equation of electrons and holes. For this model, it is used to optimize the various parameters such as thickness, the defect density of absorber layer, doping concentrations (ND and NA) of Electron Transport Material (ETM) and Hole Transport Material (HTM)

1. Introduction

The use of solar energy is recognized as a promising way to solve the global energy crisis. In recent decades, photosensitive films based on perovskite have shown promise for creating highly efficient and inexpensive solar cells. The thickness of the perovskite film is one of the key parameters for achieving high efficiency of the solar cell. The thickness should be sufficient to absorb a large number of photons but too thick a layer of perovskite creates more recombination centers and prevents the transfer of photogenerated charge carriers due to a decrease in the electric field in the perovskite, which reduces the characteristics of the solar cell. The optimal thickness should be consistent with the perovskite thickness and the diffusion length (lifetime) of the charge carriers [1]. The main stage of optimizing the structure and improving the characteristics of solar cells is modeling. This work is devoted to modeling and research of lead-based perovskite solar cells using SCAPS (Solar Cell Capacity Simulator) software. SCAPS is a program for numerical simulation of solar cells developed at the Department of electronics and information systems (ELIS) of the University of Ghent (Belgium) [2]. Promising for photovoltaic applications, ORGANOMETALLIC perovskite is a material with a crystal structure (ABX_3) similar to the crystal lattice of calcium titanate ($CaTiO_3$). The main advantages of lead methylammonium iodide ($CH_3CN_3PbI_3$) perovskites are the absence of energy-intensive and complex technological processes when applied to various substrates (crystallization temperature of 100-150 °C), as well as their characteristic forbidden zone (1.5-2.3 eV) with direct transitions, high mobility of charge carriers (0.1-10 cm2v-1C-1), relatively large diffusion length (0.1-1.5 microns) and high absorption coefficient (more than 5×104 cm-1) [3]. Modeling is an important and necessary stage in the development and manufacture of new semiconductor devices, in particular solar cells, which reduces the cost of conducting experimental research and optimizing parameters.

2. Experimental part

The output characteristics of a perovskite solar cell depend significantly on the thickness of the absorbing layer of perovskite. The solar cell model in the SCAPS program consisted of $TiO_2/CH_3NH_3PbI_3$/spiro-OMeTAD layers, where $TiO_2$is a conductor for electrons, $CH_3NH_3PbI_3$ is a photosensitive (absorbing) layer, and spiro-OMeTAD is a conductor for holes [4]. The thickness of TiO_2 and spiro-OMeTAD layers were set constant, and the thickness of the photosensitive layer varied from 100 to 500 nm for the first experiment and from 100 to 900 nm for the second experiment, respectively. This model is calculated based on the diffusion-drift system of semiconductor equations, which includes stationary differential continuity equations for determining electron and hole concentrations, as well as the Poisson equation for calculating the potential in perovskite [5]:

$$D_{e^-}\frac{d^2n(x)}{dx^2} - \mu_{e^-}\left(\frac{dn(x)}{dx}\frac{d\varphi(x)}{dx} + n(x)\frac{d^2\varphi(x)}{dx^2}\right) + G(x) - R(x) = 0, \quad (1)$$

$$D_{h^+}\frac{d^2p(x)}{dx^2} + \mu_{h^+}\left(\frac{dp(x)}{dx}\frac{d\varphi(x)}{dx} + p(x)\frac{d^2\varphi(x)}{dx^2}\right) + G(x) - R(x) = 0, \quad (2)$$

$$\frac{d^2\varphi(x)}{dx^2} = \frac{q(n(x)-p(x))}{\varepsilon\varepsilon_0}, \quad (3)$$

where n and p - the concentrations of electrons and holes; D_{e^-} and D_{h^+} - the diffusion coefficient of electrons and holes; μ_{e^-} and μ_{h^+} - the mobility of electrons and holes; φ - the electrostatic potential; q – the elementary charge; ε – the relative dielectric permittivity; ε_0 – the electric constant, G - the optical generation rate of electron-hole pairs; R - the recombination rate of electron-hole pairs. The rate of charge carrier generation was determined in the spectral range of perovskite absorption based on the booger-Lambert law and the approximation of the solar spectrum AM1.5 by the spectrum of thermal radiation at a temperature of 300 K (Planck's formula) [5, 7]:

$$G = \eta\alpha(1 - T)\Phi e^{-\alpha x}, \quad (4)$$

where η - the photon generation coefficient of electron-hole pairs; α - the perovskite absorption coefficient; T - the reflection coefficient from the front surface; Φ - the photon flux density in the perovskite absorption spectral range:

$$\Phi = \int_{280}^{800} K\left(\frac{R_S}{r_0}\right)^2\left(\frac{2\pi c}{\lambda^4}\frac{F}{e^{hc/\lambda kT}-1}\right)d\lambda, \quad (5)$$

λ – the wavelength; h - Planck's constant; c – the speed of light in a vacuum; F – the shading coefficient; R_S – the radius of the Sun; T – the temperature of the Sun; K – the coefficient reflecting the influence of the earth's atmospheric mass on the solar radiation power density for the AM1.5 spectrum.

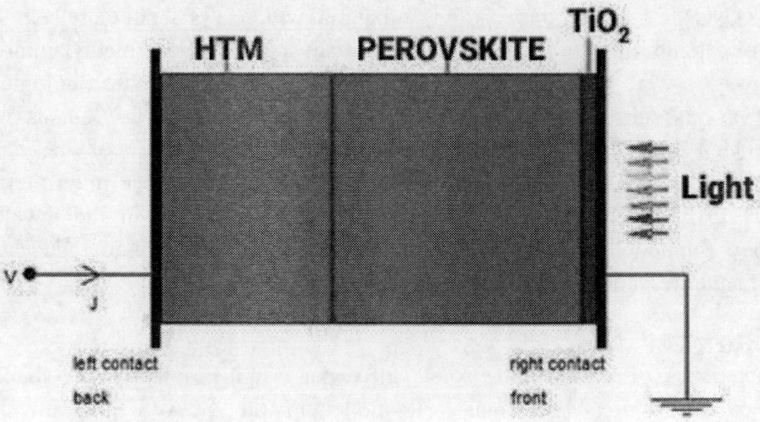

Figure 1. Model of study of the Sollar element.

The parameters obtained from the articles [6-9] were used for modeling. Table 1 shows the initial parameters of each of the layers of the studied solar cell, which were taken into account in the modeling process.

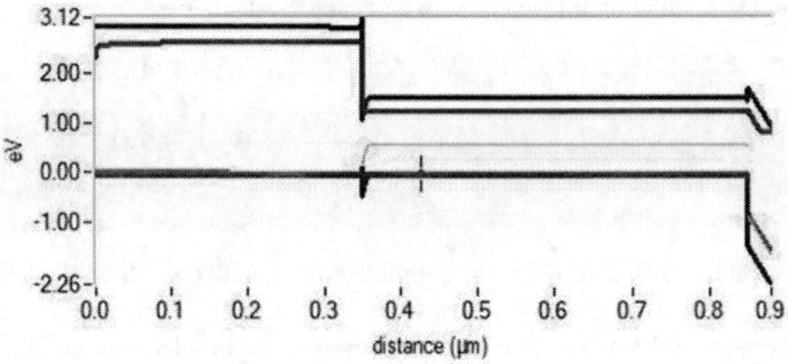

Figure 2. Zone diagram of the studied model of a perovskite solar cell.

Table 1. Initial parameters of the solar cell under study.

Parameters	HTM	$CH_3NH_3PbI_3$	TiO_2
thikness (nm)	350	variable	30
bandgap (eV)	3	1.55	3.2
electron affinity (eV)	2.45	4.5	4.26
dielectric permittivity (relative)	3	10	9
CB effective density of states (1/cm^3)	$2.2{\cdot}10^{+18}$	$2.2{\cdot}10^{+18}$	$2{\cdot}10^{+18}$
VB effective density of states (1/cm^3)	$1.8{\cdot}10^{+19}$	$1.8{\cdot}10^{+19}$	$1.8{\cdot}10^{+19}$
electron thermal velocity (cm/s)	$1{\cdot}10^{+7}$	$3{\cdot}10^{+7}$	$1{\cdot}10^{+7}$
hole thermal velocity (cm/s)	$1{\cdot}10^{+7}$	$3{\cdot}10^{+7}$	$1{\cdot}10^{+7}$
electron mobility (cm²/Vs)	$2{\cdot}10^{-4}$	$1.6{\cdot}10^{+1}$	$2{\cdot}10^{+1}$

hole mobility (electron mobility (cm²/Vs)	$2 \cdot 10^{-4}$	$2 \cdot 10^{+1}$	$1 \cdot 10^{+1}$
shallow uniform donor density ND (1/cm³)	0	0	$1 \cdot 10^{+16}$
shallow uniform acceptor density NA (1/cm³)	$2 \cdot 10^{+18}$	$6 \cdot 10^{+18}$	0

The direction of solar radiation was modeled from the TiO_2 side. The simulation process took place at a set temperature of 300 K. Figure 2 shows the dependence of the quantum efficiency of the solar cell on the thickness of the perovskite layer. Quantum efficiency describes the ratio of the number of photons whose absorption led to the generation of electron-hole pairs to the total number of photons absorbed. As a result of the study, it was found that an increase in the thickness of the perovskite layer causes a significant increase in quantum efficiency. When the thickness of the perovskite layer increases from 100 to 500 nm, the quantum efficiency increases almost twice – from 45 to 90 %. This is due to the fact that more photons will be absorbed by the perovskite layer, which will lead to an increase in the generation of electron-hole pairs [5]

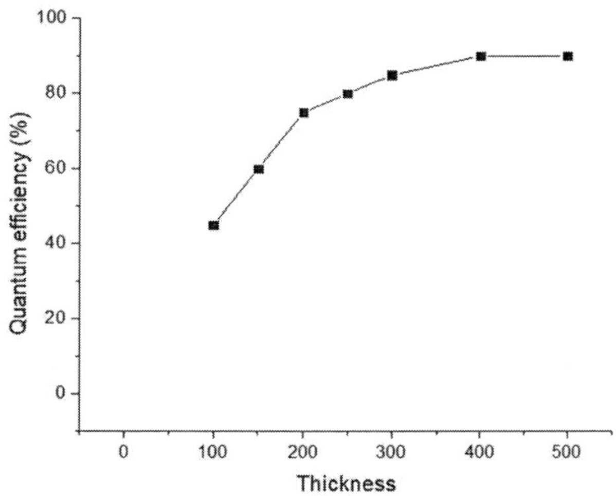

Figure 3. Dependence of the quantum efficiency of the solar cell on the thickness of the perovskite layer.

Table 2. Output characteristics of the studied SE model in the range of 100-500 nm.

QE (%)	Thickness (nm)
43	100
59	150
72	200
78	250
83	300
89	400
89	500

Figure 3 also shows that when the perovskite layer is about 400 nm thick, the quantum efficiency of the solar cell reaches a maximum of 89%. Therefore, a further increase in the perovskite layer does not lead to an increase in quantum efficiency.

Further, the estimation of the change in the efficiency of the solar cell with an increase in the thickness of the perovskite layer was made. The thickness of the photosensitive layer varied from 100 to 900 nm.

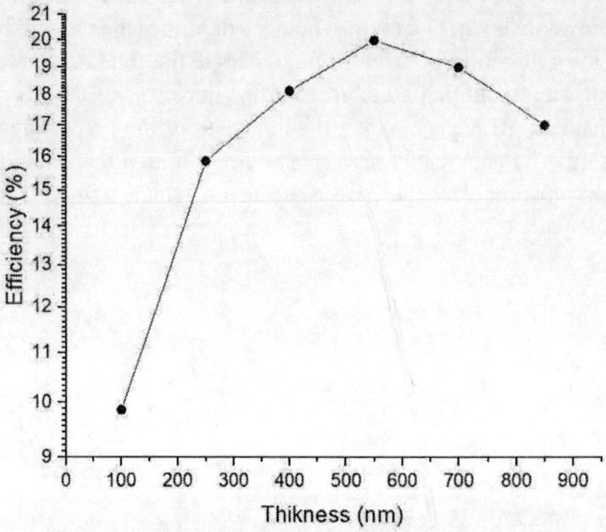

Figure 4. Dependence of efficiency on the thickness of the photosensitive layer.

Table 3. Output characteristics of the studied SE model in the range of 100-900 nm.

Efficiency (%)	Thickness (nm)
8.5	100
15.9	250
17.9	400
19.5	550
18.5	700
16.9	900

As a result, it was found that when the thickness of the photosensitive layer increases from 100 to 900 nm, the efficiency of the solar cell increases more than twice - from 8.5% to 19.5%. Comparison of the first and second graphs shows that an increase in quantum efficiency plays a significant role in increasing the efficiency at the thickness of the photosensitive layer up to 400 nm. With increasing thickness of the photosensitive layer is above 400 nm the increase in efficiency occurs only slightly and a decrease in efficiency when a thicker film, which is due to increased recombination losses (decreasing the electric field in perovskite promotes separation of photogenerated electron-hole pairs).

3. Conclusion

Thus, using the developed mathematical model of the solar cell based on perovskite, it was shown that an increase in the photosensitive layer leads to greater absorption of the number of photons and the generation of electron-hole pairs. The presence of a maximum is due to the fact that as the thickness of the perovskite film increases, the solar cell's photocurrent increases, which goes out to saturation, while the photon voltage decreases with increasing thickness due to an increase in the recombination rate. In this case, by increasing the thickness of the photosensitive layer from 100 to 550 nm, the efficiency increases more than twice - from 8.5% to 19.5%.

Acknowledgments

The research was carried out with the financial support of the Ministry of Science and Higher Education of the Russian Federation in the framework of the state task in the field of scientific activity no. 0852-2020-0015

References

[1] Malyukov S P, Saenko A V and Ivanova A V 2016 Numerical simulation of perovskite solar cells with a flat structure *IOP Conference series: Material Science and Engineering* **151** 2016 012033

[2] *Electronics and Information Systems (ELIS)* (Dept: University of Ghent) http://scaps.elis.ugent.be/ (accessed: 10.02.2020).

[3] Gratzel M 2014 *Nature Materials* **13** 838

[4] Sinshu sun, Reza Asadpour, Wanyi ne, et al. 2015 *IEEE Journal of Photovoltaics* **5** 1389

[5] Abdelkader H, Ahmed Khalil Le K, Abdallah R, Muslem Ben, Abderrahman K Imad K 2019 Modeling and optimization of CH_3 NH_3 PbI_3 based on inverted planar heterojunction of solar cells using SCAPS software *Journal. from Energetica (IJECA)* **4(1)** 56-59

[6] Du H J and Zhu J Z 2016 The device for simulation of lead-free status CH3NH3SnI3 perovskite solar cells with high efficiency *Chinese Physics* B **25** 108802

[7] Mandadapu U, Vedanayakam S V and Thyagarajan K 2017 Modeling and analysis of lead-based perovskite solar cells using SCAPS-1D *Indian Journal of Science and Technology* **10(11)** 0974-6846

[8] Ameri M, Mohajerani E and Aparcana M 2019 Investigation of this case the invisible relationship between grain size, ion defects, device physics, and performance of perovskite solar cells *Journal of Physics* D *Applied physics* **52** 125

[9] Thakur N, Mehra R and Devi K 2018 Efficient design Perovskite solar cell using parametric grading of mixed perovskite Halide and copper iodid. *Journal of Electronic Materials* **47** 6935–6942

Influence of light on electrophysical properties of thin films of mixed zinc and tin oxides

V V Petrov[1], E M Bayan[2], Yu N Varzarev[3], M G Volkova[2], V Yu Storozhenko[2], A A Rozhko[1,5] and A V Nesterenko[4]

[1]Department of Technosphere Safety and Chemistry, Institute of Nanotechnologies, Electronics, and Equipment Engineering, Southern Federal University, 2 Shevchenko str., Taganrog 347922, Russia
[2]Department of General and Inorganic Chemistry, Southern Federal University, 7 R. Zorge st., Rostov-on-Don 344006, Russia
[3]Department of Nanotechnology and Microsystems, Institute of Nanotechnologies, Electronics, and Equipment Engineering, Southern Federal University, 2 Shevchenko str., Taganrog 347922, Russia
[4]Problem Scientific and Research Laboratory "Nanobiotechnology and new Materials", Institute of Nanotechnologies, Electronics, and Equipment Engineering, Southern Federal University, 2 Shevchenko str., Taganrog 347922, Russia

[5]Corresponding author: drerozhko@mail.ru

Abstract. In this work, we studied structural and the electrical properties of thin films of mixed zinc and tin oxides obtained by low-temperature pyrolysis. XRD analysis shows that the obtained films contain phases of zinc oxide and tin dioxide and have nanocrystalline structure with crystallite size about 10-17 nm. A significant effect of daylight on the temperature dependences of the film resistance was established. The temperature dependences of the resistance of ZnO-SnO_2 films, measured in the range from room temperature to 300 °C, show the difference between the heating and cooling curves, that can be explained by two competing processes - thermal activation mechanism of conductivity and adsorption-desorption of oxygen particles on the surface of the oxide film. Using the slope of the linear section on the plots, we calculated the activation energy of conductivity, which amount is 0.7 and 1.1 eV for the samples with the different Zn:Sn ratio.

1. Introduction

Improving the efficiency of converting solar energy into electrical energy is one of the urgent tasks of our time. Currently, silicon-based solar cells are most widely used but materials based on zinc, tin oxides are also used [1]. Indium-tin oxides (ITO) as transparent conductive oxide has some disadvantages for use in electrical devices because of high cost and poor chemical stability. The nanosized films of zinc oxide, as well as films based on it, doped with various compounds, are transparent, have low resistance, are well compatible with other active semiconductor materials, which makes possible to actively use them as electrodes in solar modules [2, 3]. As electrical properties of ZnO strongly depends on stoichiometric and may vary under ambient oxygen [4], doping of ZnO films with various metals (Al, Ga, Sn) is used to improve it. Sn is the most suitable dopant for ZnO because Zn and Sn have similar ionic radii [5].

Two-component oxide system $ZnO-SnO_2$ has good electrical conductivity, high optical transmittance, good chemical stability. It can be used in a variety of applications such as energy conversion, optical materials, sensors, catalysts, etc [6]. The $ZnO-SnO_2$ films can be formed by various methods: magnetron sputtering [7], pulsed laser deposition [8], a sol-gel technique [9]. Depending on the synthesis method, the resulting materials may have different physicochemical [10] and electrophysical properties [11], therefore, one of the pressing issues of modern research is the optimal choice of method and conditions for obtaining materials. In the framework of this work, we studied the properties of nanoscale zinc and tin oxides obtained by low-temperature pyrolysis. The chemical pyrolysis is an alternative technique to previously indicated methods with its simplicity and cheapness for the formation of films on large surfaces [12]

2. Experimental

2.1. Films formation

As the main precursors for the preparation of nanosized films of zinc and tin (IV) oxides, tin tetrachloride pentahydrate, zinc nitrate hexahydrate and organic acid were used; the molar ratios of zinc and tin were 99:1, 95:5, 5:95 and 1:99. The resulting compositions of organic salts were dissolved in an organic solvent, applied to purified ceramic alumina-based substrates, and dried in air at 100 ° C. For the studied samples, 3 layers of a precursor solution were applied. Heat treatment of materials was carried out at 550 °C for one hour with an average heating rate of 10 deg/min.

2.2. Characterization

The surface morphology of the obtained films was characterized by scanning electron microscope (SEM) Nova Nanolab 600.

Structural properties of the synthesized materials were studied by X-ray diffraction (XRD) using ARL X'TRA diffractometer utilizing CuKα1-radiation with 35 kV and 30 mA.

To study the electrophysical properties of Ni, metal contacts with a thickness of 0.2-0.3 microns were formed on the films surface by vacuum thermal deposition. The temperature dependence of the resistance of the films upon heating to 300 °C and subsequent cooling was measured using Keithley 2450 sourcemeter.

3. Results and discussion

Analysis of microstructure, carried out by SEM (figure 1), exhibits that the film materials obtained by pyrolysis are uniform, dense, distribution occurs over the entire surface. The thickness of the three-layer film is about 90-100 nm.

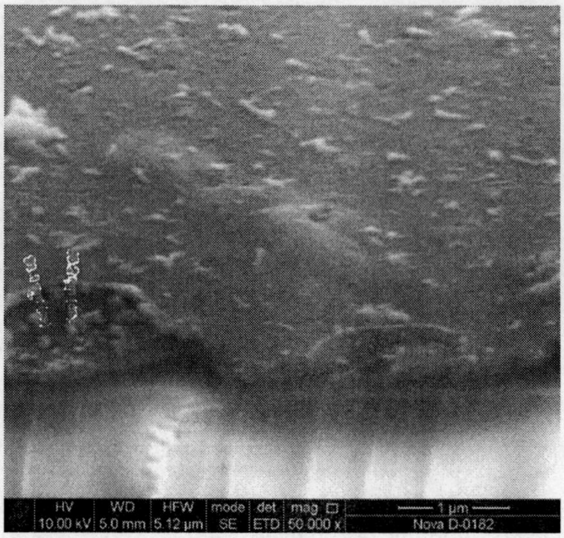

Figure 1. SEM image film ZnO-SnO_2 (99:1 in solution).

XRD analysis (figure 1) confirmed that the obtained films contain phases of zinc oxide and tin dioxide. No additional diffraction peaks were detected in the XRD patterns, indicating that thin films yield pure phase. It was shown, that obtained materials contain single-phase, hexagonal wurtzite structure ZnO according to Crystallography Open Database, COD ID 2300113 [13] (figure 1a) and single-phase tetragonal cassiterite SnO_2 structure (figure 1b) according to Crystallography Open Database, COD ID 1521419 [14]. The presence of diffraction peaks along different planes indicates that the films nanocrystalline structure is growing in different directions. It was found that signals on the XRD patterns are expanded, which is typical for nanocrystalline materials.

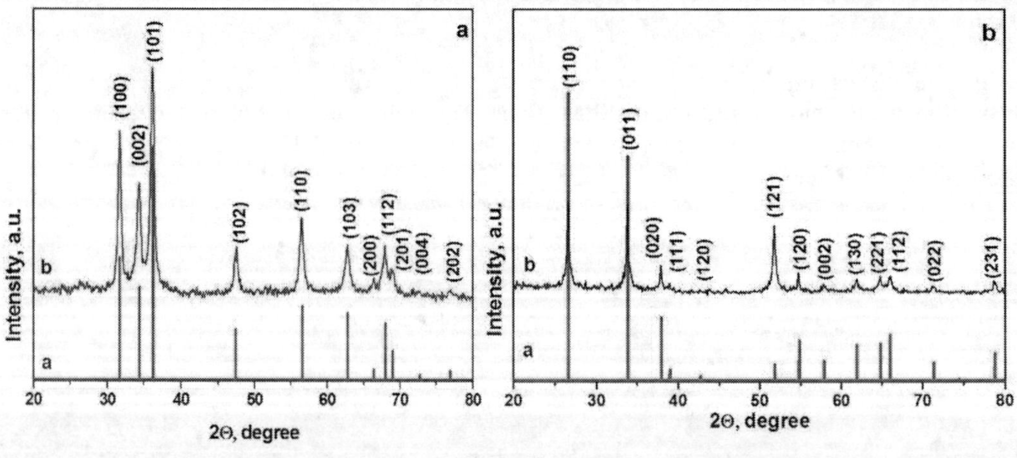

Figure 2. XRD patterns of ZnO-SnO_2 films: (a) 1 – film with Zn:Sn ratio 99:1, 2 – ZnO phase COD ID 2300113; (b) 1 – film with Zn:Sn ratio 1:99, 2 – SnO_2 phase COD ID 1521419.

The mean crystallite size D was evaluated according to the broadening of the highest intensity peak diffraction plane using the Scherrer equation [15]

$$D = \frac{k\lambda}{\beta \cos \Theta},$$

where k is the shape factor (k=0.9), λ is the X-ray wavelength (λ = 0.1540562 nm), β is the full width at the half maximum of the diffraction line and Θ is the diffraction angle. The values of β and Θ are taken for crystal plane (110) and (101) of the SnO_2 cassiterite phase and ZnO wurtzite phase, respectively [16]. According to the Sherrers equation, the particle size is about 10-17 nm.

The resistance dependence of the formed structures on the reverse temperature is shown in figure 3. Those dependences were measured after exposure to daylight and after exposure to darkness for 24 hours.

It can be seen that the effect of light on the film resistance is more pronounced for the sample with a $ZnO:SnO_2$ ratio of 99:1. The nonlinearity of the presented dependences may indicate that the influence of adsorption-desorption processes on the surface of the films is superimposed on the thermal activation mechanism of conductivity.

It is seen that during heating, the resistance of all samples first increases, then, in the temperature range 80-180 °C, a change in the nature of the temperature dependence of the resistance is observed - a smooth transition through the maximum occurs. The increase in resistance during heating can be explained by the effect of adsorbed oxygen. Oxygen molecules in the atmosphere are adsorbed on the oxide surface and become negatively charged due to the capture of electrons from the surface ZnO layers [17]. This leads to the formation of a depletion layer in the surface region of the film and a decrease in its conductivity. As the temperature rises, the thermal generation of carriers begins to exert an increasing influence on the resistance of the material, and the character of conductivity changes.

With further heating from 180 to 300 °C, a linear section is observed. This region of the temperature dependence of the resistance is well approximated by the Arrhenius equation [18]:

$$\rho = \rho_0 \exp\left(\frac{E_a}{kT}\right),$$

where ρ_0 is the preexponential factor, E_a is the activation energy of conductivity and k is Boltzmann constant. The activation energy calculated for samples with Zn:Sn ratio 5:95 and 95:5 is 0.7 eV and for samples with Zn:Sn ratio 1:99 and 99:1 is 1.1 eV. Since for oxide semiconductors, the electrical conductivity is determined by intrinsic defects, primarily oxygen vacancies, this energy corresponds to the levels of these defects in the band gap of the semiconductor.

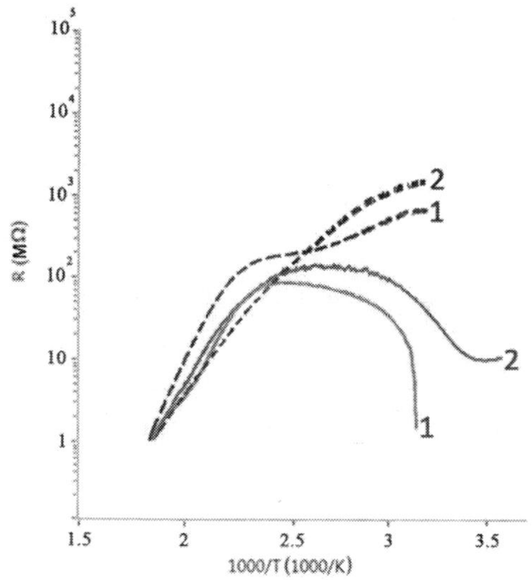

(a)

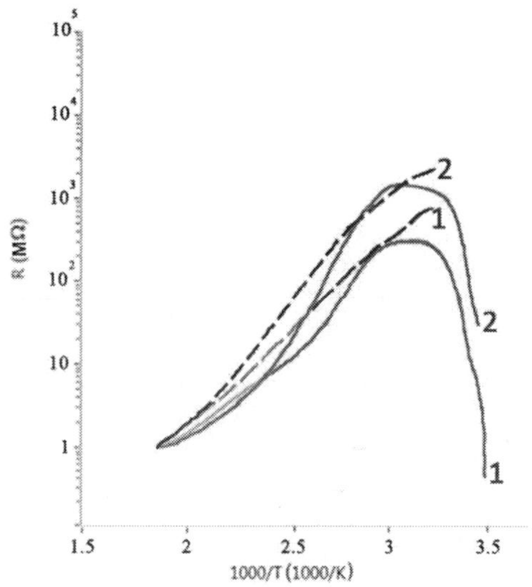

(b)

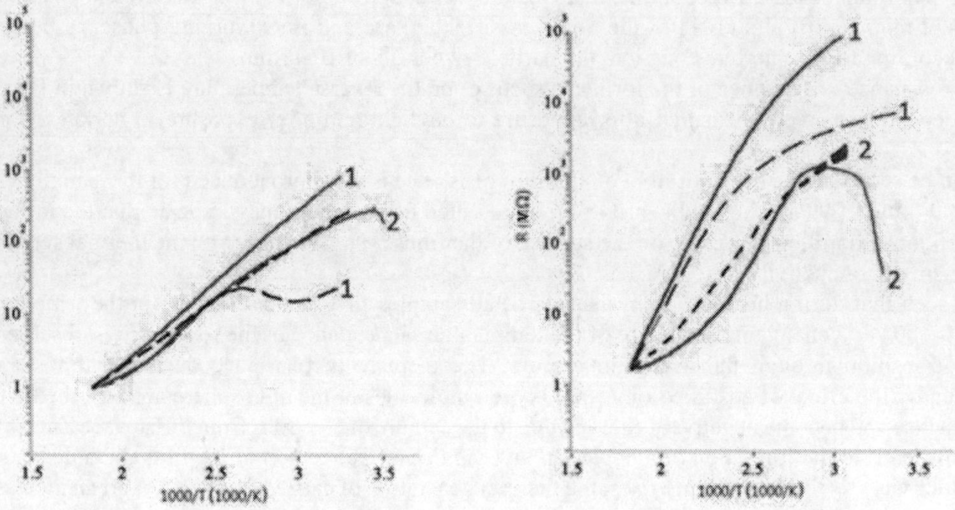

(c) (d)

Figure 3. Resistance (R) dependences on reverse temperature (1000/T) of ZnO-SnO_2 films with the Zn:Sn ratio 1:99 (a), 5:95 (b), 95:5 (c) and 99:1. (1 – after light irradiation, 2 – after dark exposure; (—) – heating, (- - -) – cooling).

At temperatures below 180 °C, the heating and cooling curves do not coincide. This difference in the course of the curves may be due to the lack of desorption of molecular oxygen ions from the surface of the films.

Thus, the change in resistance during heating and cooling is due to two competing processes - thermal generation and electron recombination on the one hand, and their capture by oxygen adsorbed on the surface on the other. Moreover, if in the first case, the equilibrium concentration is established almost instantly, then in the second it depends on the rate of the processes of adsorption and desorption. The difference in the heating and cooling curves in the temperature dependence of the resistance indicates low rates of these processes, comparable with the rates of heating and cooling.

4. Conclusions

The thin films of mixed zinc and tin oxides obtained by low-temperature pyrolysis contain phases of zinc oxide and tin dioxide and have nanocrystalline structure with crystallite size about 10-17 nm. A significant effect of daylight on the temperature dependences of the film resistance has been established. The temperature dependences of the resistance of ZnO-SnO_2 films, measured in the range from room temperature to 300 °C, show the difference between the heating and cooling curves, that can be explained by two competing processes - thermal activation mechanism of conductivity and adsorption-desorption of oxygen particles on the surface of the oxide film. Thus, taking into account the effects found, the synthesized films of zinc and tin oxides can be recommended for use in the functional elements of alternative energy.

Acknowledgements

This research was funded by the Russian Foundation for Basic Research (RFBR), grant number 20-07-00653 A.

References

[1] Lee Y J, Shin J K and Lim K-S 2006 *ECS Trans.* **1** 11-14

IOP Conf. Series: Materials Science and Engineering 1035 (2021) 012033 doi:10.1088/1757-899X/1035/1/012033

[2] Chala S, Senguoga N, Yakuphanoglu F, Rahmane S, Bdirina M and Karteri I 2018 *Energy* **164** 871-880
[3] Guo Z L, Zhuang J, Ma Z, Xia H R, Wen Q X, Luo X Y and Wen X 2019 *Thin Solid Films* **684** 1-8
[4] Wilken S, Parisi J and Borchert H 2014 *J. Phys. Chem. C* **118** 19672-19682
[5] Lee I-J, Park J, Sung N-E and Kim J 2018 *Materials Chemistry and Physics* **206** 103-109
[6] Lagariya M, Modi M, Dadhich H, Gal M, Gadani K Solanki P S and Shah N A 2020 *Physica* B *Physics of Condensed Matter* **577** 411774
[7] Cha C-N, Choi M-H, and Ma T-Y 2012 *J. Electr. Eng. Technol.* **7** 596
[8] Rakshit T, Manna I and Ray S K 2015 *J. Appl. Phys.* **117** 025704
[9] Sheikh A Kumavat Soni K and Lakshmi N 2020 *AIP Conf. Proc.* **2220** 020096
[10] Bayan E M, Lupeiko T G, Pustovaya L E and Fedorenko A G 2017 *Nanotechnologies in Russia* **12(5-6)** 269-275
[11] Petrov V V, Varzarev Yu N, Bayan E M Storozhenko V Yu and Rozhko A A 2019 *Proc. 2019 IEEE Int. Conf. Electrical Engineering and Photonics EExPolytech*. 242-243
[12] Atay F, Akyuz I and Durmaz D and Kose S 2019 *Solar Energy* **193** 666–675
[13] Heidrun S and Ahsbahs H 2006 *J. Appl. Crystallogr.* **39** 169-175
[14] Klementova M, Weiss Z and Rieder M 2000 *Int. J. Inorg. Mater.* **45** (1) 155-157
[15] Cullity B D 1956 *Elements of X-Ray Diffraction* (Addison-Wesley Publishing Company, Inc., Reading, Massachusetts)
[16] Maniv S and Zangvil A 1978 *J.Appl.Phys.* **47** 2787–2792
[17] Rambu A P, Iftimie N and Rusu G I 2012 *Mater. Sci. Eng. B-Adv* **177** 157–163
[18] Shalimova K V 1976 *Physics of Semiconductors* (Moscow: Energia)

Flow simulation of petroleum diesel fuel and rapeseed oil in the nozzle of a diesel injector

Bowen Sa[1,3], Vladimir A Markov[1], Vyacheslav G Kamaltdinov[2] and Vsevolod A Neverov[1]

[1]Power Engineering Faculty, Bauman Moscow State Technical University, 5 2-ya Baumanskaya Str., Moscow 105005, Russia
[2]Motor Transport Faculty, South Ural State University, 76 Lenin Ave., Chelyabinsk 454080, Russia

[3]Corresponding author: bowensa@yandex.ru

Abstract. The flow of petroleum diesel fuel (DF) and rapeseed oil (RO) in the nozzle of a diesel injector under operating conditions was simulated numerically. Investigated the general flow characteristics, such as mass flow rate, discharge coefficient, injection rate, averaged turbulent kinetic energy at the nozzle outlet and volume concentration of fuel vapor in the nozzle, and local flow parameters. The saturation pressure of RO is determined by summing the saturation pressure of constituting RO triacylglycerides. At a fuel temperature of 40 °C, the mass flow rate, discharge coefficient, injection velocity and turbulent intensity at the outlet of the nozzle hole for RO are significantly less than petroleum DF for all injection pressures, and the RO cavitation level (fuel vapor) in the nozzle hole is also lower than petroleum DF due to extremely slow saturation pressure of RO. However, this difference between RO and petroleum DF decreases with increasing injection pressure due to the different cavitation levels of DF at different injection pressures.

1. Introduction

In recent years, alternative vehicle fuels have been widely considered in light of energy security and environmental degradation. Vegetable oils are considered as a promising energy source to replace fossil fuels in compression ignition engines (CIEs), because vegetable oils are renewable and their direct use in CIEs reduces smoke, SO_x, CO_2 and aromatic compounds-carcinogens emissions [1].

The wide use of vegetable oil in diesel engines is hindered by the difference between the physicochemical properties of vegetable oils and those of petroleum diesel fuel (DF), especially the high viscosity of vegetable oils is 10-20 times higher than that of petroleum DF [2]. Markov et al. [3,4] studied the viscosity characteristics of two-component and multicomponent mixed biofuels based on vegetable oils and developed a method for determining their viscosity. It is noted that vegetable oils are characterized by increased viscosity, which has a negative effect on the processes of mixture formation and combustion. Increased emissions of carbon monoxide (CO) and unburned hydrocarbons (CH_x) in exhaust gases from diesel engines fuelled with pure vegetable oils have been reported in experimental studies [5-9]. This may be due to the fact that the higher viscosity of vegetable oil makes fuel difficult to atomize and evaporate, which leads to locally rich mixtures contributing to incomplete combustion and formation of CO and CH_x due to a lack of oxygen [8, 10-12]. However, the opposite results of the influence of vegetable oils on the emissions of these harmful substances have also been obtained [13-

15], which is related to the presence of oxygen in the vegetable oil molecules. Therefore, in order to clarify the effect of vegetable oils on the combustion process in the engine combustion chamber, first of all, it is necessary to study the influence of vegetable oil properties on fuel injection, atomization, evaporation and mixture formation processes, in particular, the fuel flow in the nozzle, which has a great influence on the fuel atomization, fuel evaporation and formation of a mixture of fuel and air, which may further affect the performance of diesel engines [16].

Dernotte et al. [17] experimentally studied injection rates in a non-cavitation diesel nozzle for different fuels with kinematic viscosity from 0.6 to 7 mm^2/s and density from 683 to 876 kg/m^3. At low injection pressures, an increase in fuel viscosity leads to a decrease in fuel discharge coefficient to 10%, and at high pressures, only fuel density affects mass fuel rate. However, the kinematic viscosity of most vegetable oils at a temperature of 20 °C is in the range of 50 to 130 mm^2/s [3,4,18-20]. Numerical simulation of the cavitation flow of Jatropha curcas oil [21], performed for a diesel nozzle, showed that the cavitation level of Jatropha curcas oil is much less than that of diesel fuel. However, the authors did not report the saturation pressure of Jatropha curcas oil, which is depended on oil composition and affects the occurrence and development of the cavitation process in nozzle [22].

A lot of research has been related to the flow of biodiesel fuels obtained from vegetable oils in injector nozzles [23-26], but the flow characteristics of vegetable oils have not been studied enough.

The objective of this study is to investigate the steady flow characteristics of pure rapeseed oil (RO) compared to petroleum DF in the nozzle of a diesel injector. Analyzed the distribution of pressure, flow velocity, vapor volume concentration and turbulence inside the nozzle hole at various injection pressures, as well as general parameters of the fuel flow at the outlet of the nozzle hole, such as mass flow rate, injection rate and turbulent kinetic energy.

2. Nozzle geometry and fuel properties

Flow simulation of pure RO and petroleum DF was performed in an AZPI nozzle 171.07.00 form Altai precision products plant (AZPI Russia). This nozzle is assembled in a diesel injector FDM-22 for a 4-stroke turbocharged diesel engine D-245.12S (with cylinder bore-to-stroke ratio 11/12.5). The structural scheme of the studied nozzle is presented in figure 1. Some specifications of this nozzle are given in table 1.

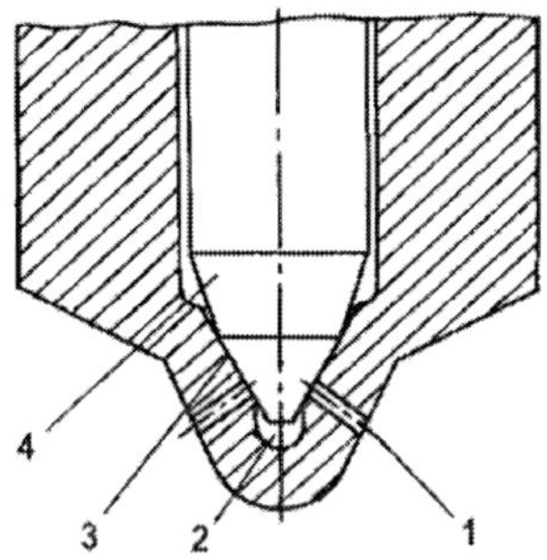

Figure 1. Structural scheme of The design diagram of the AZPI nozzle 171.07.00: 1 – the nozzle hole; 2 – the sac; 3 – the needle valve seat; 4 – the needle.

Table 1. Some specifications of the investigated nozzle.

Manufacturer, Mark number	Diameter of nozzle holes d_n, mm	Number of nozzle holes	Maximum needle lift $h_{n\ max}$, mm	Length of the nozzle hole , mm
AZPI, 171.07.00	0.35	5	0.32	1.1

Flow simulation in the flow part of the diesel injector nozzle was carried out for petroleum DF and RO. Some physicochemical properties of these fuels are given in table 2 [2, 3, 28]. The saturation pressure of RO at low temperatures was determined by summing the saturation pressure of constituting

RO triacylglycerides. And the saturation pressure of triacylglycerides can be calculated by the Clausius-Clapeyron equation [29]:

$$lg\,P(T) = \frac{-\Delta G_{\theta}^{vap}}{R\theta ln10} + \frac{\Delta H_{\theta}^{vap}}{Rln10}\left(\frac{1}{\theta} - \frac{1}{T}\right), \tag{1}$$

where, θ – the reference temperature (298.15K), R – the universal gas constant, ΔG_{θ}^{vap} – Gibbs free evaporation energy at reference temperature θ, ΔH_{θ}^{vap} – the enthalpy of evaporation at reference temperature θ. Herewith ΔG_{θ}^{vap} and ΔH_{θ}^{vap} were determined by using Fragment based-approach as the sum from each fragment of triacylglycerides - glycerol and three fatty acids [30]. Figure 2 presents the calculated results of the saturation pressure of RO depending on the temperature.

Table 2. Physicochemical properties of petroleum DF and RO.

Property	Petroleum DF	RO
Conditional composition formula	$C_{16.2}H_{18.5}$	$C_{57.0}H_{101.6}O_2$
Molecular mass	223.3	883.04
Density, kg/m^3:		
- at 20 °C	830.0	909.8
- at 40 °C	822.7	896.3
Kinematic viscosity, mm^2/s:		
- at 20 °C	3.8	75.0
- at 40 °C	2.4	41.5
Dynamitic viscosity, mPa·s:		
- at 20 °C	3.15	68.24
- at 40 °C	1.97	37.20
Saturation pressure, kPa:		
- at 20 °C	2.7	-
- at 40 °C	4.8	-

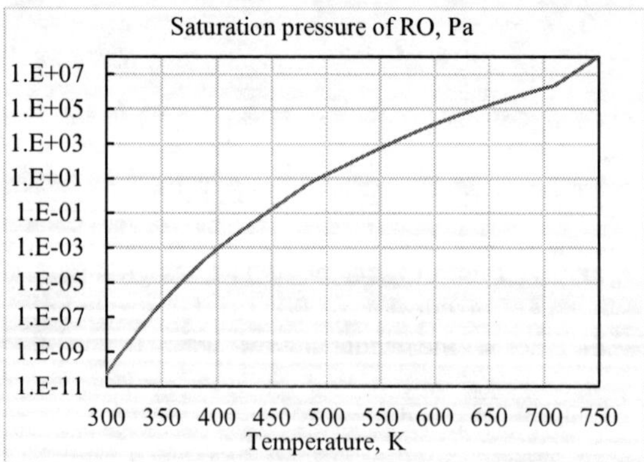

Figure 2. Saturation pressure of RO at different temperatures.

3. Numerical models

The steady flow inside the nozzle at the maximum needle lift position was simulated using the CFD program Fluent v17.2. The pressure at the inlet to the computational domain (injection pressure p_{in}) was selected to be 51.5, 40, 23 and 20 MPa corresponding to the pressure range before injector in the fuel injection system of diesel engine D-245.12S [31]. The pressure at the outlet of the nozzle hole (backpressure p_b) was 8.878 MPa, which corresponds to the working pressure in the cylinder at the start of injection [31]. The fuel temperature was set to be constant and equaled to 40 °C, which is the fuel temperature before entering the injector during high-pressure fuel pump speed of 1250 min^{-1} [31]. In order to limit the calculation time, the symmetric geometry element of the nozzle flow part with one hole was applied. Hexahedral structured grids were meshed for the computational domain using CFD ICEM and are shown in figure 3.

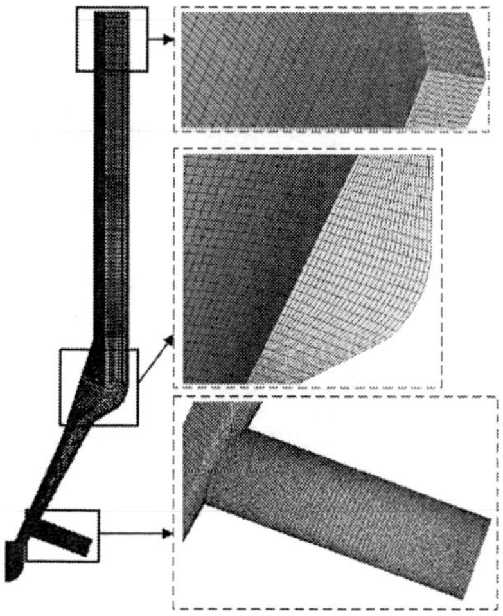

Figure 3. Hexahedral structured grids for the AZPI nozzle symmetric element with one hole.

3.1. Cavitation model

In order to simulate two-phase flow in nozzle a multiphase equilibrium model-mixture model was used. The cavitation process is described by the Schnerr-Sauer model [32], in which the mass transfer equation has the following form:

$$\frac{\partial \alpha}{\partial t}(\alpha \rho_v) + \nabla\left(\alpha \rho_v \vec{V_v}\right) = \frac{\rho_v \rho_l}{\rho}\frac{d\alpha}{dt}, \tag{2}$$

where α – the volume concentration of the vapors, ρ_v – the density of the vapor phase, ρ_l – the density of the liquid phase, ρ – the density of the mixture phase ($\rho = \alpha\rho_v + (1-\alpha)\rho_l$), $\vec{V_v}$ – the velocity of the vapor phase. The density ratio between the mixture phase and vapor phase is represented by the following equation

$$\frac{d\rho}{dt} = -(\rho_l - \rho_v)\frac{d\alpha}{dt}. \tag{3}$$

Mass transfer rate is written by the formula

$$R = \frac{\rho_v \rho_l}{\rho}\frac{d\alpha}{dt}. \tag{4}$$

Herewith, the volume concentration of the vapors can be written in the correlation to the number of bubbles per unit volume (n_b), which is represented as:

$$\alpha = \frac{n_b \frac{4}{3}\pi R_b{}^3}{1 + n_b \frac{4}{3}\pi R_b{}^3}, \tag{5}$$

where R_b – the bubble radius.

The formula for the bubble expansion velocity derived from the dynamic theory without taking into account the second-order terms has the form

$$\frac{dR_b}{dt} = \sqrt{\frac{2}{3}\left|\frac{p_v - p}{\rho_l}\right|}, \tag{6}$$

where p_v – the pressure in a bubble, which is usually taken equal to the pressure of the saturated vapors; p – the pressure in the external liquid phase.

From the abovementioned formulas (4), (5), (6), we can get the formula for describing the mass transfer rate

$$R = \frac{\rho_v \rho_l}{\rho}\alpha(1-\alpha)\frac{3}{R_\Pi}\sqrt{\frac{2}{3}\left|\frac{p_v - p}{\rho_l}\right|}. \tag{7}$$

And the bubble radius can be expressed as

$$R_b = \left(\frac{\alpha}{1-\alpha}\frac{3}{4\pi}\frac{1}{n_\Pi}\right). \tag{8}$$

3.2. Turbulent model

For describing turbulence flow in the nozzle, we chose a popular turbulent model with two equations — the k-ε model with Enhanced Wall treatment for equation ε in the near-wall boundary layers [33, 34]. In this model, the flow region is separated into a completely turbulent region and a viscosity-affected region.

In the completely turbulent layer, the conservative equations for the turbulent kinetic energy *k* and turbulent dissipation ε have the form

$$\frac{\partial}{\partial t}(\rho k) + \frac{\partial}{\partial x_i}(\rho k u_i) = \frac{\partial}{\partial x_j}\left[\left(\mu + \frac{\mu_t}{\sigma_k}\right)\frac{\partial k}{\partial x_j}\right] + G_k - \rho\varepsilon \tag{9}$$

and

$$\frac{\partial}{\partial t}(\rho \varepsilon) + \frac{\partial}{\partial x_i}(\rho \varepsilon u_i) = \frac{\partial}{\partial x_j}\left[\left(\mu + \frac{\mu_t}{\sigma_\varepsilon}\right)\frac{\partial k}{\partial x_j}\right] + C_{1\varepsilon}\frac{\varepsilon}{k}G_k - C_{2\varepsilon}\rho\frac{\varepsilon^2}{k}, \tag{10}$$

where μ – the molecular viscosity; μ_t – the turbulent viscosity, which is calculated by the formula:

$$\mu_t = \rho C_\mu \frac{k^2}{\varepsilon}; \tag{11}$$

G_k – the generation of the turbulent kinematic energy due to the velocity gradient, which is determined by the following formulas:

$$G_k = 2\mu_t S_{ij} S_{ij}, \tag{12}$$

$$S_{ij} = \frac{1}{2}\left(\frac{\partial u_i}{\partial x_j} + \frac{\partial u_j}{\partial x_i}\right). \tag{13}$$

$C_{1\varepsilon}$, $C_{2\varepsilon}$, C_μ, σ_k, σ_ε are the constants and have the values: $C_{1\varepsilon}$=1.44, $C_{2\varepsilon}$=1.92, C_μ=0.09, σ_k=1.0, σ_ε=1.3.

In the region subject to molecular viscosity, the equation for the turbulent kinetic energy *k* is the same as the equation for *k* in the completely turbulent region, but the turbulent viscosity is determined according to the method proposed by H.C. Chen [35],

$$\mu_{t,2} = \rho C_\mu l_\mu \sqrt{k}\ . \tag{14}$$

In this formula, l_μ serves as a length scale and is determined by the formula

$$l_\mu = y C_l^* \left(1 - e^{-\frac{Re_y}{A_\mu}}\right), \tag{15}$$

where y – the distance to the wall, Re_y – the turbulent Reynolds number ($Re_y = \frac{\rho y \sqrt{k}}{\mu}$), $C_l^* = k C_\mu^{-\frac{3}{4}}$, A_μ=70. Different from the equation for turbulent dissipation ε in a completely turbulent region, in this region ε is determined by the formula

$$\varepsilon = \frac{k^{\frac{3}{2}}}{l_\varepsilon}, \tag{16}$$

where l_ε – the length scale calculated by the formula

$$l_\varepsilon = yC_l^*\left(1 - e^{-\frac{Re_y}{2C_l^*}}\right). \tag{17}$$

4. Results and discussion

Using the above-presented models, numerically simulated the stationary flow of petroleum DF and RO in the AZPI nozzle. To analyze the flow of fuel characteristics inside the nozzle hole, a uniaxial coordinate system was used, the starting point of which coincides with the cross-section center of the inlet of the nozzle hole, and the coordinate axis of which is along the axis direction of the nozzle hole and has the coordinate l_{axis}. Thus, the value $l_{\text{axis}} = 0$ mm corresponds to the cross-section of the inlet of the nozzle hole, and the value $l_{\text{axis}} = 1.1$ mm - the outlet of the nozzle hole.

4.1. Pressure in the nozzle

Figure 4 presents the pressure distribution of DF and RO on the longitudinal section of the nozzle hole when the pressure at the inlet to the computational domain (injection pressure) $p_{\text{in}} = 51.5$, 40, and 20 MPa. We noted different structures of the pressure distribution between DF and RO. A zone with a small pressure in the upper inlet region of the nozzle hole, arising from a sharp change of flow direction and distortion of flow pattern, can lead to the appearance of fuel vapor (cavitation) under the condition when the fuel flow pressure is less than the fuel saturation pressure. As shown in figure 4, the zone with a small pressure decreases for RO compared with DF under the same p_{in}. The lager zone with a small pressure for DF is due to the higher saturation pressure of DF and the lower viscosity of DF [36].

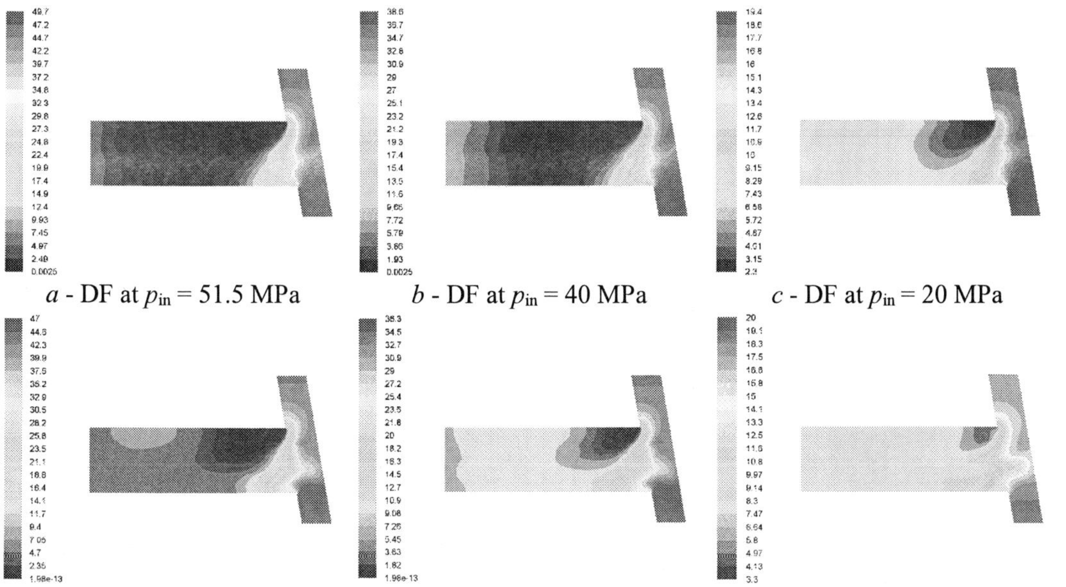

Figure 4. Pressure distribution on the longitudinal section of the nozzle hole for DF and RO at different injection pressures: a – DF at p_{in} =51.5 MPa; b - DF at p_{in} =40 MPa; c - DF at p_{in} =20 MPa; d - RO at p_{in} =51.5 MPa; e - RO at p_{in} =40 MPa; f - RO at p_{in} =20 MPa.

Figure 5 shows the curves of the variation of the pressure averaged over the cross-section of the nozzle hole for DF and RO along the flow direction. When there is no fuel vapor in the nozzle for both two fuels (see figure 7), i.e. at an injection pressure p_{in} of 20 MPa and a backpressure p_b of 8.878 MPa, the average pressure of RO in the cross-section of the nozzle hole (line 6) exceeds the average pressure of DF (line 3) in the entire nozzle hole region due to the larger pressure loss resulted from the higher viscosity of RO. And accordingly, the flow velocity of RO is slower than that of DF, as shown in figure

9 (line 3 and line 6). When the pressure difference Δp between the nozzle inlet and outlet ($\Delta p = p_{in} - p_b$) increases, cavitation starts to occur inside the nozzle hole, the average pressure of DF and RO drops in the middle region of the nozzle hole and increases in the inlet region of the nozzle hole. In addition, during the same pressure differences, the average pressure of DF in the middle region of the nozzle hole is lower than the average pressure of RO but higher in the inlet region of the nozzle hole. This is due to the higher cavitation level of DF (see figure 6). The pressure inside fuel vapors during flowing remains unchanged and equals to the fuel saturation pressure, which is much less than the pressure of liquid fuel in the zone without cavitation. Therefore, the expanded cavitation zone of DF leads to a decrease in the average pressure on the cross-section of the nozzle hole.

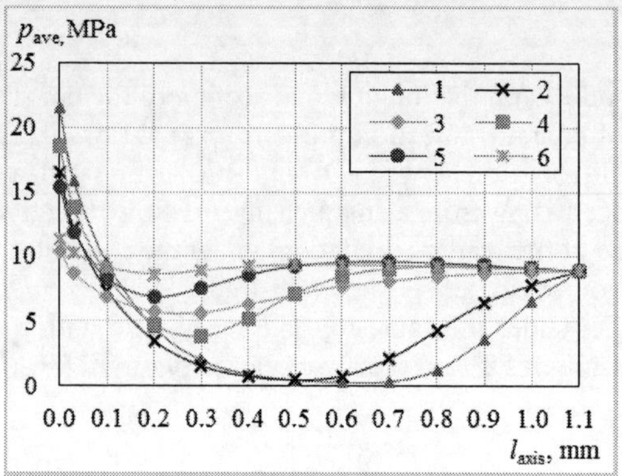

Figure 5. Curves of the average pressure of DF and RO on the cross-section of the nozzle hole along the flow direction: 1 – DF at p_{in} =51.5 MPa; 2 – DF at p_{in} =40 MPa; 3 – DF at p_{in} = 20 MPa; 4 – RO at p_{in} =51.5 MPa; 5 – RO at p_{in} =40 MPa; 6 – RO at p_{in} = 20 MPa.

4.2. Cavitation in the nozzle

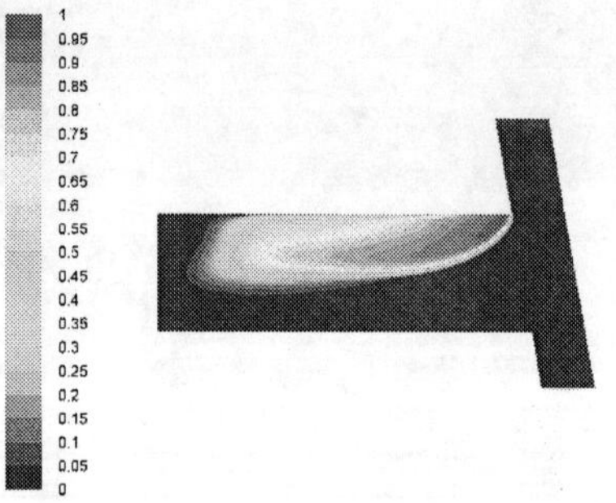

a - DF at p_{in} = 51.5 MPa

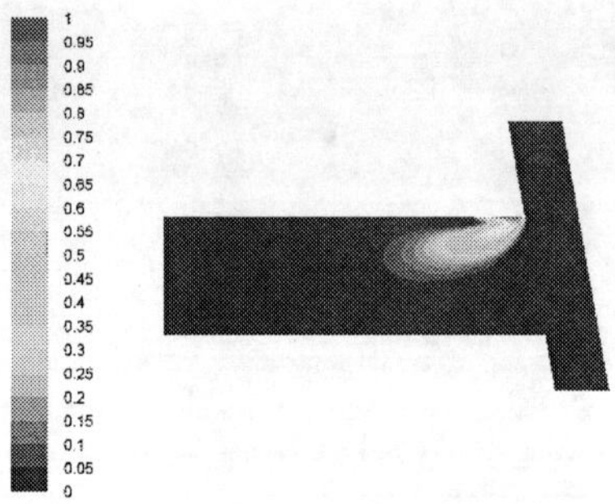

b - RO at p_{in} = 51.5 MPa

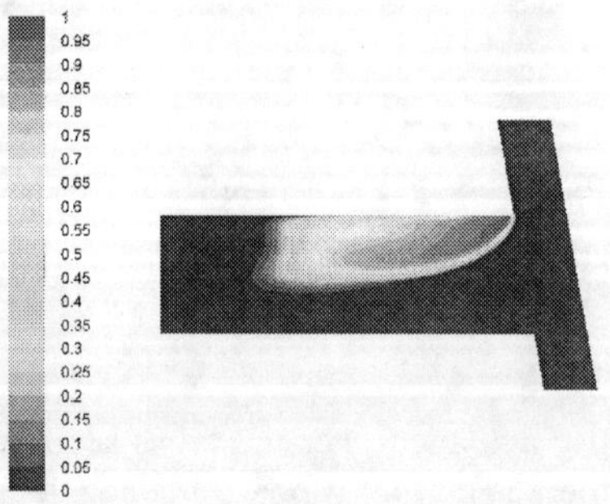

c - DF at p_{in} = 40 MPa

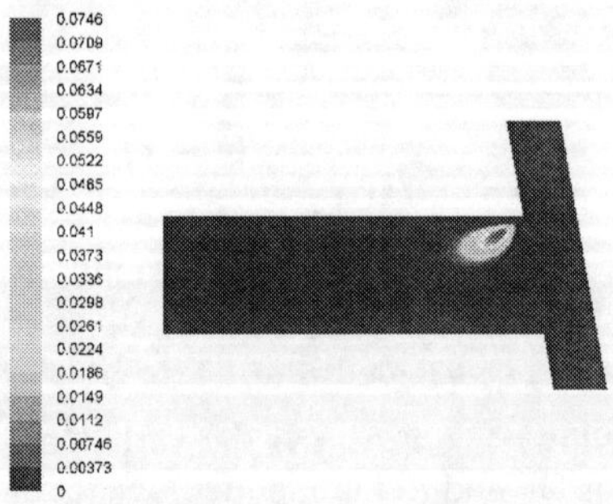

d - RO at p_{in} = 40 MPa

Figure 6. Distribution of fuel vapor volume concentration on the longitudinal section of the nozzle hole for DF and RO at different injection pressures: a – DF at p_{in} =51.5 MPa; b - DF at

p_{in} =40 MPa; c - DF at p_{in} =20 MPa; d - RO at p_{in} =51.5 MPa; e - RO at p_{in} =40 MPa; f - RO at p_{in} =20 MPa.

Figure 6 shows the distribution of the vapor volume concentration for DF and RO on the longitudinal section of the nozzle hole at injection pressures p_{in} of 51.5 and 40 MPa. These results confirm that the cavitation level of liquid RO is in a very low degree compared to the cavitation process of liquid DF, the vapor zone for the case of DF is many times larger than the vapor zone for the case of RO and even at an injection pressures p_{in} of 40 MPa the maximum vapor volume concentration of diesel fuel is approximately 12 times the maximum vapor volume concentration of RO. At the injection pressure lower 35 MPa and keeping the other conditions unchanged, cavitation does not occur inside the nozzle hole. From figure 7, which shows the dependence of the total vapor volume concentration of DF and RO in the nozzle hole on the pressure difference Δp, it is found that when the pressure difference Δp is higher than 16.122 MPa, the total vapor volume concentration of DF in the nozzle hole rapidly increases and, accordingly, the critical cavitation number for the case of DF N_{c_cr}=1.55; when the pressure difference Δp is higher than 26.122 MPa, the total vapor volume concentration of RO in the nozzle hole starts to increase slowly and is several times lower than the total vapor volume concentration of DF. For example, when the pressure difference Δp is 42.622 MPa, the total vapor volume concentration of DF in the nozzle hole is 27.5 %, and the total vapor volume concentration of RO is 5.7 %. The critical cavitation number for the case of RO is 2.94. The cavitation difference in the nozzle hole between DT and RO is mainly due to the lower saturation pressure of RO compared with DF. At a temperature of 40 °C, the saturation pressure of DF is 4800 Pa but the saturation pressure of RO is 7.74x10^{-10} Pa. Similar results about the cavitation difference for fuels with different saturation pressures were also obtained during injecting naphtha and n-dodecane through a diesel injector nozzle [22], during petroleum DF and biodiesel from the waste cooking oil flowing through an injector nozzle [25] and during the flow petroleum DF and light gasoline fuel [37].

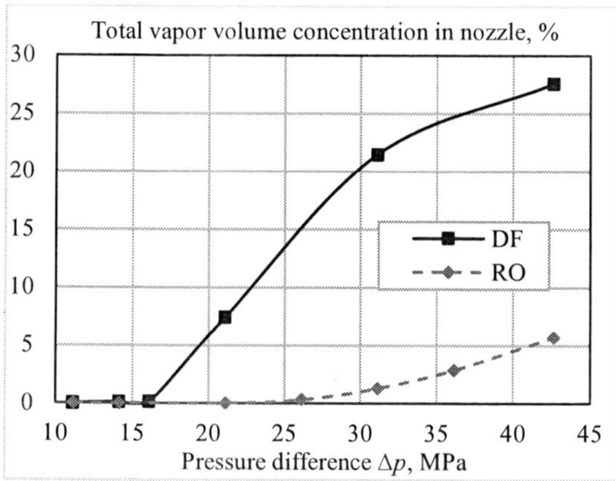

Figure 7. Total vapor volume concentration of DT and RO in the nozzle hole at various pressure differences Δp between p_{in} and p_b.

4.3. Velocity in the nozzle

Figure 8 presents the distribution of absolute flow velocity of DF and RO on the longitudinal section of the nozzle hole at a pressure at the inlet to the computational domain (injection pressure) p_{in} = 51.5, 40 and 20 MPa. It was noted that at the same injection pressure, the structure of the velocity distribution for DF and RO is different and the maximum flow velocity of DF is always greater than RO. In figure 9, the curves of the average axial flow velocity on the cross-section of the nozzle hole for DF and RO along the flow direction were plotted. It is obvious that with the same pressure difference Δp between p_{in} and p_b, the average axial flow velocity of DF is always higher than RO. At p_{in} = 51.5, 40 and 20 MPa the flow velocity of DF at the nozzle hole outlet (injection rate) is higher than the flow velocity of RO by 22.22%, 17.9% and 21%, respectively. This is due to two factors. One is that RO has an increased viscosity, which leads to an increase in the thickness of near-wall viscosity layers [22]. The other is that for the case of DF the cavitation process is stronger (more vapors generated - figure 6), which leads to a reduction of effective flow cross-sectional area and an increase of flow velocity in the no vapor region [38].

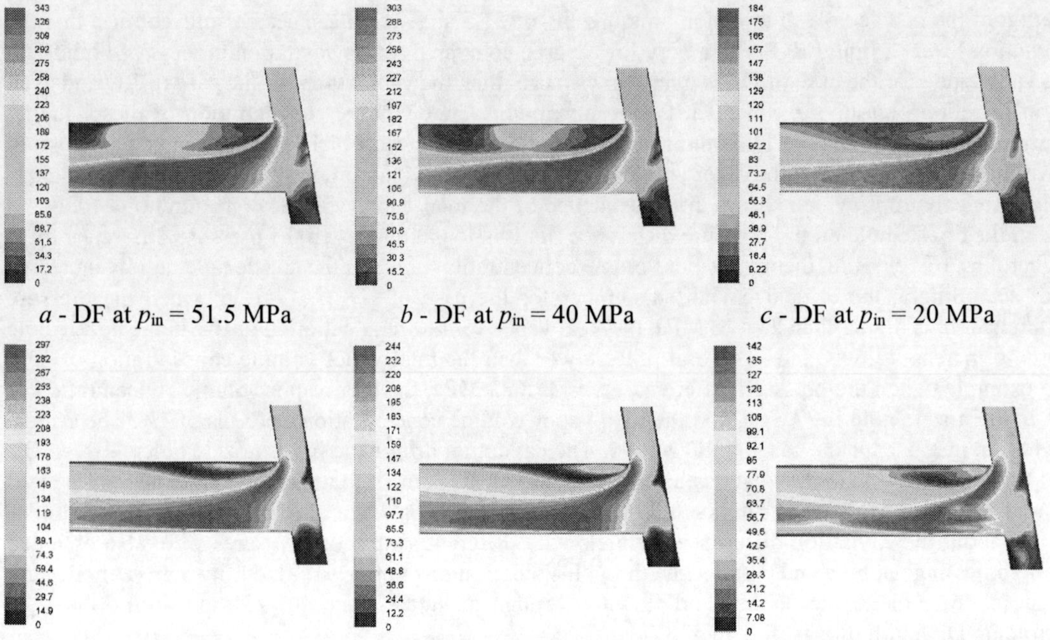

a - DF at p_{in} = 51.5 MPa *b* - DF at p_{in} = 40 MPa *c* - DF at p_{in} = 20 MPa

d - RO at p_{in} = 51.5 MPa *e* - RO at p_{in} = 40 MPa *f* - RO at p_{in} = 20 MPa

Figure 8. Distribution of the absolute flow velocity of DF and RO on the longitudinal section of the nozzle hole at different injection pressures: a – DF at p_{in} =51.5 MPa; b - DF at p_{in} =40 MPa; c - DF at p_{in} =20 MPa; d - RO at p_{in} =51.5 MPa; e - RO at p_{in} =40 MPa; f - RO at p_{in} =20 MPa.

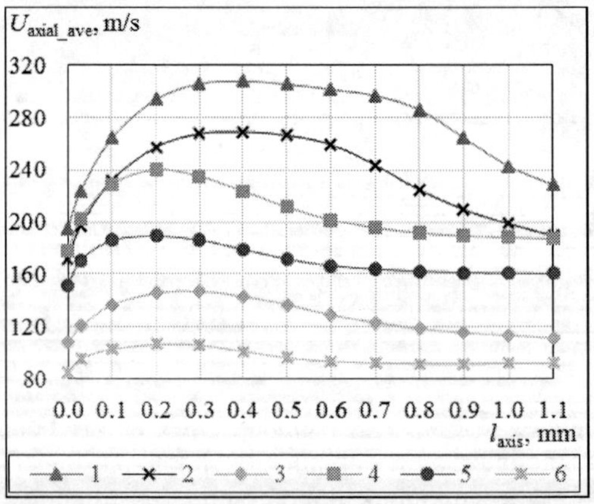

Figure 9. Average axial flow velocity on the cross-section of the nozzle hole for DF and RO along the flow direction: 1 - DF at p_{in} = 51.5 MPa; 2- DF at p_{in} = 40 MPa; 3 - DF at p_{in} = 20 MPa; 4 - RO at p_{in} = 51.5 MPa; 5 - RO at p_{in} = 40 MPa; 6 - RO at p_{in} = 20 MPa.

IOP Conf. Series: Materials Science and Engineering 1035 (2021) 012034 doi:10.1088/1757-899X/1035/1/012034

4.4. Turbulence in the nozzle

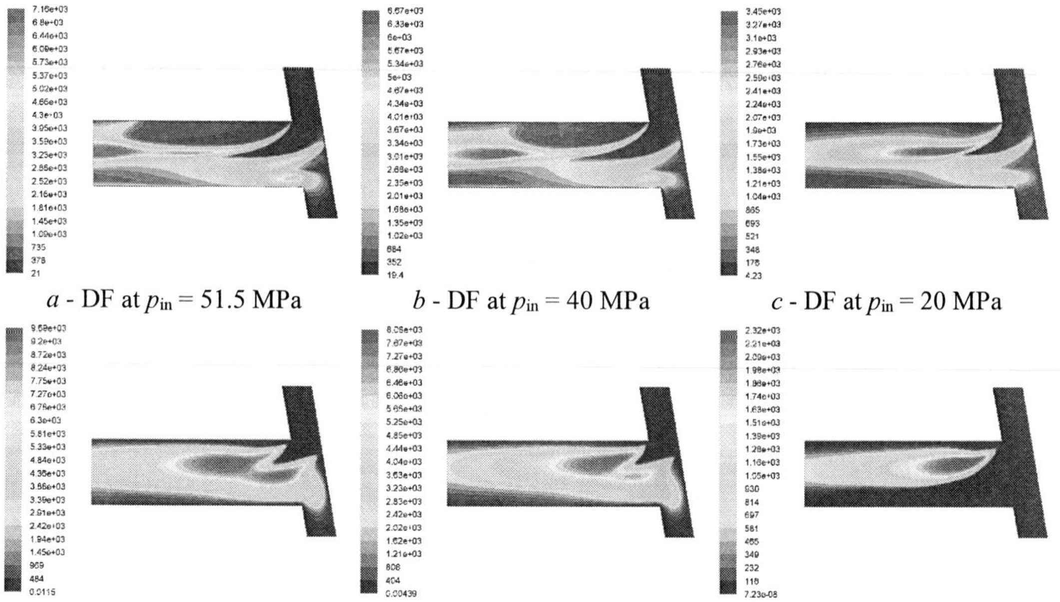

a - DF at p_{in} = 51.5 MPa; *b* - DF at p_{in} = 40 MPa; *c* - DF at p_{in} = 20 MPa

d - RO at p_{in} = 51.5 MPa; *e* - RO at p_{in} = 40 MPa; *f* - RO at p_{in} = 20 MPa

Figure 10. Distribution of turbulent kinetic energy of DF and RO flow on the longitudinal section of the nozzle hole: a - DF at p_{in} = 51.5 MPa; b - DF at p_{in} = 40 MPa; c - DF at p_{in} = 20 MPa; d -RO at p_{in} = 51.5 MPa; e –RO at p_{in} = 40 MPa; f - RO at p_{in} = 20 MPa.

To analyze the effect of RO on fuel flow turbulence, it is very useful to introduce the turbulent kinetic energy (TKE), which is characterized by the squared effective fluctuating velocity:

$$TKE = \frac{u'^2 + v'^2 + w'^2}{2}. \tag{18}$$

Figure 10 shows the TKE distribution of DF and RO flow on the longitudinal section of the nozzle hole as p_{in} =51.5, 40 and 20 MPa. In figure 11 the curves of the average TKE of DF and RO on the cross-section of the nozzle hole along the flow direction were plotted. It was found that during injecting DF the turbulent disturbance core (with the maximum turbulence level) is closer to the nozzle hole outlet and the turbulent disturbance core for the case of RO is closer to the nozzle hole inlet. Moreover, with a decrease of pressure p_{in}, the turbulent disturbance core gradually moves away from the nozzle hole outlet. The maximum TKE of RO is larger than DF at high pressures p_{in} and less than DF at low pressures p_{in}. In addition, at p_{in} = 51.5 and 40 MPa, the average TKE of RO on the cross-section with l_{axis} from 0 to 0.8 mm is also higher than this value for DF. But the TKE of RO flow at the outlet of the nozzle hole is always less than that of DF.

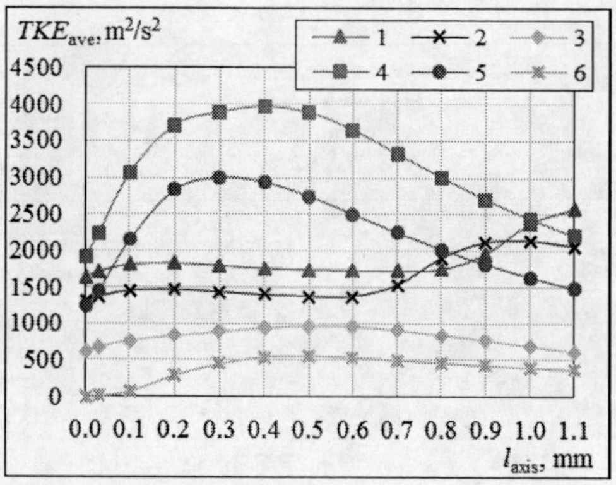

Figure 11. Curves of the average TKE of DF and RO flow on the cross-section of the nozzle hole along the flow direction: 1 - DF at p_{in} = 51.5 MPa; 2- DF at p_{in} = 40 MPa; 3 - DF at p_{in} = 20 MPa; 4 - RO at p_{in} = 51.5 MPa; 5 - RO at p_{in} = 40 MPa; 6 - RO at p_{in} = 20 MPa.

4.5. General flow parameters at the nozzle outlet

To analyze the general flow characteristics of different fuels, the fuel mass flow rate, discharge coefficient, average injection velocity and average TKE at the outlet of the nozzle hole are studied. These parameters are shown in figures 12-14.

Figure 12 shows the mass flow rate and discharge coefficient for the case of DF and RO at different injection pressures. It was shown that with the same pressure differences Δp between p_{in} and p_b, the mass flow rate and discharge coefficient of DF are higher than that of RO. Moreover, as the pressure difference Δp decreases, the discharge coefficient of DF increases at first and then almost remains unchanged when the pressure difference Δp is lower than 16.122 MPa (p_{in} = 25MPa, p_b = 8.878MPa). However, the discharge coefficient of RO at first almost does not change and then decreases. The reasons can be explained as follows. For the low-viscosity and easily evaporating DF the main role, affecting inner nozzle flow, is the cavitation, which prevents the fuel from flowing. RO is characterized by increased viscosity and low volatility so that its flow is mainly affected by viscosity.

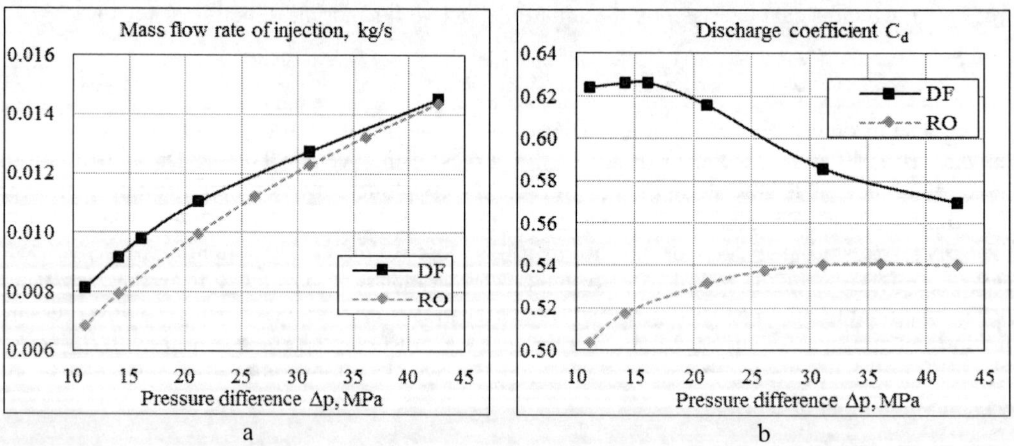

Figure 12. Mass flow rate and discharge coefficient of DF and RO at various pressure differences Δp between p_{in} and p_b: a – mass flow rate; b - discharge coefficient.

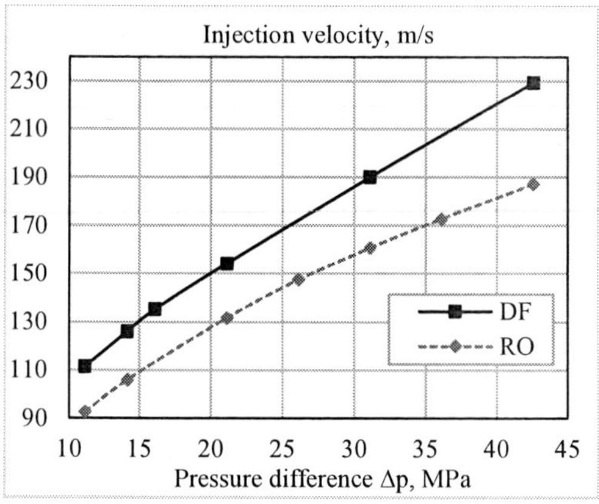

Figure 13. Fuel injection velocity of DF and RO at various pressure differences Δp between p_{in} and p_b.

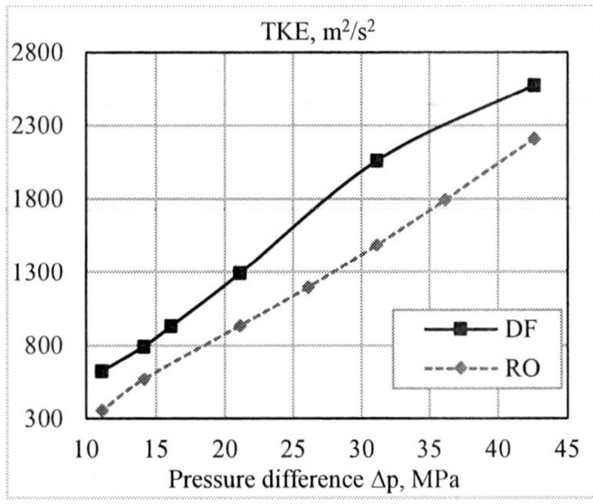

Figure 14. Average turbulent kinetic energy of DF and RO flow at the nozzle hole outlet at various pressure differences Δp between p_{in} and p_b.

Figure 13 and figure 14 show the dependence of injection velocity and the dependence of average TKE at the outlet of the nozzle hole on the pressure difference Δp between p_{in} and p_b. Obviously, with the same pressure difference Δp, both the injection velocity and average TKE of DF are higher than these of RO. The reduced injection velocity and TKE of RO might lead to a decrease of primary breakup rate and injection cone angle and an increase of spray penetration length [24-26, 37]. Based on this, using RO in a diesel engine, it is necessary to make an appropriate modification to combustion chamber geometry and the working parameters of the fuel injection system to ensure complete mixing of fuel with air and to prevent fuel spray from hitting the combustion chamber walls.

5. Conclusions

A computational study was performed to investigate the injection characteristics of petroleum diesel fuel and rapeseed oil. Important observations are as follows:

(1) The distribution structures of pressure, velocity and turbulent kinematic energy of DF and RO flow in the investigated nozzle hole are different;

(2) As the injection pressure increases, the turbulent disturbance core of RO flow tends to the nozzle hole inlet and the turbulent disturbance core of DF flow tends to the nozzle hole outlet;

(3) In the case of DF the flow velocity exceeds that for RO case. During the injection pressure p_{in}=51.5, 40 and 20 MPa, the DF flow velocity at the nozzle hole outlet (DF injection velocity) is 22.22%, 17.9% and 21% higher than the RO flow velocity, respectively;

(4) The cavitation level of DF is many times higher than RO and the critical numbers for DF and RO are 1.55 and 2.94, respectively;

(5) The DF discharge coefficient is higher than the RO discharge coefficient in all the range of injection pressure. With injection pressure decreasing, the discharge coefficient of DF grows at first and then almost stay unchanged after the injection pressure is lower than 25MPa, but the RO discharge coefficient does not change at first and then decreases.

Based on the investigation results it is necessary to point out that the direct use of pure RO in a diesel engine might lead to a reduction of injection rate and the injection turbulence level, thereby the quality of mixture and combustion process will be deteriorated. To solve this problem, an appropriate modification to combustion chamber geometry and the working parameters of the fuel injection system should be considered.

References

[1] Torres-García M, García-Martín J F, Jiménez-Espadafor Aguilar F J, Barbin D F and Álvarez-

Mateos P 2020 Vegetable oils as renewable fuels for power plants based on low and medium speed diesel engines *J. Energy Inst.* **93** 953–61

[2] Markov V A, Devyanin S N, Semenov V G, Vagrov V V and Zykov S A 2019 *Engine Fuels from Vegetable Oils* ed V A Markov (Riga: Lambert Academic Publishing)

[3] Markov V A, Devyanin S N, Zykov S A and Sa B 2017 Viscosity Characteristics of Biofuels Based on Vegetable Oils *Truck* **3** 40–6

[4] Markov V A, Zykov S A, Biryukov V V and Sa B 2017 Viscosity characteristics of multicomponent and emulsified fuels *AutoGas Fill. Complex + Altern. fuel* **16** 105–21

[5] de Almeida S C A, Belchior C R, Nascimento M V G, Vieira L dos S R and Fleury G 2002 Performance of a diesel generator fuelled with palm oil *Fuel* **81** 2097–102

[6] Wang Y D, Al-Shemmeri T, Eames P, McMullan J, Hewitt N, Huang Y and Rezvani S 2006 An experimental investigation of the performance and gaseous exhaust emissions of a diesel engine using blends of a vegetable oil *Appl. Therm. Eng.* **26** 1684–91

[7] Shah P R and Ganesh A 2016 A comparative study on influence of fuel additives with edible and non-edible vegetable oil based on fuel characterization and engine characteristics of diesel engine *Appl. Therm. Eng.* **102** 800–12

[8] Devan P K and Mahalakshmi N V 2009 Performance, emission and combustion characteristics of poon oil and its diesel blends in a DI diesel engine *Fuel* **88** 861–7

[9] Agarwal A K and Rajamanoharan K 2009 Experimental investigations of performance and emissions of Karanja oil and its blends in a single cylinder agricultural diesel engine *Appl. Energy* **86** 106–12

[10] Hebbal O D, Reddy K V and Rajagopal K 2006 Performance characteristics of a diesel engine with deccan hemp oil *Fuel* **85** 2187–94

[11] Singh P J, Khurma J R and Singh A 2010 Coconut oil based hybrid fuels as alternative fuel for deisel engines *Am. J. Environ. Sci.* **6** 69–75

[12] Yilmaz N and Morton B 2011 Effects of preheating vegetable oils on performance and emission characteristics of two diesel engines *Biomass and Bioenergy* **35** 2028–33

[13] Nwafor O M I, Rice G and Ogbonna A I 2000 Effect of advanced injection timing on the performance of rapeseed oil in diesel engines *Renew. Energy* **21** 433–44

[14] Soltic P, Edenhauser D, Thurnheer T, Schreiber D and Sankowski A 2009 Experimental investigation of mineral diesel fuel, GTL fuel, RME and neat soybean and rapeseed oil combustion in a heavy duty on-road engine with exhaust gas aftertreatment *Fuel* **88** 1–8

[15] Sonar D, Soni S L, Sharma D, Srivastava A and Goyal R 2015 Performance and emission characteristics of a diesel engine with varying injection pressure and fuelled with raw mahua oil (preheated and blends) and mahua oil methyl ester *Clean Technol. Environ. Policy* **17** 1499–511

[16] H. Q D, Q. D X, B. Z W and K. Y 2019 Spray Characteristics and Engine Performance of Vegetable Oil–Diesel–Ethanol Hybrid Fuel *J. Energy Eng.* **145** 4019011

[17] Dernotte J, Hespel C, Foucher F, Houillé S and Mounaïm-Rousselle C 2012 Influence of physical fuel properties on the injection rate in a Diesel injector *Fuel* **96** 153–60

[18] Esteban B, Riba J-R, Baquero G, Rius A and Puig R 2012 Temperature dependence of density and viscosity of vegetable oils *Biomass and Bioenergy* **42** 164–71

[19] Ramadhas A S, Jayaraj S and Muraleedharan C 2004 Use of vegetable oils as I.C. engine fuels—A review *Renew. Energy* **29** 727–42

[20] Yilmaz N 2011 Temperature-dependent viscosity correlations of vegetable oils and biofuel–diesel mixtures *Biomass and Bioenergy* **35** 2936–8

[21] Cui H, Luo F, Wang Z and Dong S 2013 Simulation analysis of flow characteristics of Jatropha curcas oil in diesel injector *Trans. Chinese Soc. Agric. Eng.* **29** 63–71

[22] Torelli R, Som S, Pei Y, Zhang Y and Traver M 2017 Influence of fuel properties on internal nozzle flow development in a multi-hole diesel injector *Fuel* **204** 171–84

[23] Wei M, Gao Y, Yan F, Chen L, Feng L, Li G and Zhang C 2017 Experimental study of cavitation

formation and primary breakup for a biodiesel surrogate fuel (methyl butanoate) using transparent nozzle *Fuel* **203** 690–9

[24] Hwang J, Bae C, Patel C, Agarwal A K and Gupta T 2017 *Near Nozzle Flow and Atomization Characteristics of Biodiesel Fuels* (SAE Technical Paper)

[25] Yu S, Yin B, Jia H, Wen S, Li X and Yu J 2017 Theoretical and experimental comparison of internal flow and spray characteristics between diesel and biodiesel *Fuel* **208** 20–9

[26] Ishak M H H, Ismail F, Che Mat S, Abdullah M Z, Aziz A and Idroas M Y 2019 Numerical analysis of nozzle flow and spray characteristics from different nozzles using diesel and biofuel blends *Energies* **12** 281

[27] Hoang A T and Pham V V 2019 A study of emission characteristic, deposits, and lubrication oil degradation of a diesel engine running on preheated vegetable oil and diesel oil *Energy Sources, Part A Recover. Util. Environ. Eff.* **41** 611–25

[28] Markov V A, Kamaltdinov V G, Zykov S A and Sa B 2019 Optimization of the Composition of Blended Biodiesel Fuels with Additives of Vegetable Oils *Int. J. Energy a Clean Environ.* **20**

[29] Cruz-Forero D-C, González-Ruiz O-A and López-Giraldo L-J 2012 Calculation of thermophysical properties of oils and triacylglycerols using an extended constituent fragments approach *CT&F-Ciencia, Tecnol. y Futur.* **5** 67–82

[30] Zong L, Ramanathan S and Chen C-C 2010 Fragment-based approach for estimating thermophysical properties of fats and vegetable oils for modeling biodiesel production processes *Ind. Eng. Chem. Res.* **49** 876–86

[31] Markov V A, Devyanin S N, Zykov S A and Gaidar S M 2016 *Biofuel for Internal Combustion Engine* (Moscow: Science Research Center Engineer)

[32] Schnerr G H and Sauer J 2001 Physical and numerical modeling of unsteady cavitation dynamics *Fourth Int. Conf. on Multiphase Flow* vol 1 (ICMF New Orleans)

[33] Wolfshtein M 1969 The velocity and temperature distribution in one-dimensional flow with turbulence augmentation and pressure gradient *Int. J. Heat Mass Transf.* **12** 301–18

[34] Jongen T 1992 *Simulation and Modeling of Turbulent Incompressible Flows* (EPF Lausanne)

[35] Chen Hc and Patel V C 1988 Near-wall turbulence models for complex flows including separation *AIAA J.* **26** 641–8

[36] Liu W and Zhao J 2016 3-D Numerical study on the effect of variant injection pressure in a diesel injector with cavitation formation *J. Adv. Veh. Eng.* **2** 174–81

[37] Yu W, Yang W and Zhao F 2017 Investigation of internal nozzle flow, spray and combustion characteristics fueled with diesel, gasoline and wide distillation fuel (WDF) based on a piezoelectric injector and a direct injection compression ignition engine *Appl. Therm. Eng.* **114** 905–20

[38] Zhao J, Liu W, Zhao J and Grekhov L 2020 Numerical investigation of gas/liquid two-phase flow in nozzle holes considering the fuel compressibility *Int. J. Heat Mass Transf.* **147** 118991

IOP Conf. Series: Materials Science and Engineering 1035 (2021) 012035 doi:10.1088/1757-899X/1035/1/012035

Remote leakage method in heat supply systems out flow

S A Sazonova[1,4], S N Korablin[1], V F Asminin[2] and A V Zvyagintseva[3]

[1]Department of Technosphere and Fire Safety, Voronezh State Technical University, 84 20-letiya Oktyabrya Str., Voronezh, 394006, Russia
[2]Department of Life Safety and Legal Relations, Voronezh State Forestry University named after G.F. Morozova, 8 Timiryazeva Str., Voronezh, 394087, Russia
[3]Department of Chemistry and Chemical Technology of Materials, Voronezh State Technical University, 84 20-letiya Oktyabrya Str., Voronezh, 394006, Russia

[4]Corresponding author: ss-vrn@mail.ru

Abstract. This paper describes the method of remote sensing of the existence or absence of leaks in functioning heat supply systems developed on the basis of the application of a two-alternative hypothesis. To achieve this goal, we proposed to apply a method for determining the location and volume of leaks, based on the use of energy equivalents. We proposed to check for leakage between static assessments of the state of heat supply systems according to data from a remote manometric survey, with the help of which a simultaneous poll of pressure sensors in the end nodes of the system in question is performed. In the event of sudden changes in the estimated system parameters, it is concluded that there is a leak. If the results of two neighboring evaluations are relatively stable, then a conclusion is made about the absence of leakage. The considered problems are based on mathematical models of flow distribution for a heat supply system. The results of a computational experiment are compared with the results of calculating the parameters in the system based on the results of gauge surveying.

1. Introduction

Work [1] examined in detail the task of static assessment of the state of heat supply systems (SHS) (and for other hydraulic systems [2, 3]) developed on the basis of mathematical models of flow distribution for SHS [4], obtained using functional (energy) equivalent [5].

The problem of determining the magnitude and location of leaks for water and gas supply systems based on energy equivalent equipment is formulated and implemented in detail in [6, 7]. In these works, the problem of the existence of a leak in hydraulic systems was not considered. It is necessary to comprehensively solve three problems of leak diagnostics: its location, its volume, and the fact of the existence of leaks, which will allow for technical diagnostics remotely using metering devices on the system, gauge data and processing information on a computer in real time with using the developed mathematical models, to recreate complete information about the studied object and promptly provide information, for the decision maker, about the presence in the complex system of leaks or unauthorized selections of the target product. As an independent task, ensuring the reliability of hydraulic systems and allowing disconnecting emergency sections and connecting backup ones, it is necessary to consider the backup problem for heat supply systems [8].

Let us consider the problem of fixing the existence of a leak for heat supply systems. First of all, it is necessary to agree that we use the two-alternative hypothesis in assessing the presence or absence of a leak. Checking for the presence or absence of leakage is supposed to be performed between two static evaluations. The value of the time interval between two evaluations is supposed to be determined practically, depending on the mode of functioning and system features. Such a time interval can be several seconds for extremely important objects (gas supply systems, oil pipelines, etc.) or hours, for new or small systems. In the selected time interval, it is much more likely that one leak can occur, and not several. Therefore, a two-alternative hypothesis was initially proposed, but we are not yet considering a multi-alternative hypothesis, it can be the basis for further developments.

2. Materials and methods

To solve the problem of the presence of an ordinary leak in the SHS, we use the detection theory [9], which assumes a statistical approach to solve it. To detect a natural leak, we apply a scientific analogy. We apply the filtering process of signals arising from the background of noise. In this case, a signal $\alpha s(t)$ with an unknown amplitude α is considered to be a leak in the system, which we can fix right after the occurrence of noticeable changes in the operation mode of the SHS (amplitude surge), recorded according to the results of a manometric survey between two state assessments. Here, $s(t)$ is a function of time given to solve the problem.

By noise, we mean relatively stable, having an average amplitude, consumption mode of the target product by users of the hydraulic system. We propose to apply an approach with filtering signals against a background of noise. Leaks comparable in amplitude with such noise will not be significant compared to possible significant leaks, therefore, the proposed approach can be effective for large leaks and, accordingly, SHS accidents that cannot be visually detected promptly.

When solving this problem, the reason for changing the state of the system must be determined.

This change can be caused either by a leak or due to an accumulated error (hindrance) of the information recipient of information (dispersion σ_1) and with random consumption of the target product due to interference in the heat supply system (dispersion σ_2).

The presence of a leak in the system, as for a statistical problem, must be found using a sample X_n

$$X_n = \lambda \alpha S_n + \Xi_n, \tag{1}$$

where n – the sample size; λ - the parameter in the binary hypothesis: at λ=1 leakage occurs with the probability p_1, at λ=0 leak does not occur with the probability p_2 ($p_1 + p_2 = 1$); S_n - the multiple possible values of fixed leakage consisting of elements s_i; Ξ_n - the multiple values of uncorrelated interference in the heat supply system and in the information receiver, consisting of the elements ξ_i.

The most common and, in fact, the most important problem of detecting unknowns is that two parameters are assumed α and σ_2, having a given range of change $\Delta\alpha$ and $\Delta\sigma_2$. The maximum likelihood method is most rationally used to solve this problem. Hypothesis testing using the maximum likelihood method involves determining the relationship in the form

$$\Lambda(X_n \mid \alpha^*, \sigma_1^{2*}, \sigma_2^{2*}) = \left[\frac{\sigma_2^{2*}}{\sigma_1^{2*}(\alpha^*)}\right]^{n/2} = \left\{\frac{\sum_{i=1}^{n} x_i^2}{\sum_{i=1}^{n} x_i^2 - \frac{1}{E}\left(\sum_{i=1}^{n} x_i s_i\right)^2}\right\}^{n/2}, \tag{2}$$

where: $E = \sum_{i=1}^{n} s_i^2$; symbol "*" is used to indicate the estimated parameters in the system under study. The ratio $\sigma_2^{2*} / \sigma_1^{2*}(\alpha^*)$ should be compared with a threshold value $C(X_n)$. When the condition is met

$\sigma_2^{2*} / \sigma_1^{2*}(\alpha^*) \geq C(X_n)$. It is customary to conclude that there is a leak in the system. The converse condition means that the leak does not exist. The threshold value $C(X_n)$ determined from the relationship

$$C(X_n) = \left(\frac{g_{21} - g_{22}}{g_{12} - g_{11}} \frac{p_2}{p_1} \frac{\Delta\alpha}{\sqrt{\pi}} \sqrt{h_0^*} \right)^{\frac{2}{n-2}}, \tag{3}$$

where: $h_0^* = E / 2\sigma_1^{2*}(\alpha^*)$ - the signal to noise ratio.

In relation (3), the value g_{ij} represents the loss function in the case of making the *j*-th decision for the *i*-th situation.

The probabilities of situations p_1, p_2 in (3) are defined as

$$p(h^* \mid 0) = \begin{cases} 0 & for\ h^* < 0, \\ \dfrac{1}{\sqrt{\pi h^*}} \exp(-h^*) & for\ h^* \geq 0. \end{cases} \tag{4}$$

And for λ=1

$$p(h^* \mid 1) = \begin{cases} 0 & for\ h^* < 0, \\ \dfrac{1}{2\sqrt{\pi h^*}} \left\{ \exp\left[-\left(\sqrt{h^*} - \sqrt{h}\right)^2 \right] + \exp\left[-\left(\sqrt{h^*} + \sqrt{h}\right)^2 \right] \right\} & for\ h^* \geq 0. \end{cases} \tag{5}$$

where $h = \alpha^2 E / 2\sigma^2$ - the real signal to noise ratio.

Statistical properties when detected by determining the probability of a false alarm using the expression

$$F = \int_C^\infty p(h^* \mid 0) dh^* = 1 - \int_0^C \frac{1}{\sqrt{\pi x}} \exp(-x) dx = 2\left[1 - \Phi(\sqrt{2C})\right], \tag{6}$$

where $\Phi(...)$ - the probability integral.

The probabilistic leak detection, which is called correct, is determined from the relation

$$D = \int_C^\infty p(h^* \mid 1) dh^* = 1 - \int_0^C \frac{1}{\sqrt{\pi x}} \left\{ \exp\left[-\left(\sqrt{x} - \sqrt{h}\right)^2 \right] + \right.$$

$$\left. + \exp\left[-\left(\sqrt{x} + \sqrt{h}\right)^2 \right] \right\} dx = 2 - \Phi\left(\sqrt{2C} - \sqrt{2h}\right) - \Phi\left(\sqrt{2C} + \sqrt{2h}\right). \tag{7}$$

In addition to the aforementioned mechanism, which is based on testing a two-alternative hypothesis on the initial sample, it will be necessary to consider a consistent and recurrent analysis procedure in order to consider the decision-making problem in the event of leaks in heating systems.

The proposed approach will quickly detect a leak, eliminate it, thereby increasing the level of security of the system and provide cost savings for its repair. The safety of various systems has been dealt with in a large number of works, for example [10, 11, 12], which can be used by analogy to ensure the safety of hydraulic systems. When performing the study, the works [13 - 26] were used.

Natural leaks are understood to mean leaks arising due to the ageing of the system during its operation or in the event of an accident from a force acting on the system. In practice, unauthorized selection of the environment may also occur due to illegal consumption of the target product from the system in the personal interest. Such environmental selections using the proposed approach to solving the problem can also be fixed, thereby preventing theft of the target product.

3. Computational experiment results

To confirm the operability of the proposed method, a computational experiment was performed in accordance with the selected design scheme of the SHS (figure 1). Based on the results of a computational experiment, for a leak in node 28 of the SHS, a graph was plotted (figure 2) for the distribution of deviations according to the results of gauge surveys from the calculated values.

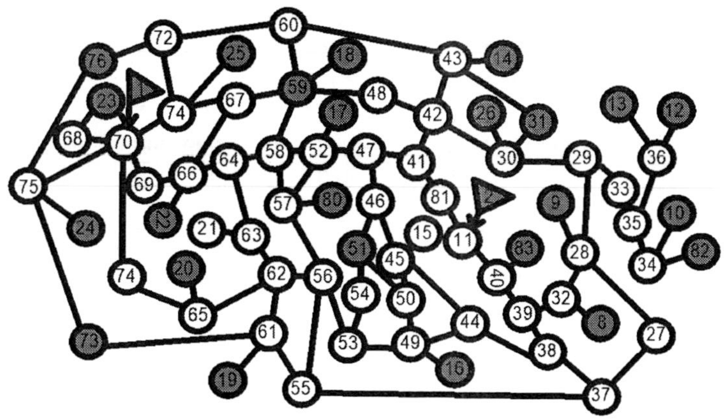

Figure 1. Calculation scheme of SHS.

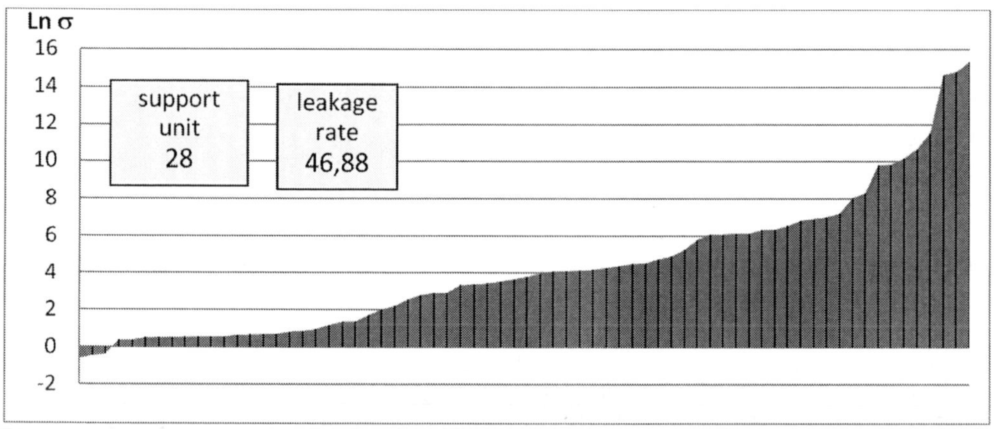

39	37	27	30	34	42	48	13	55	14	59	49	47	54	51	72	63	17	73	20	64	66	68
40	44	29	35	10	8	9	45	41	11	60	16	19	46	50	71	65	76	70	74	15	22	23
38	32	33	28	36	31	12	26	43	56	53	61	62	57	67	58	52	18	75	24	25	69	21

Figure 2. The graph of the distribution of deviations according to the results of gauge survey from the calculated values.

In figure 2, in the place of σ_j, natural σ_j logarithm applies. The table below the graph shows the numbers of the system nodes. The greatest likelihood of a leak will be in the node located in the far left upper position of the table. When moving within the table from left to right, as well as from top to bottom, the calculated value of the standard deviation will increase.

4. Conclusions

Based on the results of the research, the following conclusions can be drawn.

1. A method for remote sensing of the existence or absence of leaks in functioning heat supply systems was developed.

2. We propose to apply a scientific analogy, precisely, to apply the process of filtering signals that arise against a background of noise. It is proposed to take a leak in the heat supply system as a signal with an amplitude exceeding the noise level. In this case, by noise, we propose to mean consumer selection of the target product from the system.

3. We propose to implement the developed method for remote checking of leakage in the heat supply system based on the use of flow distribution models, the method of detecting the location and volume of leakage, and static estimation of the current state developed on the basis of the use of energy equivalents.

4. The efficiency of the proposed method is confirmed using the performed computational experiment for the selected design scheme of the heat supply system.

References

[1] Sazonova S A 2011 Solution of the problem of static assessment of heat supply systems *Bulletin of Voronezh State Technical University* **7(5)** 43-46

[2] Kvasov I S, Panov M Ya and Sazonova S A 2000 Statistical assessment of the pipeline systems state based on functional equivalents *News of Higher Educational Institutions Building* **4(496)** 100–105

[3] Kolodyazhny S A, Sushko E A, Sazonova S A and Sedaev A A 2013 The solution of the problem of gas supply systems static assessment *Scientific Bulletin of Voronezh State University of Architecture and Civil Engineering Construction* **4(32)** 25–33

[4] Sazonova S A 2011 Results of the development of mathematical models for analysis of flow distribution for heat supply systems *Bulletin of Voronezh State Technical University* **7(5)** 68-71

[5] Kvasov I S, Panov M Ya, Shcherbakov V I and Sazonova S A 2001 Energy equivalence of large hydraulic systems for the life support of cities *News of Higher Educational Institutions Building* **4(508)** 85–90

[6] Kvasov I S, Panov M Ya and Sazonova S A 1999 Diagnostics of leaks in pipeline systems during loose gauge surveys *News of Higher Educational Institutions Building* **9(489)** 66-70

[7] Sazonova S A 2010 Models for assessing the disturbed state of a heat supply system *Engineering Physics* **3** 45–46

[8] Sazonova S A, Nikolenko S D and Osipov A A 2020 Simulation of a transport standby for ensuring safe heat supply systems operation *IOP Conference Series: Materials Science and Engineering* vol 753 (Vladivostok: IOP Publishing) p 052004

[9] Repin V G and Tartakovsky G P 1977 *Statistical Synthesis with a Priori Uncertainty and Adaptation of Information Systems* (Moscow: Sovetskoe Radio) p 432

[10] Zvyagintseva A V, Sazonova S A and Kulneva V V 2019 Modeling of fugitive emissions of dust and gases into the atmosphere in open pits mining and processing plants, and improving measures to improve working conditions *Proc. of the Seventh Int. Environmental Congress* (Samara-Togliatti: ELPIT) pp 212–226

[11] Manokhin V Ya, Sazonova S A, Nikolenko S D and Zvyagintseva A V Flare emissions from asphalt plants *Int. Scientific Conf. on Energy, Environmental and Construction Engineering "EECE-2019": Lecture Notes in Civil Engineering* (St. Petersburg) in press

[12] Zvyagintseva A V, Sazonova S A and Kulneva V V Measures to improve working conditions and reduce dust and gas emissions in the quarries of the mining and processing plant *Int. Science and Technology Conf. "FarEastCon-2019" IOP Conference Series: Materials Science and Engineering* (Vladivostok: IOP Publishing) in press

[13] Nikolenko S D, Sushko E A, Sazonova S A, Odnolko A A and Manokhin V Ya 2017 Behaviour of concrete with a disperse reinforcement under dynamic loads *Magazine of Civil Engineering* **7(75)** 3–14
[14] Asminin V F, Druzhinina E V, Sazonova S A and Osmolovsky D S 2019 Development and application of a portable lightweight sound suppression panel to reduce noise at permanent and temporary workplaces in the manufacturing and repair workshops *Akustika* **34** 18–21
[15] Osmolovsky D S, Asminin V F and Druzhinina E V 2019 Reducing noise from round woodworking machines by applying vibration damping friction pads between the saw blade and the clapmping flange *Akustika* **32** 138–140
[16] Boldyreva O N and Zvyagintseva A V 2009 Technological risk regulation through optimization of equipment maintenance program *Bulletin of Voronezh State Technical University* **5(12)** 76–78
[17] Zvyagintseva A V, Boldyreva O N and Usov Yu I 2004 Construction of models for managing environmental parameters of technological processes *Engineer. Technologist. Worker* **12** 31–33
[18] Zvyagintseva A V 2008 Interaction peculiarities of hydrogen and NI-B galvanic alloys *NATO Science for Peace and Security Series C: Environmental Security* F2 437–442
[19] Zvyagintseva A V and Kravtsova Y G 2007 The problem of hydrogen permeation into the Boron doped electrodeposited nickel films *NATO Science for Peace and Security Series A: Chemistry and Biology* **2007** 661–664
[20] Zvyagintseva A V 2006 The effect of boron on the hydrogenation of nickel films *Alternative Energy and Ecology* **5(37)** 85–86
[21] Zvyagintseva A V 2015 The ability of materials based on nickel of the nanoscale range to hydrogen storage *Alternative Energy and Ecology* **21(185)** 150–155
[22] Zvyagintseva A V and Artemeva A O 2017 Modern hydrogen storage based on hybrid functional materials *The Bulletin of VSTU* **13(5)** 133-138
[23] Zvyagintseva A V 2017 Hybrid functional materials forming metal structures with optimal defectiveness for storing hydrogen in hydride form *Alternative Energy and Ecology* **16-18(228-230)** 89–103
[24] Yakovlev D V and Zvyagintseva A V 2012 Construction of an interbranch integrated geographic information system of the Voronezh region *Izvestia of Samara Scientific Center of the Russian Academy of Sciences* **1-3** 923–930
[25] Zvyagintseva A V and Kravtsova Y G 2007 Hydrogen permeation and nickel films structure correlation *NATO Science for Peace and Security Series A: Chemistry and Biology* **2007** 665-669
[26] Savchenko V S, Yushchenko K A, Zvyagintseva A V, David S A and Vitek J M 2007 Investigation of structure and crack formation in welded joints of single crystal Ni-base alloys *Welding in the World* **51(11-12)** 76–81

IOP Conf. Series: Materials Science and Engineering 1035 (2021) 012036 doi:10.1088/1757-899X/1035/1/012036

Investigation of the influence of gravitational forces on the thermal-hydraulic processes of power plants

A A Sataev[1], V V Andreev, A V Duntsev, D I Novikov and A A Berezin

Department of Nuclear Reactors and Power Plants, Nizhny Novgorod State Technical University n.a. R.E. Alekseev, 24 Minina Str., Nizhny Novgorod, 603950, Russia

[1]Corresponding author: sancho_3685@mail.ru

Abstract. Today, improving the safety of nuclear power plants is one of the main problems. Obviously, heat and mass transfer processes have a great influence on complex phenomena occurring in the reactor plant. Currently, special software systems are being developed that can describe these processes (Computational Fluid Dynamics (CFD)). However, they do not take into account the influence of external gravitational forces. They occur when the spatial position of an object changes. In the case of ship-based nuclear power plants, the development of technical specifications should include a mandatory requirement for trouble-free operation when affected by pitching and rolling. The paper describes the study of the effect of shipboard pitching on the mixing processes of non-isothermal flows in a small-scale model. Based on the analysis of the results obtained, it can be concluded that the external dynamic force has a negative effect on the mixing of non-isothermal flows.

1. Introduction

Progress in the development of nuclear submarines and surface fleets, stationary nuclear power, and the creation of promising facilities with nuclear power plants is impossible without comprehensive analysis, complex testing and research. It is advisable to conduct such research on ground-based large-scale prototype stands.

Conducting experiments at high parameters or close to full-scale ones is sometimes associated with significant costs. However, many phenomena occurring in reactor installations can be modelled with the necessary degree of representativeness on small-scale models. A significant amount of information can be obtained from such small-scale models [1].

Currently, in our country, and especially abroad, software tools are being developed that allow describing complex heat and mass transfer processes (Computational Fluid Dynamics (CFD). The relevance of CFD programs is confirmed by a large number of scientific publications on this topic and a number of specialized scientific conferences devoted to this issue.

The most significant activities in this area are those coordinated by the OECD/NEA (OECD / Nuclear Energy Agency) in Europe, and by the CASL (Consortium for Advanced Simulation of Light Water Reactors) program in the USA. Thus, since 2002, the OECD / NEA has been holding annual CFD4NRS (CFD for Nuclear Reactor Safety) conferences dedicated to the tasks and problems of using CFD programs in the nuclear power industry [2].

The most significant achievements in this field were made by scientists from Germany (using the Rossendorf Coolant Mixing Model (ROCOM) [3]). No less significant results were also obtained at

IOP Conf. Series: Materials Science and Engineering 1035 (2021) 012036 doi:10.1088/1757-899X/1035/1/012036

the Vattenfall installations (USA), the Fortum PTS installation (Finland), and the stand of OKB GIDROPRESS (Russia) [4].

Obviously, heat and mass transfer processes have a great influence on complex phenomena occurring in the reactor plant. One of these processes is the mixing of non-isothermal (as well as isothermal, differing in the concentration of impurity [5]) flows, which occurs in the main equipment of a nuclear power plant (NPP) (reactor, steam generator, volume compensator, etc.).

All the above-mentioned models described mixing processes only in stationary modes. However, when developing technical specifications for ship-based nuclear power plants, a mandatory requirement must be provided for failure-free operation when external dynamic forces are applied to the nuclear power plant (short-term periodic tilts of the ship up to 45° and long-term tilts up to 15° on any side) [6].

High-amplitude pitching restricts the use of certain types of reactors (for example, with a boiling coolant). It can cause oscillations in reactor power due to the occurrence of a change of density in local regions of the active zone. The greatest impact of pitching is on reactor installations with natural circulation.

2. Materials and methods

To research the processes of mixing non-isothermal flows, a reduced hydraulic model of the reactor installation was made. It is represented by a single circulation loop. The reactor core is modelled by a hole sheet. It is part of the extraction unit. In this paper, the circulation loop is understood as the hydraulic path from the model inlet (inlet pipe), the lowering section, the annular gap, the pressure manifold, the core simulator (made in the form of hydraulic resistance — hole sheet), and the outlet pipe.

Table 1. Main parameters of the experimental model.

Parameter	Amount
Simulated reactor plant	KLT - 40 – 1:5
The number of simulated circulation loops	1 (plan-4)
The number of measuring probes in the simulator	17
The number of measuring probes in the annular gap	4-12
The number of control probes for the preparation of media injections	5
Flow sensors	3
The number of level sensors	3
Elements of valves	15
Injection temperature	10-90°C

As the main research method, the study of temperature distribution fields in characteristic mixing regions was applied. This is achieved by constructing a layer-by-layer picture of the temperature distribution. This is done by changing the vertical coordinates of sensors located on the path of mixing flows [7], [8].

IOP Conf. Series: Materials Science and Engineering 1035 (2021) 012036 doi:10.1088/1757-899X/1035/1/012036

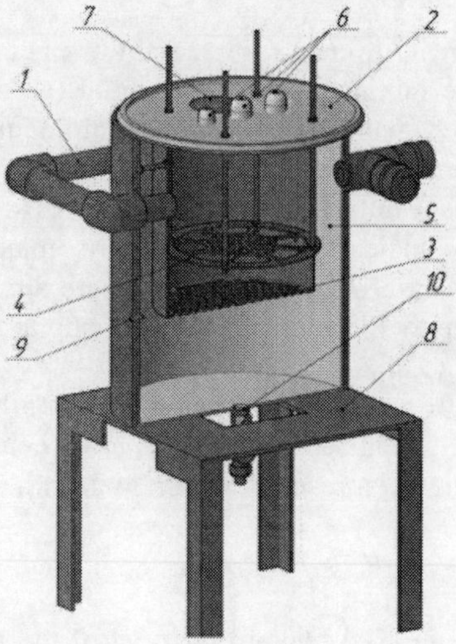

Figure 1. General view of the experimental model
1-Circulation loop; 2-Cap; 3-Reactor core simulator; 4-Sensor matrix; 5-Ring gap; 6-Hermetic inlets; 7- Viewing window; 8-Platform; 9-Temperature sensor; 10-Drain Valve.

To obtain a 3D field of temperature distribution, sensitive temperature elements (thermal sensors) were installed in a certain way. Altogether, this makes up the sensor matrix. To receive data and further process it on a PC, a high-speed program was compiled for the survey of thermal sensors.

To simulate the effect of changes in spatial orientation, it was necessary to determine the vibrations and forces acting on the ship (and, accordingly, the reactor).

The pitching is called the oscillatory movements that the ship makes near its equilibrium position. Regarding this, all types of vibrations can be divided into:

- vertical - the oscillations of the ship relative to the vertical representing a progressive periodic movement;
- shipboard pitching - the oscillations of the ship in the plane of the frames, which are angular displacements;
- keel - ship oscillations in the diametrical plane, also represent angular movements [9].

Usually, all three types of vibrations occur in various combinations, or all at the same time.

This article will look at the impact of shipboard pitching.

The left figure 2 shows the ship in equilibrium; in the right figure 2 - a ship with a roll to starboard.

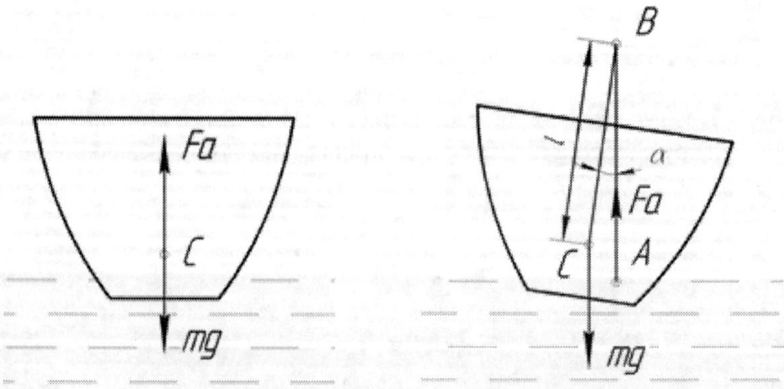

Figure 2. Forces acting during shipboard pitching.

Let us write down the equation of the dynamics of the rotational movement of the ship under the action of shipboard pitching:

$$J \frac{d^2 \alpha}{dt^2} = M , \tag{1}$$

$$J \frac{d^2 \alpha}{dt^2} = -mgl \sin \alpha , \tag{2}$$

where J - is the moment of inertia of the body relative to the longitudinal axis; α - the angle of inclination (roll) of the ship.

The resulting equation is a nonlinear differential equation of the 2nd order, that is, the body makes nonlinear oscillations [10].

Accordingly, an experimental stand with a swinging platform was made to simulate these oscillations. The swinging platform is rigidly fixed with a movable axis, which is held on support bearings attached to the frame. The electric motor, through the transition coupling, transmits the rotating force to the movable axis, which in turn drives the model under study. By setting a specific law of motion, the model can be made to oscillate periodically relative to its vertical position.

The model can oscillate in a single plane like a mathematical pendulum, according to the law:

$$\phi(t) = \phi_{\max} \cdot \sin(\frac{2 \cdot \pi \cdot t}{T}), \tag{3}$$

where $\varphi_{\max}$ - the oscillation amplitude, radian; T- the period of oscillation, s; t – the current time, s.

Figure 3. Swinging platform.

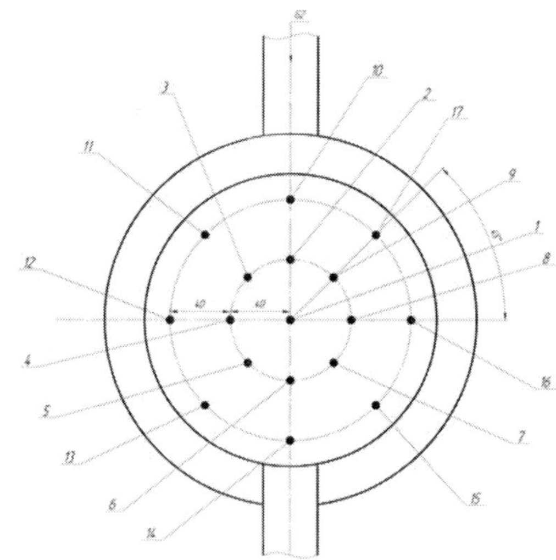

Figure 4. The location of sensors (1-17) in the section of the model.

3. Results

Figures 5 and 6 show the temperature fields at the input and output of the model core simulator.

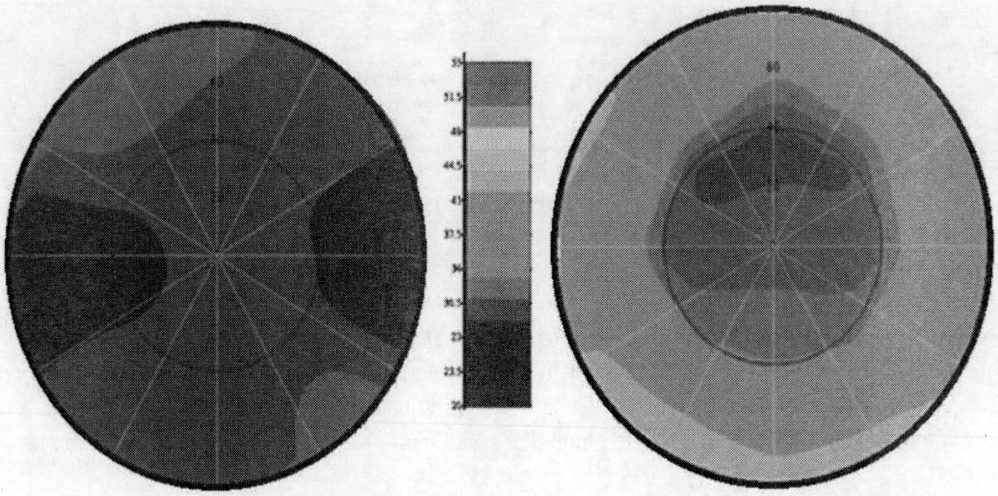

Figure 5. Visualization of the distribution of the temperature field in the area of the entrance to the core simulator: on the left - the experimental model taking into account the external dynamic force, on the right the experimental model in the stationary mode.

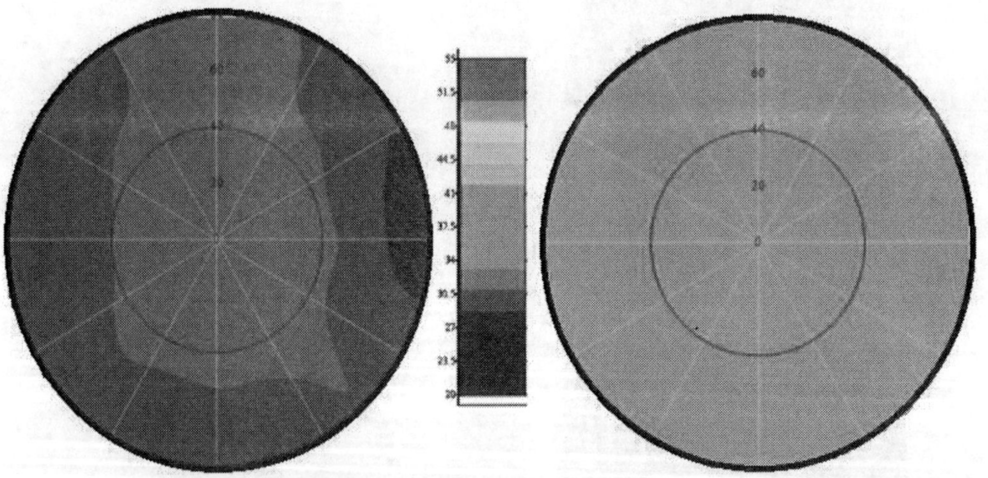

Figure 6. Visualization of the distribution of the temperature field in the area of the exit from the core simulator: on the left the experimental model taking into account the external dynamic force, on the right the experimental model in the stationary mode.

4. Discussion

For the stationary mode, when analyzing the temperature field at the entrance to the core simulator, the maximum temperature is observed on the sensors T2 and T4 located on the periphery. With vertical movement, the value of the parameter T2 weakens, and T4 increases (the peripheral zone of 3 quarters). This indicates the presence of a turbulent vortex passing through this area. The steady temperature value is observed on average for 250-280 seconds after the start of injection.

It is also possible to draw a conclusion about the uneven mixing and formation of a stagnant zone in the region of the 1st quarter (figure 5 - sensors T1 and T3). The temperature gradient is on average from 4-10 ^0C.

Under the influence of periodic oscillations on the model for the temperature field at the entrance to the core simulator, the maximum temperature is observed on the sensor T1 and T6. Note that for the static mode, the maximum temperature was observed on the sensors T2 and T4 located on the periphery. Therefore, we can assume that the inertial forces turn the turbulent vortex and shift the zone of intense mixing to the region of 2 quarters at the inlet pipe. The steady temperature value is observed on average for 300-320 seconds, which is more than for static mode.

Separately, the mathematical processing of graphs for the presence of temperature fluctuations associated with the laws of oscillatory motion was carried out. To do this, the graph for the T1 sensor was divided into sections. For each section, a linear approximation was used to determine the linear component of the temperature change from the mixing time.

Accordingly, to find the vibrational component, the graph of the equation of the approximating linear line was subtracted from the graph of the original function. However, in addition to the vibrational component of the temperature (associated with the model oscillation), there are temperature pulsations associated with vortex formations in the flow (turbulent pulsations). However, this kind of pulsation is characterized by small amplitude (no more than 0.05-0.1^{0}C) and is transient in time (1-2 seconds). The studied ripples, which occur due to oscillations in the model, have a much longer period and amplitude.

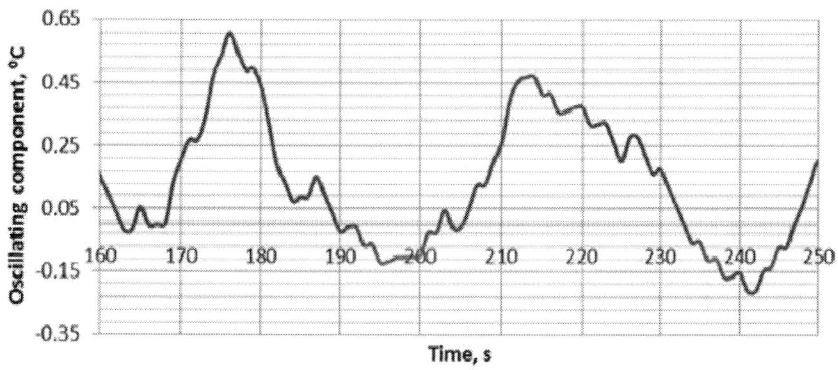

Figure 7. Graph of the dependence of the oscillatory component on time.

5. Conclusion

The main goal of our research was to find differences in the mixing processes of non-isothermal flows in stationary and dynamic modes and to evaluate them. Experiments showed that this effect significantly affects the mixing of flows. The results obtained (models, methods, and computational algorithms for describing complex mixing processes of non-isothermal flows) will allow us to optimize existing reactor plant designs and justify safety parameters.

References

[1] Krapivtsev V G and Solonin V I 2019 Model studies of inter-loop mixing heat carrier in the VVER-1000 in reactor pressure path *Atomic Energy* **125(5)** 271-77

[2] Bolshukhin M A, Budnikov A V, Fomichev V I, Sveshnikov D N and Romanov R I 2013 Actual issues of development of experimental database for verification of CFD programs when using them in nuclear power engineering *Proc. of Nizhny Novgorod State Technical University Named after R. E. Alekseev* **2(99)** 117-25

[3] Höhne T and Kliem S 2006 Coolant mixing studies of natural circulation flows at the ROCOM test facility using ANSYS *ANSYS-CFX CFD4NRS 2006* **23**

[4] Lisenkov E A, Bezrukov Yu A, Ulyanovsk V N, Saliy L A, Ulyanovsk D V, Zaitsev D V, Sergeev S G, Bykov M A, Zaitsev S I 2008 Research of mixing of the heat carrier in the

lower chamber of the reactor *Questions of Nuclear Science and Technology Series: «Ensuring the Safety of Nuclear Power Plants»* **23** 3-17
[5] Kuopanportti J 2015 Behavior of the VVER-440 reactor core at Loviisa NPP with incomplete mixing of boron solution during shutdown under natural circulation conditions *Ensuring the Safety of Nuclear Power Plants with VVER* ***9***
[6] Kudinovich I V 2019 Justification of the nuclear and radiation safety of an atomic vessel under external influences *Proc. of Krylov State Scientific Center* **1(387)** 131-42
[7] Sataev A A, Duntsev A V, Vorobev D A and Krasavin N A 2018 Study of mixing processes of nonisothermal flows on a single-loop model of a reactor installation *Modern High Technology* **3** 96-101
[8] Sataev A A and Duntsev A V 2018 Investigation of the processes of mixing non-isothermal flows on a model of a ship's nuclear power plant *Bulletin of ISEU* **5** 26-32
[9] Chaynikov K N 1971 *General Device of Courts* (Leningrad: Shipbuilding)
[10] Wulfson I I 2017 *Short Course in the Theory of Mechanical Oscillations* (Moscow: VNTR)

IOP Conf. Series: Materials Science and Engineering 1035 (2021) 012037 doi:10.1088/1757-899X/1035/1/012037

Integrated information system of life cycle support of networks and electrical network equipment

V V Safronov[1], V F Barabanov, A M Nuzhnyy and N I Grebennikova

Department of Automated and Computing Systems, Voronezh State Technical University, 14 Moskovsky Avenue, Voronezh 396066, Russia

[1]Corresponding author: vitolik@bk.ru

Abstract. Taking into account the concept of seamless integration, this paper presents the structural and functional scheme of information systems' integration. The proposed scheme is specified for typical life cycle control systems of networks and electrical network equipment in information systems, with the dominant use of domestic software products. In the proposed scheme, the functional features and limitations of the module for controlling and correcting drawings, as well as the module for digitizing archive documentation, are clarified to form a unified system for digital prototyping of products in the context of a continuous lifecycle. We present the developed structure of the integrated information system for lifecycle support of networks and electrical network equipment. An integrated information system for lifecycle support implies the implementation of software in the form of a series of specialized platform-independent software intermodular interfaces, which allows for quick configuration of the system for the existing industry and enterprise specifics.

1. Introduction

The analysis of the existing practice of implementing digital network design systems reveals the systemic presence of a number of problems, characteristic for domestic design enterprises as a whole. This is due to the fact that the creation of the park of software products, used to automate the design process, was carried out at the time when the maintenance of systemic design data integration was technically impossible and not expected. In the course of design, the automation means from various manufacturers were applied (primarily, those of foreign origin) [1, 2]. In this connection, the existing and newly created project documentation is not strictly standardized, being presented in various, often non-integrable formats. The contemporary approach assumes providing the information support of the designed product at all lifecycle stages.

Nowadays, the Russian software market witnesses the generation of domestic software products, meeting all present-day requirements and providing the preparation of project documentation in strict conformity with industry standards. These factors demonstrate the necessity of creating integrated information systems of lifecycle support of networks and electric network equipment, focused on priority use of domestic software products.

2. Materials and methods

2.1. The research objective

The development of informational, methodological and software means of providing highly effective interaction within a set of domestic software products, included in the IISLCS of networks and electric network equipment, is aimed at achieving the following goals:

- raising the quality of the information structure of design enterprises;
- the integration of the information structure components of an enterprise into a single information space;
- providing the use of a universal data standard in the design of networks and equipment;
- handling electronic catalogues and databases of standard technical solutions;
- providing the centralized storage, processing and management of data, obtained from various elements of the information structure of the enterprise in a single database;
- providing the possibility of the effective use of archive design engineering documentation;
- providing data exchange between all elements of the information structure of the enterprise;
- improving the quality of project documentation by using the means of automatic search and correction of standard errors in project data;
- reducing the time of product manufacture due to providing the possibility of parallel execution of product design stages;
- providing the comprehensive support of the product lifecycle.

2.2. The list of procedures related to the organization of the study

The development of algorithms for data communication between the elements of IISLCS, related to:

- identification of possible areas, related to the organization of the data communication;
- the analysis of formats, methods and means by which data communication is organized;
- organization of the data communication;
- the analysis of applicable CAD systems and formats of graphic files, storing the data, related to the 3D models;
- the comprehensive analysis of methods and means of graphic data exchange in formats used;
- the analysis of archives of graphic project documentation for the purpose of detecting standard errors and graphic data elements, not subject to translation;
- the classification of standard errors and graphic data elements, not subject to translation;
- graphic data translation.

The development of software components and other means of software complex elements' integration, i.e.:

- translation programs or customization of third-party translators;
- the automated system of graphic data analysis and correction;
- software tools for the interaction of graphic systems with the formation of the integrated database of graphic data;
- software products for the organization of data communication between the elements of the IISLCS.

3. Results and discussion

3.1. The structure of the IISLCS of networks and electrical network equipment

Taking into account the comprehensive analysis, the integrated information system of life cycle support of networks and electric network equipment was formed, with priority use of domestic software products, providing the most effective solution of the following tasks:

- raising the efficiency of the information structure of an enterprise;
- the integration of the information structure components of an enterprise into a single information space;

- providing the use of a universal data standard in the design of networks and equipment;
- handling electronic catalogues and databases of standard technical solutions;
- providing the centralized storage, processing and management of data, related to all elements of the electronic structure of the enterprise;
- providing the possibility of the effective use of archive engineering design documentation;
- providing the comprehensive support of the product lifecycle.

The structure of the IISLCS of networks and equipment is presented in figure 1.

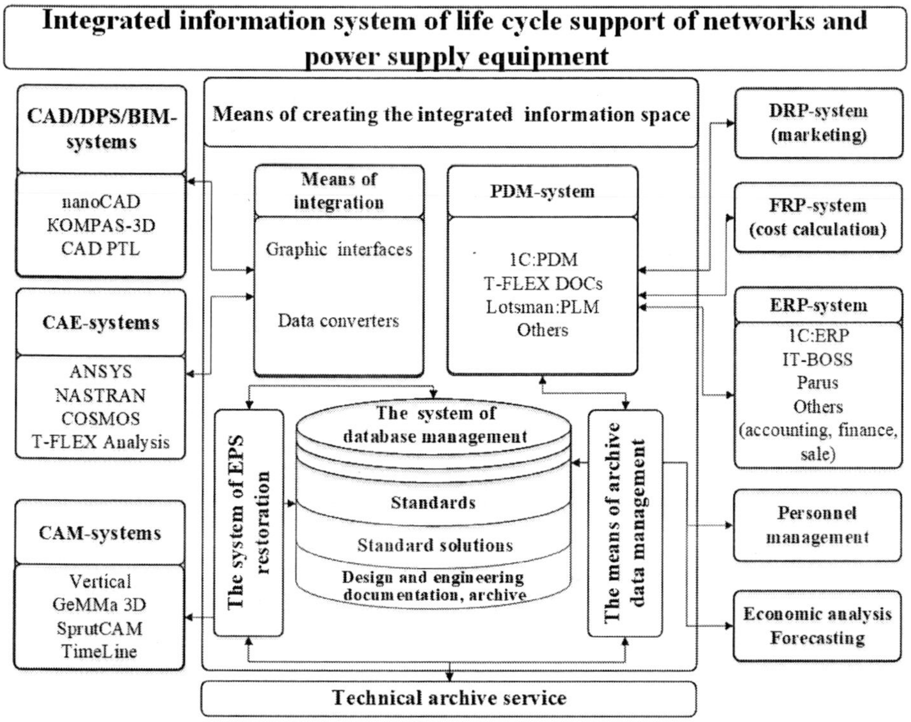

Figure 1. Structure of the IISLCS of networks and equipment.

3.2. The modules for digitizing archive design engineering documentation (DED), the analysis and correction of graphic data

Providing the effective use of the archive DED is one of the essential conditions for creating an effective system of product lifecycle management.

The reasons, hampering the reuse of the archive DED:

- the DED exists in hard copy only, which hinders the search and use of such documents;
- the DED is executed in outdated formats, not used by CAD;
- the DED is presented in the form of orthogonal projections and 3D models are missing.

All reasons, listed above, exclude the possibility of using the archive DED in modern design systems.

The same reason makes it difficult to include such DED as part of present-day corporate e-archives, storing design documents and data in the electronic product structure (EPS).

The problem of archive graphic data digitization may require the solution of the following sub-tasks.

Scanning and vectorization of drawings, existing in hard copy only. During the execution of this procedure, the following problems are most likely to occur:

- the bitmap, obtained by scanning, may not meet the requirements for systems for graphic data vectorization. Therefore, the results of scanning, presented in the form of bitmap data, must be subjected to methods of automated quality restoration and correction of corrupted images;
- the execution of a 2D drawing, obtained by the use of vectorization systems, does not comply with the enterprise standards and may prove to be inappropriate for use in the system of spatial model restoration. To solve this problem, it is required to develop the module of complex control and correction of graphic objects, allowing for automatic solution of such tasks as, elimination of open contours, rearrangement of projection types on the drawing field, the set of procedures for model-based formatting of graphic primitives (the correction of text attributes, the creation of a layer list, the layer distribution of primitives, etc.)

The conversion of data storage formats into the one, eligible for use in the module of archive DED digitization.

The importance of the problem of translating graphic formats is, as a rule, determined by the following factors:

- the use of CAD systems of various complexity levels to solve various project tasks;
- the use of various highly specialized CAD systems at certain design stages with the integration of the results at later stages.

The solution of the problem assumes the creation of the graphic data conversion system, making it possible to perform the conversion of all required formats for both 2D and 3D graphic data.

The implementation type of the module of archive DED digitization is presented in figure 2.

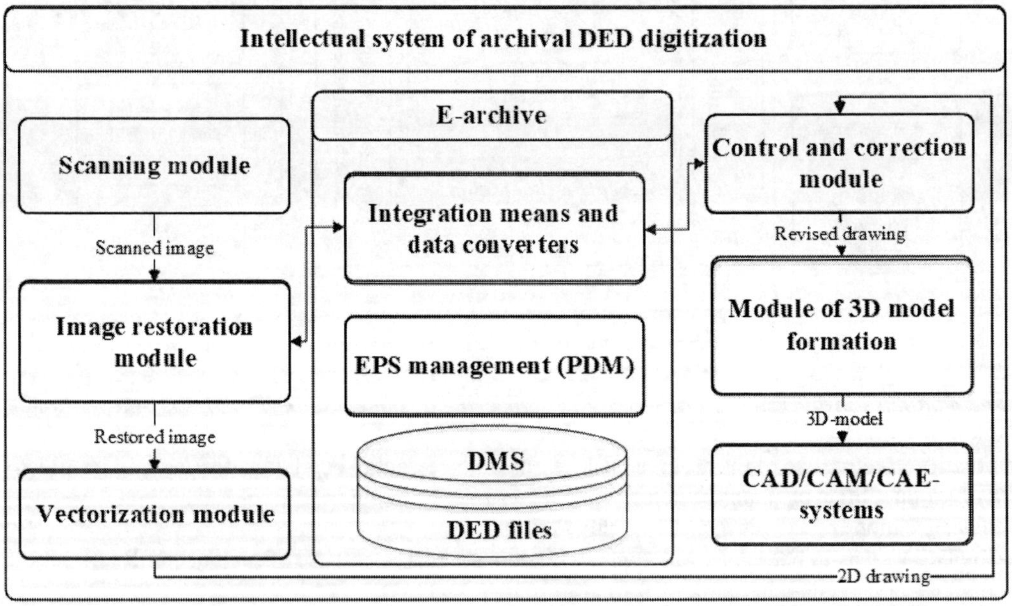

Figure 2. The structure of the archive DED digitization module.

3.3. The problem of reusing archive DED

The enterprises, engaged in project activities for a long time, possess substantial archives of design engineering documentation, the reuse of which seems problematic for the following reasons:

- the DED exists in hard copy only, which hinders the search and use of such documents;
- the DED is executed in outdated formats, not used by CAD;
- the DED is presented in the form of orthogonal projections and 3D models are missing.

All reasons, listed above, exclude the possibility of using the archive DED in modern design systems, focused on the use of 3D models, since these systems do not possess the means for the restoration of 3D models using orthogonal projections.

The same reason makes it difficult to include such DED as part of present-day corporate e-archives, storing design documents and data in the electronic product structure (EPS), formed on the base of the 3D model.

Figure 3 presents a type of a structural diagram of the module for control and correction of drawings.

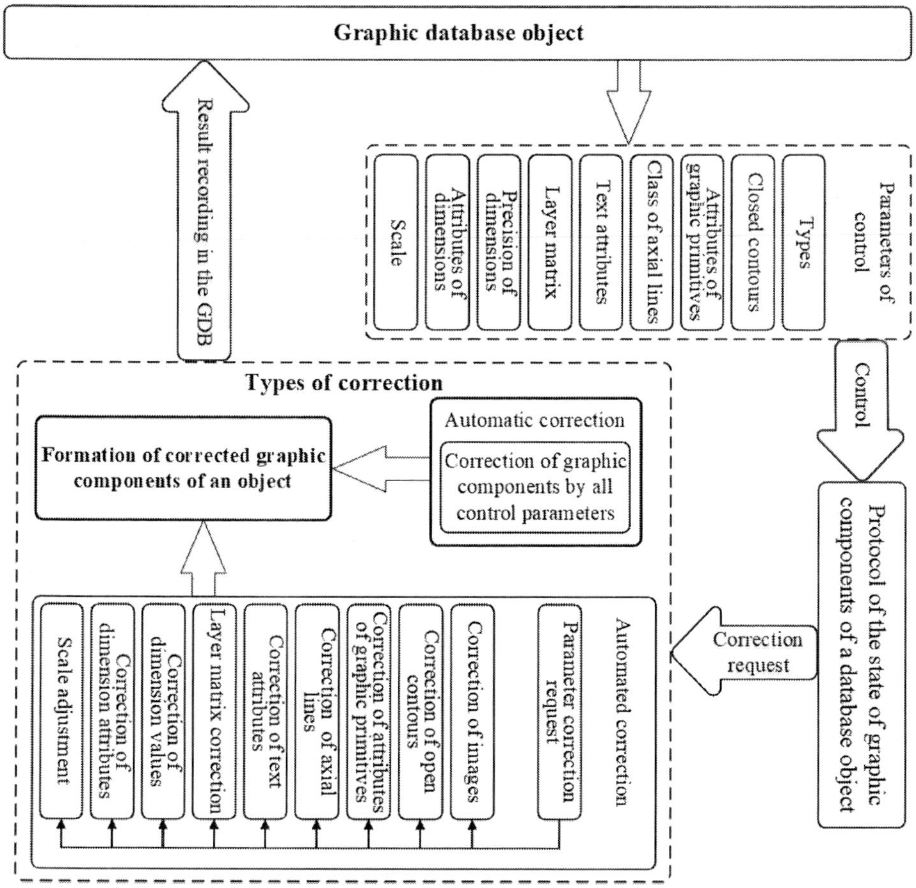

Figure 3. The structural diagram of the module for control and correction of drawings.

4. Conclusion

The offered block and function schemes of the specialized software are developed on the basis of the concept of seamless integration with support, planning and life cycle maintenance systems. They are oriented to the preservation of functional client-server model integrity. The software is realized in the form of a series of specialized program intermodular interfaces.

The proposed structure of the integrated information system is based on the use of the model of seamless integration of large software complexes in integrated lifecycle support systems, as described in [3-6], which ensures the preservation of functional integrity of the client-server model of software systems. The software is implemented as a series of specialized platform-free software intermodular interfaces.

By creation and implementation of the IISLCS of networks and electrical network equipment, it is planned to achieve the following scientific and technical results:

- the components for IISLCS creation are focused on the application of domestic software products;
- the optimal structure of the IISLCS for solving the problems of automation at an enterprise;
- the application of present-day formats and author's algorithms of data communication between graphic and non-graphic components of the IISLCS;
- the contemporary methods and algorithms of control and translation of graphic data;
- the use of software solutions for implementing seamless integration and organization of single information space for IISLCS components focused on the specific nature of an industry and an enterprise;
- providing the centralized storage, processing and management of product data, obtained from all elements of the information structure of an enterprise;
- standardization and unification of the DED due to the use of domestic software products;
- providing the possibility of the effective use of archive DED;
- providing the information support of networks and equipment throughout their lifecycle;
- reducing the cost of network and equipment design due to the use of domestic software products.

References

[1] Kenin S L, Barabanov V F, Nuzhnyy A M and Grebennikova N I 2013 *Bulletin of Voronezh State Technical University* **3(1)** 4-8

[2] Nuzhnyy A M, Safronov V V, Barabanov A V and Gaganov A V 2013 *Bulletin of Voronezh State Technical University* **6(1)** 23-27

[3] Nuzhnyy A M, Grebennikova N I, Barabanov A V and Povalyaev A V 2013 *Bulletin of Voronezh State Technical University* **6(2)** 25-31

[4] Safronov V V, Barabanov V F, Prudnikov E E and Kovalenko S A 2015 *Bulletin of Voronezh State Technical University* **4** 20-22

[5] Safronov V V, Podvalny S L, Barabanov V F and Nuzhnyy A M 2016 *Izvestia of Samara Scientific Center of the Russian Academy of Sciences* **4(3)** 646-654

[6] Safronov V, Barabanov V, Podvalny S, Povalyaev A and Achkasov A 2019 *Lecture Notes in Networks and Systems* **60** 309-322

IOP Conf. Series: Materials Science and Engineering 1035 (2021) 012038 doi:10.1088/1757-899X/1035/1/012038

Detection of damage sites on overhead power lines with a voltage of 110 kV or higher

V N Kuryanov[1], V G Kul'kov, E M Skopova, Yu V Firsov and D S Norov

Department of Electric Power and Electrical Engineering, Volzhsky branch of National Research University Moscow Power Engineering Institute, 69 Lenina Avenue, Volzhsky 404110, Russia

[1]Corresponding author: vek077@yandex.ru

Abstract. The article is devoted to problems in the electric grid complex related to ageing of components of power lines, reduction of reliability and increase the number of failures. It shows the need of development of scientific and methodical bases of perfection of management and organization of the system of technical maintenance and repair of equipment in power network systems, taking into account the current state of development of the power industry. Problems of disconnections of high-voltage power lines occur periodically in many countries and the search for options for their early elimination or pre-termination is a very urgent task. Researching and developing methods and devices for early detection of fault locations will minimize the time for troubleshooting and thus reduce downtime and economic losses for all parties, both the grids organization and the electricity consumer. The paper presents the results of research and development of devices used to detect places of power line damage in the conditions of "digital transformation" of the energy complex. The implementation of the developed proposals will have a positive impact on reducing losses from under-supply of electricity and will reduce the time for identifying fault locations associated with determining the locations of short-circuit currents.

1. Introduction

Development of the energy sector of the Russian Federation is connected with the implementation of the concept and state programs for the transition to modern effective technological and economical aspects of managing the country's energy system. The implementation of the national project "Digital economy" in the concept of energy development until 2035 forms new approaches for energy companies to implement digital transformation programs. Existing models for managing power facilities need to be adapted in accordance with technical solutions to improve the reliability of operation of electric networks in the framework of digital transformation [1,2] (during operation, repairs, reconstruction, etc.) in the conditions of digitalization, which currently does not always provide high efficiency, and the introduced digital technologies often relate only to the line of newly introduced new equipment. However, the current equipment of the country's energy industry (more than 60%) operates with a service life of more than 25 years and can still be effectively used for the production and transfer of energy resources. During power supply interruptions, many production facilities stop completely, and today the urgent task is to identify and eliminate various incidents as quickly as possible with the resumption of normal operation.

In the Russian Federation, electricity transit is carried out mostly on grids of 110 kV and higher, in addition, the main consumers can be powered from substations, in this case, the disconnection of the supply over the head line by the consumer is felt more acutely. Overhead power lines run through various terrain and therefore are exposed to all possible factors: ice load, wind load, atmospheric pollution, thunderstorms, falling branches and trees, human activity, and so on. The impact of these factors leads to permanent, repeated, or single-time damage. When disconnecting, with a successful re-activation (especially once), it is often not possible to identify the place of damage during a normal walking inspection. In this case, it is necessary to disconnect the line and perform a thorough riding inspection of several supports and spans between them, which leads to the disconnection of the consumer and significant material costs for the operating organization.

A number of devices are used to determine the location of the damage. Basically, these are devices installed at substations that determine the distance from the place of installation of the device to the place of damage by various characteristics. In practice, taking into account the network modes and measurement errors, it takes quite a long time to find the location of a short circuit on the power line, even if there are special fixing devices installed at substations. Devices for determining the location of the damage, installed directly on the support, are not widely used, which is due to some shortcomings of existing modifications.

The purpose of the work is to develop a device for detecting short-circuit currents in the elements of overhead power lines with a voltage of 110 kV and higher, which are installed directly on the support and allow one to determine quickly and accurately the location of damage without disconnecting the equipment. To achieve this goal, a device was developed that meets the following requirements: it is installed onto a support without placing additional weight on it; it is possible to install and disassemble it without disconnecting the air line and in a short time; installation, disassembly and maintenance are performed by maintenance personnel, without additional involvement of specialists; there is a light indicator to facilitate search at night.

The paper analyzes the accumulated experience and scientific developments [3] in this direction: the most likely paths of short-circuit currents on overhead power lines with a voltage of 110 kV and higher are determined; the method of signaling when the device is triggered is determined; the characteristics of the device components and parts are determined.

The result of the research is the development of devices for detecting damage sites on overhead power lines with a voltage of 110 kV and higher, followed by digitalization: event recording, transmission, accumulation, and analysis of data arrays in decision-making centers

2. Analysis of the reasons for overhead line failures

The variety of causes of damage and failures of overhead power lines makes it difficult to systematize them. Failures that occur for each of the reasons listed below are not the same at different stages of operation of the overhead line. All detected manufacturing and installation defects remain inside the products and are hidden foci of future failures in the operation of high-voltage power lines. Operational failures – the ones that occur for reasons related to a violation of the established rules or operating conditions. For example, failures due to non-compliance with the actual working conditions of the project; contamination of insulators and surfaces of overhead lines, etc. However, the analysis of retrospective data on the operation of overhead lines allows us to get high-quality views on the localization of failures by overhead line elements, on their distribution by time groups.

Over the past decades, there have been serious deviations in the operation of equipment, which led to its damage due to errors of personnel and production instructions. This fact indicates that special attention should be paid to the elimination of shortcomings in the work of the operating personnel, on whose work the safety and reliability ultimately depend. All failures related to operation can be divided into two groups: "eliminated" and "unavoidable". The first group includes failures that can be prevented by organizational and technical measures. The second group includes failures that can only be prevented by replacing equipment or its individual parts. Production (technological) failures often occur for reasons related to the imperfection or violation of the established process of manufacturing or installation of components and parts of the overhead line. The most common failures are those associated with a violation of the connection process of overhead line elements. The share of failures due to operational shortcomings has increased sharply in recent years [4]. Modern repairs performed at a high technical and technological level and in full accordance with the normative and technical documentation and diagnostic results ensure reliable operation of the equipment. In practice, repairs are not always carried out in good quality, so it is not uncommon for equipment to fail soon after repairs. Such failures are usually referred to as repair defects if they occurred within the first two months after the end of repair work or have a pronounced repair defect and after a longer period of operation. This situation leads to the need to carry out a significant amount of work on defecation of overhead line elements during planned repairs and, as a result, to increase their duration.

It is known that the materials for manufacturing wires have a dependence of long-term strength and creep on the operating time [5, 6]. Accordingly, the overhead lines resource has specific limits, exceeding which will cause damage and failures. When determining the reasons for failures, it is not always possible to clearly determine the reason for disconnecting the power line. In the practice of overhead lines operation, it often happens that design flaws are hidden behind other circumstances that are taken as the reasons for failures or are replaced by existing violations of operating modes. The measures developed in such cases are aimed at eliminating the side causes of regime violations, while the main design flaws remain unrecorded, which leads to repeated, repeated failures. In addition, for a large percentage of overhead lines failures, it is not possible to find out the true reasons for failures.

The main causes of accidents are the following factors: damage to the wire and lightning protection wire (8%), thunderstorm activity (11%), disconnections as a result of birds' activity (18%), overlapping of insulation and the air gap between the wire and the support (along the insulation) (56%), and other reasons. When covering the insulation (without destruction) or the air gap between the wire and the support (along the insulation), there are often very small traces in the form of thin glass flakes on the upper part of the insulator and small specks - marks of short-term thermal effects of an electric arc on the traverse and the wire, which are not always possible to see from the ground, even with binoculars. In most cases, after the overlap of the polymer insulator, insulators of glass or porcelain insulators, the performance of the suspension is preserved. The reasons for power outages over a five-year period are shown in figure 1.

Due to the development of scientific and technological progress in the past few decades, new information and communication technologies have been rapidly developing in order to improve reliability, prevent or reduce the number of failures in the Russian Federation's power grid complex. Questions of reliability of technical objects and their reflection in modern standardization are considered in detail in [1, 7, 8].

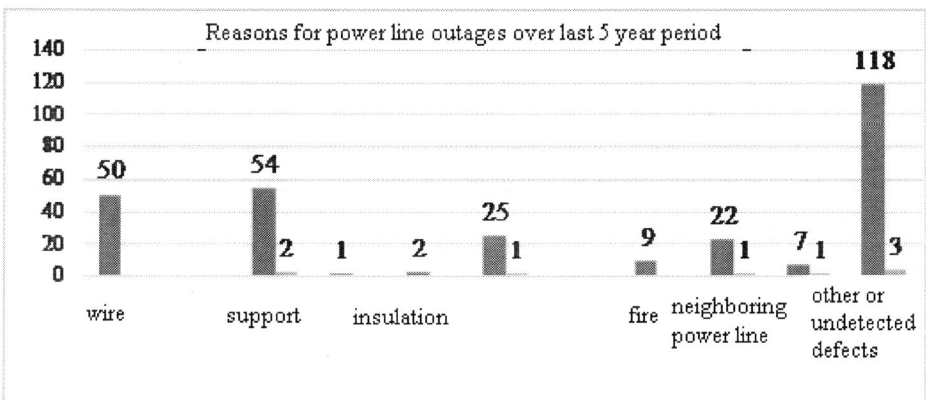

Figure 1. Reasons for power outages for a five-year period.

3. Remote detection of damage sites

To determine the location of damage, power lines are equipped with various types of devices. As a rule, these devices are installed at substations and their area of operation covers several lines, including substation equipment.

The operation of such devices is aimed at determining the distance from the place of a short circuit to the place of their installation, using the following basic methods.

Pulse method, which is based on the measurement of time intervals when electromagnetic waves propagate through overhead lines. This method is divided into location-based and wave-based.

The location method is based on measuring the time interval between the moment when the probing pulse is sent and the moment when the pulse reflected from the site of damage comes to the beginning of the line. That is, the time of the double pulse run to the point of damage is measured. In this case, the distance to the short-circuit location is defined as the product of the propagation speed and the interval, divided in half.

The wave method is based on measuring the time interval between the moments when the ends of the line are reached by the fronts of electromagnetic waves that occur at the point of damage.

The induction method is based on fixing the parameters of the magnetic field of currents that flow through the wires and in the soil along the route of the overhead line.

The above methods, as well as other methods, are based on determining the distance from the short-circuit location to the place where they are installed, and therefore their operation is negatively affected by a number of factors, such as:

- condition of overhead line wires (the reliability of the length, the reliability of the cross-section of the wires and repair inserts; the presence of damage with a decrease in the cross-section, corrosion, especially relevant for wires made of aluminum steel alloys, etc.);
- condition of wire connections (voltage drop, reliable contact, etc.);
- equipment status of intermediate substations;
- modes of operation of electrical grids, and a number of other factors.

Taking into account the above factors, as well as the "passport-permissible" error, the readings of damage detection devices installed at the substation may differ from the actual values by several hundred meters to several kilometers. In cases where there are no obvious traces of a short circuit, it is necessary to unscheduled output to repair the overhead line and perform a riding inspection of several supports, and sometimes several dozen supports, which leads to a decrease in grids reliability, and, on end overhead lines, to disconnect end consumers while searching for short-circuit locations.

4. A new device for determining the places of passage from short-circuit currents

Analyze the existing devices for solving the problem of detecting short-circuit locations that installed directly on the overhead line support. A number of inventions (signaling devices, indicators) were considered. However, the invention is not widely used, perhaps this is due to some non-modifications, the main of which are the complexity of installation, high cost, design flaws, and the inability to reuse. During the operation of polymer insulators, it was found that when the insulator is overlapped, the screen rings located at both ends of the insulators are damaged. Such damage occurs from minor meltdowns to visible burnouts.

Based on these observations, a new design of the digital device shown in figure 2, indicating the overlap of insulators on overhead power lines, which is installed directly on the support and allows you to quickly and accurately determine the location of damage without disconnecting the equipment and meets the following requirements:

- mounted on a support without increasing the weight load;
- installation and dismantling is carried out in the shortest possible time;
- possibility of mounting and dismounting without disconnecting the power line;
- allows service personnel to accurately determine the location of a short circuit both remotely and from the ground;

- installation, disassembly and maintenance is performed by the service personnel, without additional involvement of specialists;
- light indication, for easier search at night time.

This research shows aspects of scientific and methodological bases for improving the management and organization of the system of diagnostics, maintenance and repair of equipment for power grid complexes, taking into account the current state and "digital transformation" of the energy grids complex.

A number of devices and devices have been developed to determine the path of short-circuit currents directly in place (on the support). The research on this problem is still underway, which indicates the urgency of the problem. A model of the insulator overlap indicator, which is a digital signaling device with a fuse insert, is proposed. According to the developed model, the signaling device consists of several parts connected to each other, as shown in figure 2. This indicator also allows one to determine the insulator that requires maintenance during a visual inspection from the ground, as well as sends a signal from a digital device to the dispatcher's console. Digital module 3, short-circuit current detector is a compact protected digital device that is installed next to the visual short-circuit current detector. The device is programmed and configured for correct operation. Figure 3 shows the appearance of the microprocessor electronics on the board.

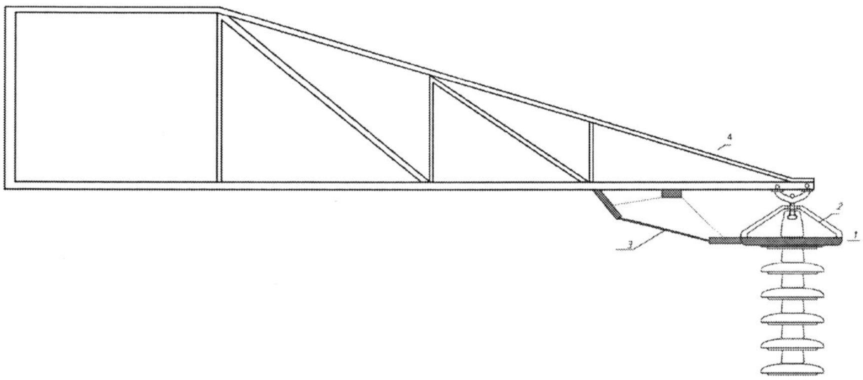

Figure 2. Digital device for determining the location of short-circuit currents in power transmission lines. 1-visual short-circuit current detector; 2-mounting spokes; 3-digital module short-circuit current detector; 4-elements of the power line support.

Figure 3. Implementation of microprocessor electronics on the board.

The device is mounted in a small box that is easy to install. The device is designed to detect places where short-circuit currents pass through overhead power lines and the result of the device operation is a file with data on determining the places where short-circuit currents pass through overhead power lines. The digital device transmits a signal, i.e. informs the power system dispatcher about the current operation by sending text messages.

5. Conclusions

The following solutions are proposed for detecting damage sites on overhead power lines with a voltage of 110 kV and higher.

1. A visual and digital signaling device for the passage of short-circuit currents of power lines passing through insulators on the traverses of power transmission line supports was developed.

2. We developed a digital GSM device for transmitting information about the operation of the short-circuit current detector to the automated workplace of the power system dispatcher, in order to timely detect fault locations when disconnecting the overhead line and reduce the time of troubleshooting.

The implementation of the developed proposals, starting from the design stage of new power transmission lines and installation on all operated overhead lines, will have a positive impact on reducing losses from under-supply of electricity and will reduce the time for identifying fault locations associated with determining the locations of short-circuit currents.

Acknowledgements

The research is funded by the Russian Federation public contract №0720-2020-0025 "Technique development and method analysis for ensuring power system object security and competitiveness based on the digital technologies".

References

[1] Voropay N I, Kovalev G F, Kucherov Yu N et al. 2013 *The Concept of Ensuring Reliability in the Electric Power Industry* (Moscow: Energiya) p 212

[2] Mayorov A V 2019 Technical solutions for improving the reliability of electric distribution networks in the framework of digital transformation *Electricity. Transmission and Distribution* S3 **14** 2-6

[3] Dzyubin A S 2019 The indicator of breakdown of the insulator, patent no. 187995 U1, IPC G01R 31/08 (2006.01), SEC G01R 31/08 published 26.03.2019

[4] Fedotova G A, Voropai N I and Kovalev G F 2015 Reliability of technical objects. Standardization issues *Reliability and Safety of Energy* **4(31)** 2-6

[5] Kuryanov V N, Shvets E S, Timashova L V and Fokin V A 2017 Application of domestic innovative high-temperature ASW wires for 110 kV overhead lines and their efficiency *Energy of the Unified Network* **5(34)** 12-18

[6] Kuryanov V N, Fokin V A, Kushch L R and Merzlyakov A S 2018 Optimal solutions for power grid development problems *Electricity Transmission and Distribution* Special issue 34-37

[7] Grabchak E P 2015 Problems of normative and technical regulation in the electric power industry *Reliability and Safety of Energy* **4(31)** 7-13

[8] Kuryanov V N, Kushch L R, Sultanov M M and Fokin V A 2016 Research of technical losses of electric energy and economic efficiency of innovative load-bearing cable application in the contact network of railways *Alternative Energy and Ecology* **1-2** 50-58

Development of a training complex of elements of power systems and metering of electric power with application of VR technologies

A V Strizhichenko[1]**, E G Zenina, D S Kochergin and V V Efimov**

Department of Power Engineering, Volzhsky branch of National Research University Moscow Power Engineering Institute, 69 Lenina Avenue, Volzhsky 404110, Russia

[1]Corresponding author: strijichenko@yandex.ru

Abstract. The paper describes the results of simulation of virtual power equipment of a substation and an electric power metering system for the purpose of training and practicing the actions of personnel during scheduled switching and emergency situations. An experimental model is presented that demonstrates the basic functionality of the simulator, which allows for rapid switching on a virtual substation model. The virtual model will help the trained operational personnel gain operational experience, as well as eliminate the possibility of electric shock and equipment failure. Due to the flexible functionality of the program that builds the project, it is possible to implement additional blocks of the simulator, having previously tested the innovation separately from the main project. This feature guarantees compatibility of new versions of the project with the old installed versions. It also allows the client to update the product without reinstalling the program. The model was created in the Unreal Engine 4 program. This product allows you to create projects for Windows, Lunix, and Mac OS operating systems. Unreal Engine 4 is compatible with the virtual reality system. Due to compatibility, it is possible to create interactive projects, in particular, an interactive project in the field of energy.

1. Introduction

According to the statistics of JSC "SO UES" for 2018, and the results of the investigations into the causes of accidents in electric grid equipment in the UES of Russia, 16% are due to erroneous actions of personnel, and in the electric grid sector of Russian Railways. The share of erroneous actions of operational personnel accounts for 6% of accidents.

The main violations are the lack of verification of removal of portable grounding or disconnection of grounding knives after completion of work, the lack of verification of compliance of dispatch names of the required switching equipment.

In this regard, we find relevant the issues of the use of simulators based on virtual reality systems for educating students, as well as the training and retraining of personnel of energy enterprises, which allow you to work out the actions of personnel during switching at the commissioning of substation equipment for repair and equipment inspections, its repair and replacement.

2. Development of the simulator

The advantage of using virtual reality systems (VR) in the electric power industry is the ability to create a virtual environment for specific technical situations. A virtual substation is not only a detailed simulator with the ability to conduct training without the need for access to real equipment but also a tool for testing knowledge. The power equipment of the 110/10 kV substation was chosen as the object under study to create a virtual switching simulator [1]. Figure 1 shows the type of outdoor switchgear-110 kV.

Figure 1. 110 kV outdoor switchgear.

In this simulator, it is possible to conduct switching simulations for putting into repair and putting into operation power and measuring transformers, power switches, disconnectors, power lines (figure 2).

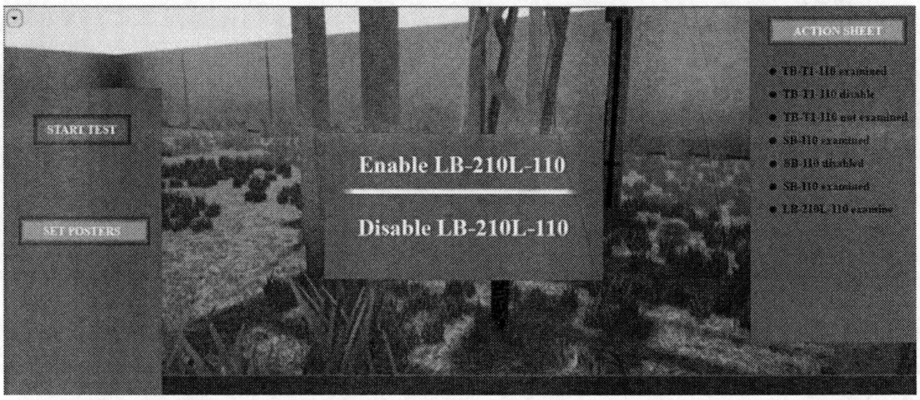

Figure 2. Menu of operations with a linear disconnector.

Possible emergency situations are modeled: emergency shutdown of the transformer due to breakdown of insulation, disconnection of the transformer under the condition of the gas relay being triggered, failure of the disconnector due to damage to the mechanical part, disconnection of the line due to a short circuit, etc. When working with the simulator, the operator has the ability to perform the following actions: inspection of equipment, switching devices, installing grounding, hanging posters (figure 3).

Figure 3. Setting a Poster.

If the operator performs erroneous actions, depending on the degree of criticality of the error, visualization of the consequences is provided with the possibility of correcting the situation or terminating the test task (figure 4).

Figure 4. Message about the task.

In addition to modeling the power unit, we made it possible to simulate the work with electricity metering systems. The description of the electrical counters is given in [2]. The simulator allows you to perform the following actions: installing a meter, simulating actions that ensure the safety of work when installing and replacing an electric meter, connecting phases in a certain order using a device for determining phases [3]. It is possible to conduct meter inspections in the simulator in order to detect deformations, damage to the case, check electrical insulation, check the correct operation of the counting mechanism in the meter, confirm the absence of “self-propelled”, check the sensitivity threshold of the apparatus, etc. (figure 5).

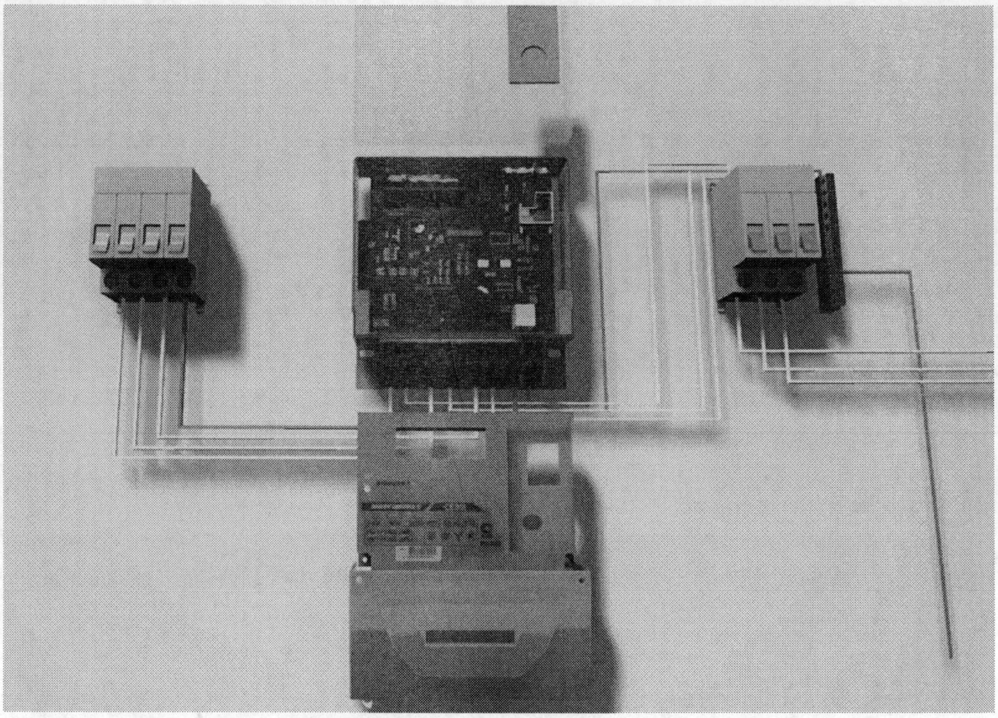

Figure 5. VR-model of the electric meter.

It is also envisaged to practice the inspection and repair of the meter in case of the following possible malfunctions: core wear during long-term operation, mechanical damage to the device and its parts due to vibration, corrosion of metal parts of the device.

3. Conclusion

When training specialists in the field of "Power Engineering and Electrical Engineering", an important aspect of training is the implementation of a practice-oriented approach. If it is impossible to gain experience on the real equipment of an energy facility due to a lack of qualifications and potential danger to the student, there is a need for the use of virtual reality simulators, which allows you to work out the actions of personnel during planned switching and emergency situations, test the student's knowledge and simulate various modes of energy equipment.

Acknowledgements

The research is funded by Russian Federation public contract № 0720-2020-0025 "Technique development and method analysis for ensuring power system object security and competitiveness based on the digital technologies".

References

[1] Kochergin D S and Strizhichenko A V 2019 Development of a Virtual substation simulator for training and retraining of specialists *Improving the Energy Efficiency of Electricity Production and Transmission: Student Conf. as Part of the RusHydro Group Corporate Championship Program According to WorldSkills Standards* (Volzhsky) pp 14-18

[2] Akimov E G 2011 *Three-phase Electronic Meters* (NCR Rukont) p 124

IOP Conf. Series: Materials Science and Engineering 1035 (2021) 012039 doi:10.1088/1757-899X/1035/1/012039

[3] Kochergin D S and Strizhichenko A V 2020 Development of the Virtual Counter simulator for training and retraining of specialists *Energy and digitalization - factors of stable development of a modern city: Int. Scientific and Practical Conf.* (Volzhsky) pp 46-48

Vertical axis wind power plant

P Yu Belyakov[1], Yu V Pisarevsky[2], A V Tikunov[2,3], A Yu Pisarevsky[2] and T E Chernykh[2]

[1]Department of Electric Power, International Institute of Computer Technology, 29 B Solnechnaya Str., Voronezh, 394026, Russia
[2]Department of Electromechanical Systems and Power Supply, Voronezh State Technical University, 14 Moskovskiy Avenue, Voronezh 394026, Russia

[3]Corresponding author: tikunov_a@list.ru

Abstract. The issue of developing technologies using alternative renewable energy sources today is a very urgent task both around the world and in Russia, which is explained by economic, technological, and, most importantly, environmental factors. The climatic and geographical features of our country create all the necessary conditions for the use of such areas as solar energy, small hydro energy, wind energy, etc. But the most promising in terms of energy potential is wind energy, as many regions of our country are located in areas with high average wind speeds. However, there are a number of problems that impede the development of wind energy in Russia, the most important of which is the almost complete absence of wind generators designed and manufactured in our country that could compete with foreign models. The product proposed in the article is an attempt to solve this problem through the use of a new design of a wind wheel and an electric generator, as well as the use of magnetic suspensions.

1. Introduction

Today, energy conservation issues around the world are becoming more and more important; the reasons for this are the constant increase in energy prices and the complexity of relations between countries, and, most importantly, the issue of depleting easily accessible deposits of organic fuels (oil, coal, gas). In this regard, energy conservation is currently very relevant, and alternative, in particular, renewable energy sources and technologies for their use are in the spotlight.

Unfortunately, the Russian energy sector is seriously lagging behind European countries in the introduction of alternative technologies, and if in a number of developed and developing countries wind energy, photovoltaic and hydroelectric power plants produce up to 50% of all electric energy, in Russia this indicator does not reach 1% [1].

In recent years, the issue of the necessity of developing alternative energy in Russia is being discussed at all levels, but, unfortunately, at this point in time, this direction is seriously hindered by the almost complete absence of competitive domestic units produced in series; all we have is either experimental constructions or ones produced in a single copy.

Talks about the need to develop this energy sector are regular at all levels, but government investments are negligible, and the private sector has little interest in projects with long payback periods.

Nevertheless, interest in wind power plants in our country is increasing due to the resumption of projects for the development of remote areas, and a number of enterprises of various profiles are developing in this direction [2-5]. One of such enterprises is the Baltic Machine-Tool Plant (St. Petersburg), by order of which, in 2010-12, the Department of Electromechanical Systems and Power Supply of Voronezh State Technical University, together with the Department of Electric Power of International Institute of Computer Technologies (Voronezh), put into practice the project of wind-driven electric plants (WDEP) with a capacity of 10 kW. The orientation to the indicated power was adopted based on the analysis of the world practice of power supply of autonomous objects based on wind turbines with capacities from 1 to 15 kW [6].

2. The concept of a wind-driven electric plant

A reliable and inexpensive wind-driven electric plant for use in areas with severe climatic conditions should have the following qualities: simple design of the aeromechanical part with a minimum number of moving elements (absence of regulation of angles of attack and pointing to the wind), bearing units with long service life and minimal maintenance, a simple rotation frequency control system depending on wind speed. These requirements are met by a WDEP with a vertical axis of rotation, which allows one to use unconventional solutions for the generator and bearing units. In particular, the main shaft of the WDEP carries a generator rotor, and to reduce friction losses and increase the service life of bearings, a magnetic suspension of the rotating part can be used, which receives most of its weight.

To verify the technical solutions adopted in terms of the magnetic suspension and the automatic control system, we made a working model (figure 1) with a power of 1.5 kW, developed at the calculated (most often repeated in the region of possible use) wind speed.

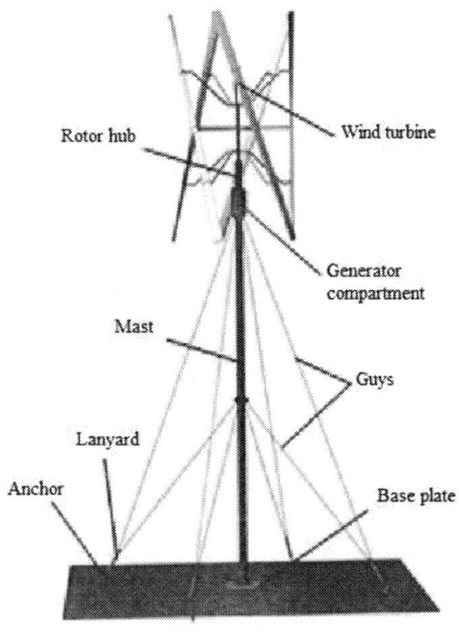

Figure 1. Wind-driven electric plant design.

To convert wind energy into mechanical energy, we used a specially developed and patented wind turbine with a vertical axis of rotation (figure1) [9].

In accordance with the requirements specification for the project, when developing a prototype, we used consumers that have low requirements for most of the quality indicators of electric energy, namely, accumulator batteries and illuminators based on incandescent lamps.

To reduce material costs at the first stage of the development of the WDEP, as an electric generator, we decided to use a 4MP 31.5 gear motor based on the AIR 100L6E2 asynchronous electric motor with a built-in electromagnetic brake with a capacitor excitation, the rotation frequency of which is practically unchanged and is a multiple of the 50Hz network frequency. The selected geared motor has a vertical design and a flange on the upper bearing shield, which makes it easy to mount it in the generator compartment of the WDEP (figure 2). A feature of the selected gear motor is the presence of an integrated mechanical brake, which makes it possible to stop and fix the wind wheel during a WDEP shutdown.

Figure 2. Mounting the gear motor.

In accordance with the theory of wind engines, the maximum efficiency of converting wind energy into the mechanical energy of a rotating shaft corresponds to a certain ratio of specific speed and wind speed. For WDEPs with known dimensions of a wind turbine, it is fair to talk about a certain ratio of rotational speed and wind speed. The main feature of the developed wind wheel is the impossibility of regulating the torque and frequency of rotation during fluctuations in wind speed and generator load using aerodynamic means. To regulate the rotational speed of the wind wheel, on which the output voltage of the generator also depends, we made it possible to adjust the current in the load, which allows us to influence the braking torque on the shaft of the wind turbine, which is created by the electric generator [8].

3. The electrical part of a prototype WDEP

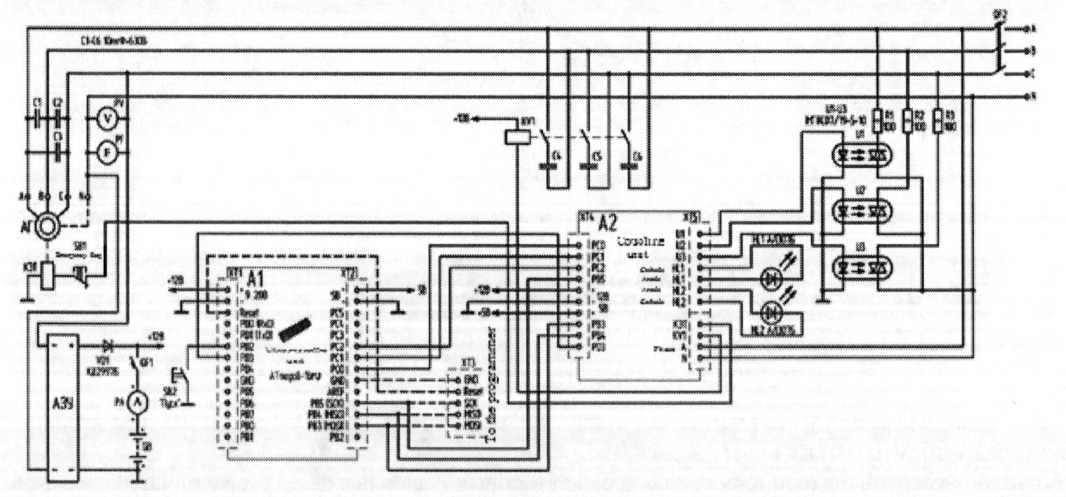

Figure 3. Scheme of the electrical part of the WDEP:
AG – the asynchronous generator; KET – the electromagnetic brake; C1 - C3, C4 - C6, respectively, the main and additional excitation capacitors; KV1 – the contactor for connecting additional capacitors; R1-R3 – the unit of ballast resistors; U1 - U3 – the opto-symistors; QF2 – the load switch; AC – the automatic charger; GB – the accumulator batteries.

To control the load current during the operation of the plant, we developed an automatic control system (ACS) that implements the principle of phase control of the load current distributed between the active payload, accumulator batteries, and ballast resistors [8]. Thus, the electric part of the prototype WDEP consists of two parts: the power circuit and ACS (figure 3).

ACS consists of a microcontroller unit - A1, and a unit for interfacing the microcontroller with the power elements of the plant - A2. For visual control of the current state of the WDEP, the circuit used control and measuring devices PV, PF PA, as well as LED indication H1, H2.

To ensure the temperature regime of equipment mounted in an electrical cabinet, and, in fact, to cool opto-symistors and ballast resistors, we provided radiators and fans, which are not shown on the electrical circuit diagram for simplification (figure 4).

The bench tests fully confirmed the operability of the WDEP automatic control system.

Figure 4. General view and internal structure of the WDEP control cabinet.

4. Magnetic levitation suspension

An important part of the work on creating a promising WDEP was the development of magnetic levitation units that take up most of the vertical load on the WDEP shaft bearings in order to reduce mechanical friction losses and increase durability.

The developed permanent magnet levitation suspension (PMS) consists of several magnetic modules of the same design, the number of which varies depending on the amount of force that needs to be compensated (figure 5). This constructive solution is explained by the fact that the overall dimensions of the module are limited by the inner diameter of the hub into which it is mounted, as well as by the diameter of the axis of the wind wheel. In addition, in many respects, the geometric dimensions of the modules are determined by the dimensions of the permanent magnets and also depend on the distance between the bearing supports [11].

The proposed design of the magnetic suspension uses three magnetic modules (figure 5 b), each of which consists of two magnetic platforms. The upper platform (position 2) is fixed on the tubular axis of the wind wheel 4, the lower platform 3 is on the inner surface of the hub housing, which transmits torque from the wind wheel to the generator.

Magnetic platforms, in turn, consist of steel shells 9 and 10, on which permanent magnets of the brand N342/1.32/876/955 are placed (position 11). For ease of installation, the magnets were made in the form of segments, which allows us to assemble them in the form of a ring inside the shell (figure 5 a), each prototype platform contains 24 magnets.

In the course of the research, we found that in the places of contact of two adjacent magnets the uniformity of the magnetic field is violated, which results in a pulsation of the magnetic flux inside the module, which inevitably leads to eddy currents and, consequently, to additional losses. In order to

IOP Conf. Series: Materials Science and Engineering 1035 (2021) 012040 doi:10.1088/1757-899X/1035/1/012040

reduce the effect, a steel magnetic strip 14 was added to the platform design over a ring of magnets. In addition, platform cover 15 acts as a damper when the magnetic flux of the magnetic module changes since it is made of an aluminum alloy with high electrical conductivity.

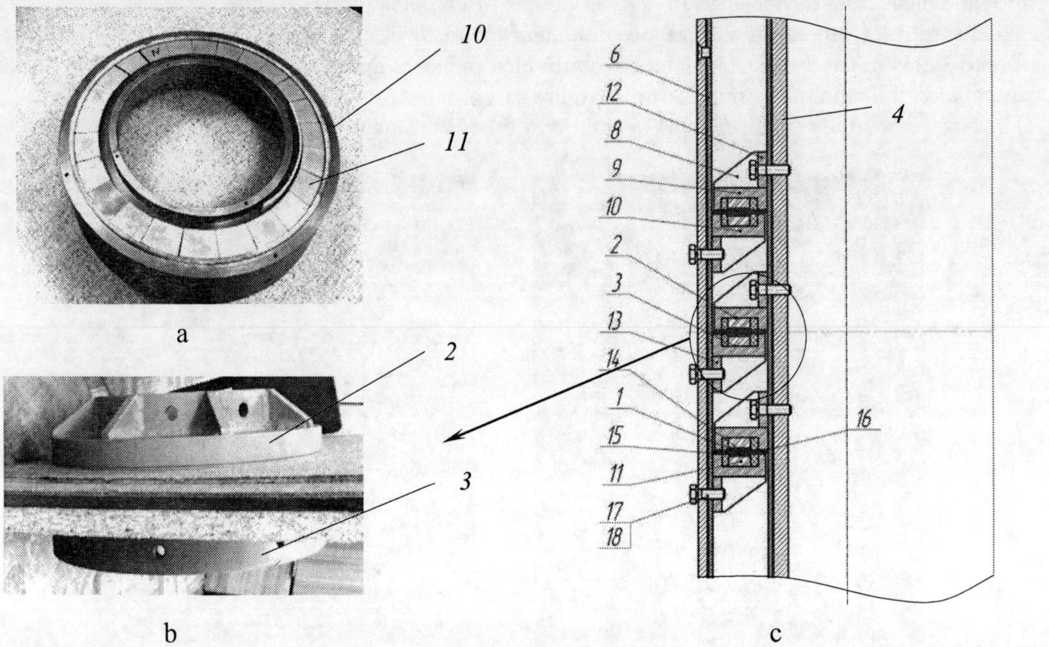

Figure 5. Magnetic suspension design:
a - lower platform with magnets installed during the assembly process; b – module assembled from unlike type platforms, prepared for transportation;
c - design of the WDEP with magnetic suspensions.

The proposed module design reliably protects the permanent magnets from various external influences, including moisture. The possibility of condensation in the inner cavity of the hub housing is not excluded. In this regard, all the steel parts of the module are painted, the final painting is carried out after assembling the elements of the magnetic module, and to ensure complete proofness, the shell of the module is filled with epoxy compound.

"To ensure reliable fastening of the magnetic modules, inner ring 12 and gussets 8 are welded to shell 9, and outer ring 13 and gussets 8 are welded to shell 10. Upper platform 2 is attached to inner fiberglass cylinder 5 with screws 17. Similarly, the lower platform is attached to outer fiberglass cylinder 5 with screws 17. In turn, inner fiberglass cylinder 5 is attached to the axis of the rotor 4 with bolts 21. Additional bond strength is provided by bolts 23. Outer fiberglass cylinder 6 is attached hub housing 7 with bolts 18. Bolts 25 provide additional bond strength.

In order to reduce the scattering magnetic fields and increase the interaction force between the magnetic supports, the construction of the supports includes fiberglass cylinders that create a non-magnetic gap between the hub housing and the magnetic module.

A distinctive feature of the proposed PNMS is the possibility of using various combinations of magnetic modules, some of which work on attraction, and some on repulsion, which makes it possible to reduce the total mass of applied permanent magnets "[11].

5. Special electric generator

As the studies showed, the realization of all the advantages of a vertical-axis wind turbine mentioned above can be fully achieved by using a special electric generator integrated into the design of the plant itself. An analysis of the available information showed that a synchronous generator with excitation from permanent magnets having a disk design can be most effective [7] for the realization of the tasks set.

In the course of the research, we developed and subsequently patented [10] a design scheme of a disk synchronous generator. Figure 6 shows a sketch of the layout of the developed generator used in the research.

The rotor design of the developed generator consists of two magnetic conductors mounted on a shaft. Magnetic conductors are in the form of disks, permanent magnets are assembled on one of them round the edge. As the research results showed, the cylindrical shape of the magnets is the most optimal, but it is possible to use the shape of the segments, as is the case with a magnetic suspension. The generator shaft is made in such a way that it can be rigidly connected to the wind wheel shaft, which makes it possible to exclude bearing assemblies from the generator design.

A distinctive feature of the proposed generator is that there is no steel magnetic conductor on the stator, it is a monolithic winding, which is made by pouring coils with a compound.

During the research we found that for this generator we can use different types of windings, the choice of winding depends only on the required parameters of the machine, and the best performance is given by the use of a single-layer coil or two-layer loop windings [12].

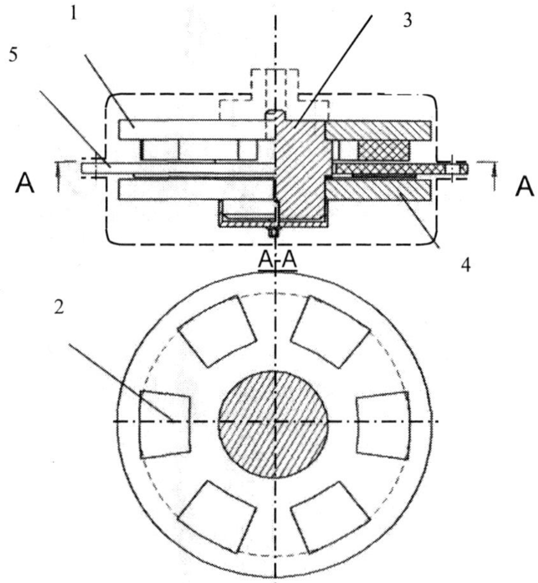

Figure 6. Design of the synchronous disk generator:
1 - the upper magnetic conductor of the rotor; 2 - a permanent magnet; 3 - shaft;
4 - lower magnetic conductor of the rotor; 5 - stator winding.

Thus, the proposed generator in comparison with machines of a similar purpose has a lower mass (due to integration into the design of the WDEP), a simpler structural scheme, and therefore it is easier to service and repair, and most importantly, it has a low cost. In addition, an important feature of the adopted concept is the modular principle of a magnetic suspension and generator, which allows one to build a range of WDEPs for a number of capacities by using a different number of generator blocks.

6. Conclusion

Thus, the complex of research and development work allowed us to formulate a promising concept of a vertical-axis WDEP for a power range from 1 to 15 kW. We tested a number of solutions and implemented them in the form of existing prototypes, for example, manufactured at the Baltic Machine-

Tool Plant (St. Petersburg) (figure 7). Unfortunately, it was not possible to carry out full tests and refinement of the product based on their results, since the work was suspended due to financing problems.

Figure 7. Prototype WDEP.

References

[1] 2011 La production d'électricité d'origine renouvelable dans le monde. Collection chiffres et statistiques. Treizième inventaire. EDF – Observ'ER http://www.observ-er.org
[2] Official website of SRC "Vertical" http://www.src-vertical.com/information/scientist
[3] Official site of CoLtd "Electroveter" http://electroveter.ru.
[4] Official website "Energieforum.ru" http://www.energieforum.ru
[5] Official website of the company ATMEL www.atmel.com
[6] All small wind turbines; all the world's small wind turbines in one overview http://www.allsmallwindturbines.com
[7] Chernykh T E, Pavlov A S, Belyakov P Yu, Pisarevsky Yu V and Tikunov A V 2012 Study of an end synchronous direct-drive generator for vertical-axis wind turbines *Electrical Technical Complexes and Control Systems* **2** 11-16
[8] Belyakov P Yu, Panov R M, Pavlov A S and Tikunov A V 2012 Electrical part and control system of vertical-axis wind-driven electric plant *Electrical Technical Complexes and Control Systems* **1** 69-77
[9] Belyakov P Yu and Tikunov A V 2012 Wind turbine with the vertical axis of rotation, Patent 2470181 of the Russian Federation, IPC F03D3/00 No 2010136214/06
[10] Pisarevsky Yu V, Belyakov P Yu, Pisarevskii A Yu, Tikunov A V and Chernykh T E 2011 Generator, Patent 111365 of the Russian Federation, IPC H02Л21/24 No 2011133630/07
[11] Pisarevsky Yu V, Belyakov P Yu, Pisarevskii A Yu and Tikunov VA 2014 Development of a magnetic levitation bearing units for use in vertical-axis wind-driven electric plants of low power *Electrical Technical Complexes and Control Systems* **3** 53-56
[12] Chernykh T E, Belozorov S A and Tikunov A V 2018 The choice of disk windings of a synchronous generator with excitation from permanent magnets *Bulletin of Voronezh State Technical University* **14(5)** 84-94

Methodological problems of replacing low-voltage AC networks with DC networks

L N Titova[1,3] and V A Sergeev[2]

[1]Department of Electromechanical Systems and Power Supply, Voronezh State Technical University, 14 Moscovskiy Avenue, Voronezh, 394026, Russia
[2]Joint-Stock Company "MEL", 88 Peshe-Streletskaya Str., Voronezh, 394038, Russia

[3]Corresponding author: tln64@mail.ru

Abstract. The paper discusses the prospects for replacing low-voltage AC power grids of industrial frequency with DC power grids in order to introduce energy-saving equipment and technologies. The modern achievements in the field of energy-saving electric drives are analyzed in detail, which make it possible to introduce constant voltage electric networks attractive for investments. The generalized analysis can be performed using the synergetic criterion of the technical and economic efficiency of innovation. To use such a criterion, all indicators to be analyzed must be presented in dimensionless form. We propose to consider separately the technical and operational groups of the generalized technical characteristics of AC and DC electric drives. The selected subset of groups of parameters of electric drives is justified from the point of view of regulatory and technical documents currently in force. This methodology seems to be more objective than the existing industry methodology for assessing product quality based on functional analogues of products being developed. With a synergetic approach, any characteristic is invariant relative to the others, which confirms the objectivity and reliability of the comparison method. The system of indicators is necessary and sufficient, both for technical assessment and for the creation of new designs and the development of directions for the design of electric drives. In conclusion of this work, we give examples of the implementation of such an approach by comparing the achieved indicators of quality and operational reliability of electromechanical energy converters produced by domestic enterprises.

1. Introduction

In modern conditions, the implementation of commercially successful energy-saving measures is entirely dependent on the depth of elaboration of design decisions and the assessment of existing risks. In addition to the topics already discussed in previous works [1–4], we propose for discussion in the professional community a new methodological approach to the use of direct current power electric drives for general industrial mechanisms instead of alternating current electric drives.

For further analysis, let us consider the general requirements for innovative design solutions.

Firstly, design decisions should be aimed at improving the electrical safety of consumers' electrical units. This is achieved not only by reducing the voltage of the unit to a safe ultra-low level (not more than 48 V) but also by undertaking special measures of protection against electric shock. The list of such measures is well known [5] and does not need additional description and justification for use.

Secondly, design decisions should be energy saving. The possibilities of energy-saving technologies are realized both when reducing losses in the transmission of electric energy and when reducing the power consumption of electric power receivers.

Thirdly, design decisions should ensure the maintenance of the necessary and sufficient level of unification of wiring accessories and electric energy metering devices.

Fourthly, design decisions should increase the reliability of power supply, related both to protection against probable erroneous actions of the consumers, and to protection of electrical equipment from switching and atmospheric interference. For example, the quenching of a constant voltage arc requires special measures, the illiterate use of which can negate all the benefits of energy saving.

Finally, design decisions in the field of DC voltage supply should have the lowest possible implementation cost. An example of cost reduction of, for instance, electrical work is the reduction of the consumption of conductive material while reducing the length of electrical networks, as well as reducing the cost of insulation (DC mains can be performed in the form of bare buses) and others. The absence of special protective grounding conductors PE in the extra-low voltage circuits is also a factor in reducing the cost of electrical work.

A feasibility study for the use of new energy-saving design solutions can be performed according to the criterion of the total cost of ownership of electrical networks and electrical installations. [4].

2. Problem statement

In this paper, we consider the issues of the reasonable replacement of a controlled AC electric drive based on asynchronous electric motors with a squirrel-cage rotor (AC electric drive) by a direct current electric drive (DC electric drive) based on non-contact direct current electric motors (NCDCM) controlled from an industrial frequency AC network through controlled three-phase rectifiers (CR).

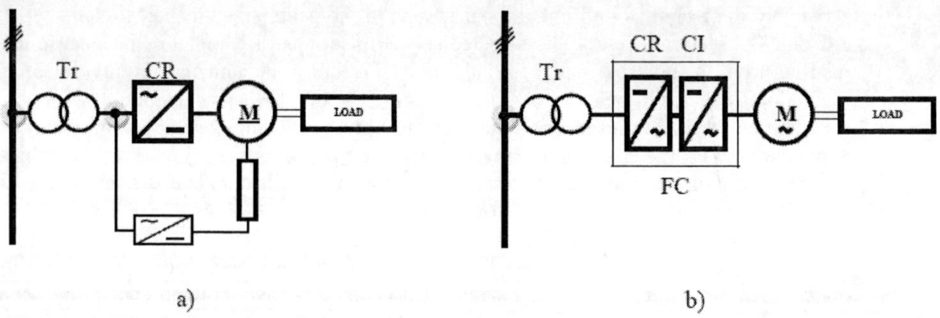

Figure 1. Block diagrams of electric drives: a) direct current with controlled rectifier; b) variable frequency AC drive with pulse-width regulator (PWR); Tr – power transformer; M – electric motor; CR – controlled rectifier; CI – current inverter; FC – frequency converter; CI - current inverter; FC - frequency converter.

From the wide variety of circuit solutions for converters and electric drive control systems, we selected two most used circuits for analysis:

DC electric drive based on NCDCM with excitation from rotor permanent magnets and CR based on a three-phase thyristor voltage regulator (TVR) without filter capacitor with parametric feedback by EMF (in case of controlling the rotation frequency of the NCDCM);

AC electric drive based on an asynchronous squirrel-cage rotor motor controlled by a frequency converter with the double conversion of the current type (vector control with a PWR current inverter and a CR with a low-pass filter).

Comparison indicators that determine the choice of a particular type of an electric drive, as it was previously indicated in [6], are divided into two large groups of generalized characteristics: technical and operational.

IOP Conf. Series: Materials Science and Engineering 1035 (2021) 012041 doi:10.1088/1757-899X/1035/1/012041

Since 2012, a standard in the field of energy efficiency has been in force for asynchronous electric motors in Russia [7], which was developed in Federal Law No. 261-FZ from 11.23.2009 "On Energy Efficiency ...". A similar document regarding electric motors and DC drives has not yet been created.

We classified the following as the generalized technical characteristics:

1.1 The electromagnetic power of the electric motor P_{EM};

1.2 The efficiency coefficient (η - for a DC electric drive) and (the product $\eta' = \eta \cos\varphi$ - for an AC electric drive);

1.3 The maximum developed (long-term permissible) electromagnetic moment of M_{MAX} at nominal load and supply voltage;

1.4 The starting moment, which determines the short-term overload capacity of M_S;

1.5 The regulation range of rotation frequency of the output end of the shaft D;

1.6 The rigidity of the mechanical characteristics γ;

1.7 The overload capacity in short-term operation K_m;

1.8 The electromechanical time constant T_{EM};

1.9. The moment of inertia of the rotating parts, reduced to the moment of inertia of the rotor J_{Σ};

1.10 The specific mass per unit of power (or the maximum developed electromagnetic moment).

We attributed the following to the operational characteristics:

2.1 The assigned resource (running hours) in the nominal operating mode according to GOST IEC 60034-I;

2.2 The service life, including the shelf life, as well as the maintainability indicator in accordance with GOST 27.410;

2.3 The resistance to the impacts of multiple mechanical (GOST 17516.1) and climatic (GOST 15543.1) external influencing factors during operation;

2.4 The performance according to the degree of protection of the electric motor outer shell against the penetration of water and solids according to GOST IEC 60034-5;

2.5 The execution according to the installation method in accordance with GOST 2479;

2.6 The level of own vibrations and noise in accordance with GOST 11929 and GOST 20815;

2.7 The heat resistance of electrical insulation in accordance with GOST 8865;

2.8 The indicators of interproject unification of the main units (assembly units), determined in accordance with GOST 23945;

2.9 The indicators of electromagnetic compatibility of AC and DC electric drives with consumer technical equipment in accordance with GOST 32144;

2.10 The cost of manufacturing a unified product, thousand of rubles/kW [8].

3. Justification of the method for solving the problem

The methodology for solving this problem depends entirely on the accuracy of the presentation of the source data.

The operational characteristics of the products are set in the relevant state standards. As the achieved indicators for both types of electric drives, one can use indicators included in the maps of the technical level of products compiled by the engineering and technical services of Russian electric machine-building enterprises at the stages of development work on the creation of innovative products. The procedure for compiling and confirming the indicators included in these documents is regulated by the requirements of GOST 2.116 and is available to any user of the state standard information system (SSIS) of Rosstandart.

An objective review requires the development of uniform criteria for a variety of electric drives. We examined the operational characteristics of electric drives with a power of 1 to 100 kW, powered by three-phase electric networks with a voltage of 380/220 V and a frequency of 50 Hz.

The situation is different with the technical characteristics of the products being compared. There is no point in discussing the generally diverse designs of electric machines without being tied to the requirement indicators, as well as operating modes and conditions of safe operation.

IOP Conf. Series: Materials Science and Engineering 1035 (2021) 012041 doi:10.1088/1757-899X/1035/1/012041

We propose the following conditions for evaluating the generalized technical characteristics of electric drives:

1. An analysis should be carried out for electric drives of the same purpose, expressed in the identity of the operating mode. This article analyzes electric drives of continuous operation (S1 according to GOST IEC 60034-1).

2. When replacing one type of electric drive with another, the parameters of their power source (transformer substation), as well as the length and cross-section of the power transmission lines should not be changed.

3. Design decisions related to the replacement of multi-engine electric drives with a single-engine electric drive or vice versa are not considered.

4. The operating conditions in terms of exposure to factors harmful or dangerous to human health are considered to be constantly valid regardless of the type of electric drive selected.

5. Reliability indices when replacing one type of electric drive with another should not deteriorate. For this reason, DC collector electric motors of all types and designs, including electric motors with electromagnetic excitation, the speed control range of which can be increased by weakening the field of the inductor, are excluded from consideration.

6. Mechanical losses, including friction losses in the bearings, as well as the outer surface of the rotor against the air, will be considered equal for both types of electric drives.

7. The power supply is considered to have an infinitely large power. The voltage of the power source, including CR, which is part of both devices, is assumed to be constant independent on the operating mode of both electric motors and valves. The valves are supposed to be ideal, allowing a switching frequency lying above the industrial frequency range.

8. The highest harmonics of the current and voltage arising in the power supply system are neglected for reasons not related to the operation of CR and CI.

4. Theoretical research results, practical value

The list of generalized technical indicators of both types of the electric drives should be expressed in relative units. Only in this case, it will be possible to operate the list for the analysis of electric drives, regardless of power, rotation frequency, moment, and other parameters.

By means of analysis, let us choose the mechanical characteristic *M=f(s)*, for which we extend the concept of "sliding" to a DC electric drive. Despite the different physical interpretations of the results obtained using this concept for AM and NCDCM, the generality of the mathematical expressions allows us to obtain reliable quantitative results that are subject to further analysis.

First, we obtain analytical expressions for the mechanical characteristics of drives of both types. Let us consider the following basic models:

a) an NCDCM of "classical" design with excitation from permanent magnets;

b) a symmetric three-phase AM with a squirrel-cage rotor, powered by a three-phase network;

c) a symmetric three-phase AM, powered by a single-phase alternating current network and using a reactive element for starting;

d) an asymmetric two-phase AM powered by a single-phase AC network.

Figure 2 shows the mechanical characteristics of the motors of the basic models, constructed in relative units at the same magnitude of the starting electromagnetic moment m_S.

The expression of the mechanical characteristic for the NCDCM has the simplest form

$$M^{NCDCM} = m_S - \left(m_S - 1\right)\omega_* = m_S - m_S(1-s) = m_S S, \quad (1)$$

where $s_n = \frac{\omega_0 - \omega_H}{\omega_0} = 1 - \frac{m_S - 1}{m_S} = \frac{1}{m_S}$ - the nominal slide of the NCDCM in relative units;

ω_* - the rotor angular velocity (relative to the nominal angular velocity ω_n).

To obtain an expression of the mechanical characteristics of AM in a dimensionless form, we use the well-known Kloss formula:

$$\frac{M}{M_{\max}} = 2\frac{1+\rho s_{kp}}{s/s_{kp} + s_{kp}/s + 2\rho s_{kp}}, \qquad (2)$$

where s_{kp} – the critical slide, r.u.; $\rho = R_1 / C_1 R_2'$ - the constant coefficient depending on motor parameters.

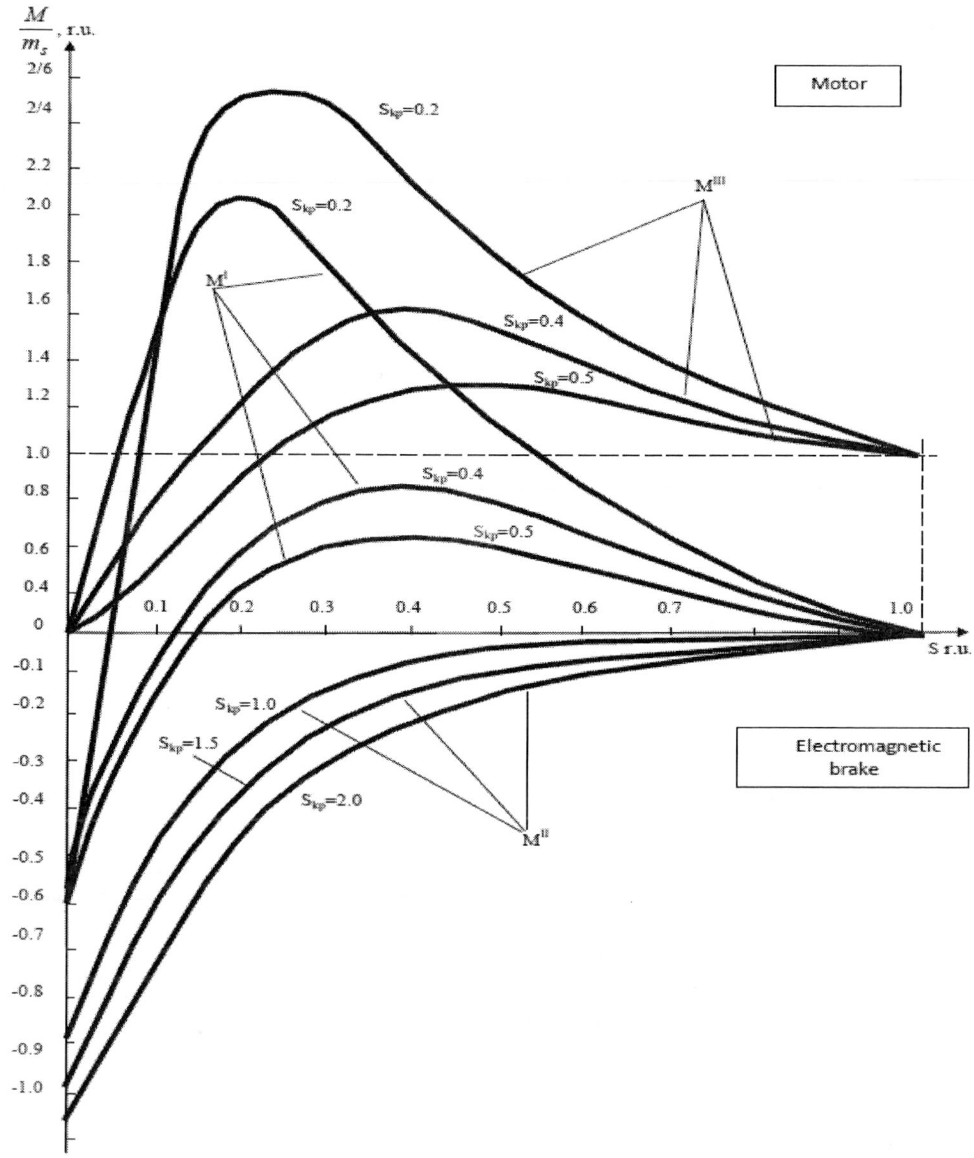

Figure 2. Mechanical characteristics of AM - three-phase (M^{III}), two-phase (M^{II}) and single-phase (M^{I}), at different critical slide values.

Additionally, let us suppose that for $s_{kp} \leq 0.3$, the quantities ρ and s_{kp} are mutually independent and the term ρs_{kp} in (2) can be neglected. Then, for a symmetric three-phase AM we get:

$$M_*^{III} = \frac{m_s (1 + s_{kp}^2) \cdot s}{s^2 + s_{kp}^2}. \tag{3}$$

As previously noted, the value C_1, which determines the relative voltage drop in the winding of the AM stator for low-power motors (P_2 <1 kW), does not exceed 1.2, and therefore, the product $(2\rho \cdot s_{kp})$ will be less than 0.5 for $s_{kp} \leq 0.4$.

A single-phase AM has, in addition to the moment from the direct sequence of the magnetic field M_d, the moment of the reverse rotating field M_r, determined when the rotating field slides backward s_r: $s_r = 2 - s_d$, where: s_d is the sliding of the direct field.

A two-phase AM differs from a single-phase one in that the moment from the reverse field at 0<s<1, which corresponds to the condition for the absence of self-running. In this case, the resulting moment will obviously be negative in this region, the term $2\rho \cdot s_{kp}$ will be more than one, and it cannot be neglected in (2) since motors of this type have critical sliding values $s_{kp} \geq 1$.

For further analysis, we will use formulas (1), (3).

Figure 2 shows the mechanical characteristics of AM in all three basic models in the region $0 < s < 1$.

We now turn to the consideration of generalized technical characteristics that are directly related to the mechanical characteristics of NCDCM and AM:

4.1. Rotation frequency control range

Unlike asynchronous motors with a nominal rotation frequency that is a multiple of the synchronous frequency of 3000 rpm, an NCDCM can be designed with a nominal rotation frequency in the range from 100 to 10000 rpm for both general and special applications. As a rule, AM with the number of pole pairs $2p \geq 10$ are not commercially available. According to [9], an NCDCM of modern series can have the number of pole pairs $2p \geq 20$ because modern magnets made of iron alloys with rare-earth elements have noticeably decreased in size compared with previously widely used iron-nickel-cobalt magnets made of Alnico alloys.

Depending on the size, DCM can have an area of "safe" work 2.5-3 times more than AM. The limitation of the control range upward from the nominal rotation speed is determined by the mechanical strength of the rotor, and downward from the nominal rotation speed by poor cooling conditions at low and ultra-low ("creeping") rotation speeds.

The limitation of the range of regulation of AM is associated with a decrease in the maximum moment of an asynchronous motor inversely proportional to the square of the supply voltage and a decrease in overload capacity and multiplicity of the starting moment m_S.

Figure 3 shows the ranges of regulation of DCM in comparison with three-phase AM in relative units.

4.2. The rigidity of a mechanical characteristic

By this we mean the following expression:

$$\gamma = \frac{M_{\max} - M_o}{s_{kp}}, \tag{4}$$

where M_{max} – the maximum value of the electromagnetic moment developed by the motor; M_o – the synchronous operation moment (s=0).

For DCM we will have: $M_{max} = m_S$; $M_O = 0$; $s_{kp} = 1$, and γ_{NCDCM}, equal to m_S, is a constant value independent of other motor parameters.

For AM, it is necessary to take the derivatives in (3) with respect to s and equating them to zero, determine the values M_{max}. It is easy to show that the maximum value of the moment in all cases is achieved at $s=s_{kp}$.

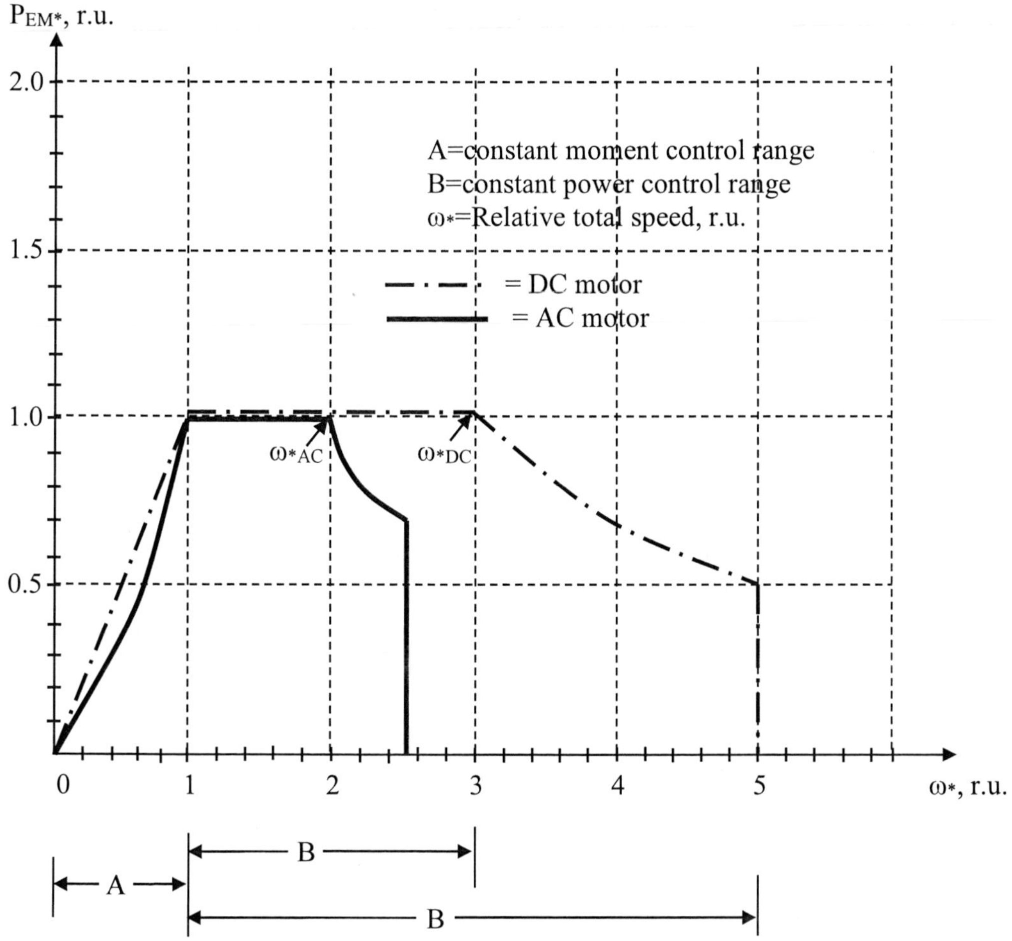

Figure 3. Comparison of the areas of "safe" operation of electric motors of direct (DC) and alternating current (AC).

4.3. Overload capacity
By overload capacity we mean the ratio:

$$K_M = \frac{M_{\max}}{M_n}, \qquad (5)$$

- for NCDCM powered by CR on the basis of TVR $K_M = m_S$,
- for AM controlled by a PWR with a current inverter, the overload capacity should obviously be estimated by the ratio of the maximum moment to the starting moment.

Overload capacity depends not only on the parameters of the motor but, to a large extent, on the characteristics of transistors included in the FC or thyristors included in the TVR. The wider the range of rotational speeds in which the motor should provide maximum power, the better the NCDCM or AM should be adapted to processes that require a constant moment in the entire speed range.

4.4. The electromechanical time constant

It is the most important characteristic that determines the dynamic properties of an electric drive [10]. In the system of units adopted in this article for both types of drives we have:

$$T_{EM} = \frac{\omega_n}{\varepsilon_n (m_s - 1)}, \quad (6)$$

where $\varepsilon_n = \frac{M_n}{J_n}$, 1/s^2- nominal angular acceleration of the NCDCM or AM with the moment of inertia *J*, including elements of the reducer and rotating parts of the electric machine.

Obviously, at the same angular velocities, the outer diameter of the NCDCM rotor is significantly smaller than the diameter of the AM rotor, which means that the electromechanical time constant of the NCDCM will be much less than the time constant of the AM.

It is also clear that for motors of the same nominal power, the value of the multiplicity of the starting moment of the NCDCM will always be higher than that of the AM of the same overall dimensions. In the general case, the losses in the anchor at the same values of the nominal slide will be greater in the NCDCM.

4.5. Rotor moment of inertia J_a.

DC motors have significantly lower rotational axis heights and the mass of both the rotor and the machine as a whole than asynchronous motors and, therefore, have a lower rotor inertia J_a, which is a significant advantage for the use of automatic control systems in electric drives, such as servo systems of all types, as well as in reversible drives, operating in both motor and brake modes.

4.6. Specific mass per unit of power (or the maximum developed electromagnetic moment)

It also gives the advantage of NCDCM over AM in electric drives of the same functional purpose. It is difficult to formalize this advantage in the form of any kind of dependence because such calculations are strongly influenced by design decisions on the body parts of both types of motors.

In addition, it should be noted that at present, thanks to the progress of modern materials science, a significant drawback of all NCDCM has been overcome - the high cost, associated primarily with the cost of magnetic materials. Also, the service life of the machine increases when using reliable radio-electronic elements and heat-resistant insulation.

For operation in electrical installations located in areas of humid tropical climates, explosive media, as well as rooms with increased fire or chemical hazards, asynchronous motors are still indispensable.

Asynchronous electric drive has proven its effectiveness in those sectors of the industry that are characterized by aggressive environmental conditions, a high degree of contamination with a variety of chemically aggressive products, as well as in conditions of high humidity.

5. Results and discussion

The limits of the article do not allow us to dwell on the current system of state standards in Russia in the field of ensuring the operational reliability of electric motors and electric drives. We only note that standardization in the field of electric machines (class 33 of the All-Russian classifier of industrial products) reflects the requirements for a motor as an object of design in much more detail than the standards in the field of electric drive. To date, there are more than forty current state standards aimed at regulating individual requirements for electric machines, methods for their design, manufacture, and quality assurance during the life cycle. In the class of electric drives (class 42 OKP), there are only three standards of the type of general technical conditions, two of which were developed in Soviet times and are greatly outdated. This, ultimately, designates the methodology for determining operational characteristics, which is thus based on the characteristics of the electric motors that make up the electric drives.

Figure 4 presents a diagram of the so-called synergetic criterion obtained on the basis of relations (1) - (6) described in the article. The diagram complies with the criteria specified in [11].

The methodology for constructing the diagram is as follows:

1. All generalized technical characteristics are expressed in relative units and plotted as segments from the center of the diagram to the periphery on the corresponding radial lines.

2. The total length of the segment is the same for all axes of the diagram and the length of each axis is taken as a unit. The same value determines the current technical level of development.

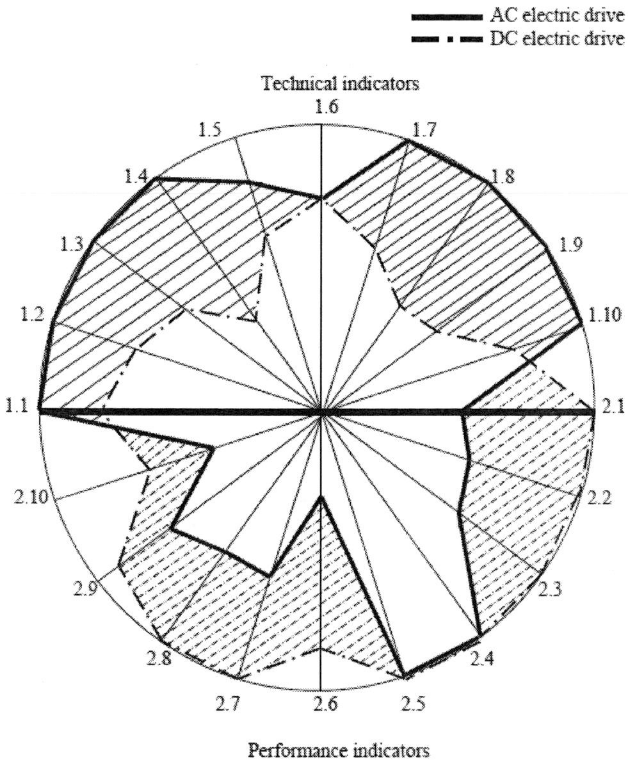

Figure 4. Synergetic criteria for selecting the type of electric drive.

3. If in various sources (catalogs of firms - manufacturers of electromechanical equipment, scientific papers, advertising materials) there were discrepancies in the definition of parameters, they were interpreted in favor of the highest indicator achieved.

4. It was assumed that all 20 indicators are equivalent, which is illustrated by the same sector area of the chart.

5. Shaded areas indicate the advantages of one type of electric drive over another in a number of ways. The diagram shown in figure 4 is built for electric motors of two manufacturers:

- Vladimir Electromotor Plant, part of the "RUSELPROM" Electrotechnical Holding. The plant is an exclusive developer and supplier of induction motors with a height of the axis of rotation of 132 and 180 mm in the power range from 0.75 to 315 kW;
- Closed Joint-Stock Company "MEL" is an exclusive developer and manufacturer of DC electric motors and electric drives based on them in the power range from 0.1 W to 22 kW.

Table 1 presents the main electromechanical parameters of two electric motors; generalized technical and operational characteristics are presented in figure 4.

Table 1. Basic electromechanical parameters of two electric motors.

Parameter name (characteristics)	Norm for type	
	4A180M2U3	DBU250-22000-R09-D11
Nominal power, kW	22	22
Nominal rotation speed, rpm	3000	3000
Supply voltage, V	220/380	440
Current consumption, A	41	30
Mode of operation	S1	S1
Starting moment, Nm	9.5	19.2
Overall dimensions (diameter x length), mm	360×450	250×280
Weight, kg	151	8.5
The source of information	www.vemp.ru	www.mel.vrn.ru
Manufacturing plant	VEMZ OJSC Vladimir	CJSC “MEL” Voronezh

6. Conclusion

The methodological approaches presented in the article and the results of their application for constructing a simple and clear graphic model (synergetic criterion) allow us to draw the following conclusions:

1. A DC electric drive in the system for automatically controlling the rotation speed or coordinates of the actuator has a significant advantage in almost all generalized technical characteristics over an AC electric drive, regardless of the power, rotation frequency, operating mode, or other parameters of both electric machine units. A DC electric drive has an advantage over an AC electric drive in nine out of ten characteristics and equality in one.

2. The control range of the rotational speed of a DC electric drive is 3-4 times higher than the similar characteristic of an AC electric drive, which opens up great opportunities for constructing automatic control systems with various properties.

3. The electromechanical time constant, which determines the losses in transient conditions, in a DC electric drive is 2.5-3 times less than the same value for an AC electric drive due to the significantly smaller overall dimensions of the rotating parts.

4. The power-to-weight ratio of the drive is 7-10 times higher for a DC drive. Such a solid “gain” is due to the improvement of the properties of magnetic materials and magnetic systems in general, achieved in recent years.

5. The performance of an AC electric drive for eight characteristics has advantages over a DC electric drive, including the technical resource, the cost of maintenance and maintainability, as well as the level of vibration and noise [12].

6. The best indicators of the interproject unification of AC electric drives are explained by a large number of existing series of electric motors and electric drives, therefore, the development and production of a new AC drive are less expensive [8].

7. There is a tendency towards the gradual reduction in the cost of manufacturing DC electric drives. Reducing the range of permanent magnets with an increase in their output is the main factor in reducing the cost of DC electric drives. In our opinion, while maintaining this trend, AC electric drives will be completely supplanted from the market of high-tech controlled electric drives in the foreseeable future.

IOP Conf. Series: Materials Science and Engineering 1035 (2021) 012041 doi:10.1088/1757-899X/1035/1/012041

References

[1] Sergeev V A, Krysanov V N and Titova L N 2017 Prospects for energy saving in the organization of the supply of residential complexes with a constant current *Energy Security and Energy Efficiency* **4** 22-28

[2] Pisarevskiy A Yu, Tikunov A V and Titova L N 2019 DC in liquefied natural gas transportation systems *Proc. of the Int. Scientific-Tech. Conf. Dedicated to the Memory of Doctor of Technical Sciences, Professor A I Zaitsev: New Technologies in Research, Design, Management, Production* (Voronezh) pp 302-305

[3] Pisarevskiy A Yu, Tikunov A V and Titova L N 2017 Ways to improve the efficiency of the electricity systems of airports when organizing the power supply with direct current *Proc. of the Int. Scientific-Tech. Conf.: New Technologies in Research, Design, Management, Production* (Voronezh) pp 114-117

[4] Astakhov R S and Titova L N 2014 Total cost of ownership of a street lamp as selection criteria of street lighting lamp *Proc. of All-Russian Student Scientific-Tech. Conf.: Applied Problems of Electromechanics, Energy, Electronics. Engineering Ideas of the XXI Century* (Voronezh) pp 104-106

[5] 2007 *Rules for Arranging Electrical Installations* chapters 1.1, 1.2, 2.4, 2.5, 6.1-6.6-7 (Moscow: NTS ENAS) p 552

[6] Volkov V D, Sergeev V A and Popov S P 2016 Algorithmizing the decision-making process for the energy management of industrial buildings *Actual Problems of Science of the XXI Century* (Voronezh) pp 105-112

[7] GOST R 54413-2011 Rotating electrical machines. part 30. efficiency classes of single-speed three-phase asynchronous motors with a squirrel-cage rotor

[8] Titova1 L N, Pertsev Yu A, Samodurova S A and Nuhin P O 2020 *IV Int. Scientific and Technical Conf.: Energy Systems* vol 791 (Belgorod: IOP Publishing Ltd)

[9] Hobbs D P 2003 *Lean Manufacturing Implementation: A Complete Execution Manual for any size Manufacture* (Fort Lauderdale, FL: J. Ross Publishing) p 264

[10] Shchepilov I N, Mamontov D A and Sergeev V A 2013 Energy-saving complex based on an asynchronous valve cascade *Proc. of the All-Russian Scientific and Technical. Conf.: Applied Tasks of Electromechanics, Energy, Electronics* (Voronezh: VSTU) pp 137-138

[11] Sergeev V A and Repin A A 2014 Power supply systems for residential and public buildings with constant voltage *Proc. of the Regional Conf.: Innovative Technologies and Materials in Airdrome and Road Construction* (Voronezh) pp 210-212

[12] Shchepilov I N, Pitolin V M and Sergeev V A 2011 The degree of detail of the initial information when building a mathematical model for controlling the electrical complex in emergency conditions *Proc. of the 16th Int. Scientific and Technical Center and the Russian Scientific School of MuiS: Systemic Problems of Reliability, Quality, Information, Telecommunication and Electronic Technologies in Innovative Projects (Innovatika-2011)* (Moscow: Energoatomizdat) pp 140-145

Methods for dealing with icing of power line wires

V G Kul'kov[1], V N Kuryanov and R A Fokin

Department of Fundamental Disciplines, Volzhsky branch of National Research University «Moscow Power Engineering Institute», 69 Lenina Avenue, Volzhsky 404110, Russia

[1]Corresponding author: vikulkov@yandex.ru

Abstract. This article describes the types of ice and snow deposits on metal wires of overhead power lines. It has a harmful effect on the functional state of power lines. Under the influence of the force of resistance to air flows in the wind and the gravity of the wires with an increased mass, accidents are possible. These include breaks, wire interconnections, and short circuits in the line. Here we give a brief overview of existing methods for preventing the formation of ice on the line wires and fighting icing by removing the ice shell. This article describes a combined method for breaking the ice coating on the wires. For this purpose, we propose a combination of two methods. One of them is to load the wires with high-frequency current. Due to the skin effect, the inner surface of the ice heats up, and a thin layer melts. This significantly reduces the adhesion of the ice to the metal. Simultaneously, low-frequency current of the order of one Hertz flows through the wires. This current causes the wire to resonance oscillations, which destroy the ice covering. The article provides an estimate of the duration of the high-frequency current exposure time, its value, and the frequency of the second current.

1. Introduction

The icing of power line wires is a very harmful phenomenon [1]. The presence of an ice shell on a metal wire leads to an increase in mass, a decrease in its strength and service life. It also changes the drag of the wind. All these can lead to broken wires and damaged supports. Effective control of wire icing is a relevant task.

The process of the appearance of the ice shell is associated with such atmospheric influences as sleet, frost, ice, and their combination [2]. Wet snow has a high degree of adhesion to metal, has a mass density of 0.12–0.3 g/cm^3. Frost is a white, crystalline, opaque precipitate, with a density of 0.02-0.3 g/cm^3. It is formed at a temperature of -10 – -40 °C. Ice is a solid, continuous translucent sediment with a density of 0.6-0.9 g/cm^3, often formed during rains and fogs. The thickness of ice on the wires and lightning protection cables can reach 60 – 70 mm, significantly weighing them down. The essence of the phenomenon of icing is the deposition and subsequent freezing of water droplets on the surface of the wires or the sticking of wet snow at a wind speed not exceeding 10 – 20 m/s. Icing occurs on the windward side of the wires if the wind is perpendicular to the transmission line, and evenly along the entire length if the wind is directed parallel to the wire. In the latter case, the ice is less dangerous for the line, since it has a more porous structure and lower density. Usually, the ice layer is not uniform in thickness. The SAG of icy wires can increase by several meters.

In the process of mathematical modeling of ice formation, it is represented as a hollow cylinder with the same wall thickness on all sides. AC-185/43 wire with a diameter of 19.6 mm and a kilometer length has a mass of 846 kg. The weight of a wire with a thickness of ice deposits of 20 mm increases by 3.7 times, with a thickness of 40 mm – by 9 times, with a thickness of 60 mm – by 17 times. The total weight of the transmission line of eight-kilometer-long wires increases to 25, 60 and 115 tons, respectively, which leads to the breakage of the wires and metal supports.

The icing of wires causes chaotic vibrations with two degrees of freedom under the influence of air flows. This leads to a violation of the adjustment of wires and cables, their breakage, and the overlap of the linear insulation of the high-voltage line. These effects also cause the pillars to collapse. Such accidents lead to significant economic damage. Repairs to the line involve disconnecting it for a period of several hours to several days.

2. Methods for removing ice from wires

The fight against icing of wires is a complex problem and includes a complex of technical means [3]. These include mechanical, electrothermal, physical and chemical, thermodynamic, and electromechanical methods.

The simplest mechanical method is the destruction of ice deposits using long poles from the car or from the ground surface. It is used on fairly short sections of power lines. This method can be used without disconnecting the line using electrical insulation tools. The disadvantages of the method are related to its long duration, the need to be present in difficult-to-reach locations of the line, and the possibility to damage it. It is assumed to use manual physical labor and attract a large number of people, which is not effective. This method is used on short-distance lines.

The mechanical method can be attributed to the effect on the deposits of ice moving along the wires of cutting devices that use the energy of permanent magnets, wind or electromagnetic field of the electric current of the line. There is a variation of the method in which rollers move along the wires, lifting and bending the wires. In these places, the ice is breaking down. Usually, the rollers are mounted on rods attached to special machines moving along the line. In this case, there is always some probability of damage to the line elements. The cutting tool has a limited service life. There are inconveniences of its placement on the wires. This method also includes the use of various pneumatic devices and robotic systems. Their work is based on the impact on the ice cover. Robots are used that are installed on a wire. They are controlled remotely from the ground. Their application was quite effective. They allow one to carry out some repairs. The disadvantages of such devices include the need to control the operator, as well as the high cost of the robot itself.

In the electromechanical method, the ice removal occurs by passing periodic current pulses through the wires. The ampere force acts between the wires. In this case, the impact occurs. There are fluctuations and further destruction of the icing when mechanical resonance appears. Various forms of current pulses and duty cycle are used. The main advantage of the method is its ability to work in a preventive mode. In this case, you do not need to destroy the formed ice. It will not appear on the wires.

The electrothermal method consists in heating the wires and melting the ice on them [4]. Various types of current are used to achieve this goal. It can be alternating or direct current. The frequency of the alternating current can be different: infra-low, industrial, or high. It can be tens or hundreds of megahertz. The current value usually exceeds the nominal value by 1.5-2 times. The duration of exposure varies. The disadvantages of this method include the need to disable the line. The operating mode of the line is close to a short circuit. This may not reflect well on the technical condition of the line. In addition, you must use separate generators, rectifiers, and electronic devices.

The physical and chemical method consists in applying special solutions to the wires that prevent water crystallization on the wires [5]. The method is based on the use of coatings with low adhesion to snow, ice, and water. Such coatings are called hydrophobic. The peculiarity of the method is not the destruction of the ice shell but the prevention of its formation [6]. The general property of such coatings is that the more effective it is, the shorter its service life. Fluoropolymer coatings have good prospects.

The intensity of icing depends on both the height of the overhead line wires and the diameter of the power line wire. As the height of the wire increases, so does the thickness of the resulting ice. When increasing the overall thickness of the wire, it twists less.

Each of these methods has its advantages and disadvantages. Combined methods are more effective for dealing with wire icing. They combine the simultaneous impact of different processes on the ice shell, which increases the probability of its destruction. Some physical aspects of one of these methods are discussed below. It consists of simultaneous action of high-frequency current and low-frequency alternating or pulsed current on the ice cover. The first factor melts the ice layer. The second method of shock loading [7] destroys the ice shell.

3. Mathematical model of the current effect on the ice shell

In the case of mechanical failure, one of these methods must be used to separate the ice layer from the metal wire. To increase the efficiency of this process, it is enough to create a liquid layer at the interface between them. This can be done by melting the layer of ice directly adjacent to the bound.

The actual situation must take into account many factors, such as the heterogeneity of ice thickness along and across the wire, its loose structure, heat exchange conditions on the inner and outer surfaces of the ice shell, the movement of the ice-water interface during the phase transition, the state of the environment, etc. Here is an estimated calculation of the melting time by passing high-frequency current till the appearance of a water layer.

It is known that at high frequencies, there is a surface effect (skin effect). It consists in the fact that high-frequency current is displaced to the surface of the wire. In fact, it flows in a certain surface layer with an effective thickness that depends on the frequency. The effective cross-section of the conductor decreases, and its resistance increases. As a result, it can be many times higher than the value determined for direct current. The ratio of resistances, in this case, is equal to

$$\frac{R}{R_0} = \frac{r_0\sqrt{\omega\mu\gamma}}{2\sqrt{2}}, \tag{1}$$

where R and R_0 are the AC and DC resistances, r_0 is the wire radius, ω is the frequency, μ is the magnetic permeability of the wire material, and γ is its specific conductivity.

First, let us estimate the required amount of current in the wire. The heat output in the wire is used to heat the wire and inhomogeneous heating of the ice shell. Some of this power is lost by heat exchange with the environment. The wire temperature in steady mode must be higher than the melting point of ice T_0. Otherwise, the ice layer will not melt. Consider the limit stationary mode when the wire temperature is equal to T_0. In this case, all the allocated power is equal to the loss of power. The value of this current is called critical. For a unit of wire length, the heat balance equation looks like

$$q = \Delta T\left(\frac{1}{2\pi k}\ln\frac{r_1}{r_0} + \frac{1}{2\pi r_1\alpha}\right)^{-1} = \frac{I_c^2 R}{l}. \tag{2}$$

Here q is the linear density of heat flow through the ice wall, ΔT is the difference in melting temperatures of ice and ambient air, k is the coefficient of thermal conductivity of ice, α is the coefficient of heat exchange between the outer surface of the ice shell and the surrounding air, r_0 and r_1 are the radii of the wire and the shell, I_c is the effective value of the critical current, l is the length of the wire. Note that

$$R_0/l = \left(\pi r_0^2\gamma\right)^{-1} \tag{3}$$

and express from (2) the critical current strength.

$$I_c = \left(\sqrt{\frac{32\gamma}{\omega\mu}}\,\frac{\pi^2 k r_0 r_1\alpha\Delta T}{r_1\alpha\ln r_1/r_0 + k}\right)^{\frac{1}{2}} \tag{4}$$

In order to get a layer of water, it is necessary to pass a current greater than the critical value of $I = nI_c$, $n > 1$ through the wire. The greater the multiplier n, the faster the layer will appear.

Now let us estimate τ – the required time of current passage. The amount of heat released in the wire goes to the following processes. 1 - inhomogeneous heating of the ice shell. 2 - melting of the cylindrical layer of thickness b. 3 - heating of the formed water layer. The specific heat capacities of ice and water are $2.1 \cdot 10^3$ and $4.2 \cdot 10^3$ J/kg·K, respectively. The specific heat of melting ice is $3.35 \cdot 10^5$ J/kg.

Non-uniform heating of ice requires an amount of heat

$$Q_1 = Km_1 c_1 \Delta T . \tag{5}$$

Here m_1 is the initial mass of the ice shell, c_1 is the specific heat capacity of the ice, and K<<1 is the coefficient that takes into account the uneven radial heating. The internal temperature of the shell reaches a value of T_0, and the external temperature reaches a value close to the ambient temperature. Heating the resulting amount of water requires energy, which is calculated using a similar (5) formula with $K = 1$. The amount of heat required to melt the ice layer is equal to

$$Q_2 = \lambda m_2 , \tag{6}$$

where λ is the specific heat of melting ice, m_2 is the mass of the molten ice layer.

Based on the given ratio of energy values, for a rough estimate, we can neglect the values (5) for the water layer and the ice shell in comparison with (6). Then the energy consumption for obtaining the water layer and heating the wire to the melting point of ice is

$$Q = \pi\lambda\rho_1 l\left((r_0 + b)^2 - r_0^2\right) + \pi r_0^2 l \rho_3 c_3 \Delta T , \tag{7}$$

where ρ_1 is the mass density of ice and b is the thickness of the water layer. The second term refers to the metal of the wires. It has high thermal conductivity. Therefore, we can assume that its heating also occurs uniformly from the ambient air temperature to a temperature slightly higher than T_0. The amount of heat released by the current I in the wire during the time τ is $Q = I^2 R\tau$. Equating it to (6) and counting $b << r_0$, we get the required warm-up time

$$\tau = \sqrt{\frac{8\gamma}{\omega\mu}} \frac{\pi^2 r_0^2 \left(2\lambda\rho_1 b + r_0 \rho_3 c_3 \Delta T\right)}{I^2} . \tag{8}$$

Further destruction of the ice shell can be carried out by passing a low-frequency sinusoidal or pulsed current along the wires. A wire with an ice shell, located between two adjacent poles of a power line, is like a stretched string with fixed ends. The lowest natural frequency of string vibrations is defined by the expression

$$\omega_0 = 2\pi\sqrt{F/l^2\rho'} , \tag{9}$$

where F is the tension force of the wire, and ρ' is the effective linear mass density of the iced wire, taking into account the water layer. After simple transformations taking into account the ratio $b << r_0$ we find the frequency of the supplied current for dropping the ice shell

$$f = {\omega_0}/{2\pi} = \left({F}/{\pi l^2 \left(r_1^2 \rho_1 + 2 r_0 b(\rho_2 - \rho_1) + r_0^2 \rho_3\right)}\right)^{1/2} . \tag{10}$$

Theoretical estimates of the natural frequency of vibrations of the wire can not be made with an acceptable degree of accuracy. The fact is that the thickness of the ice layer and its density are not the same along the wire. In addition, the shape of the shell is not cylindrical. The vibrations of the wire have components in two planes. The process is significantly affected by the surrounding air masses. In [8], it is proposed to introduce a coefficient that takes into account the change in the natural frequency of vibrations of a wire with ice frozen on it relative to the natural frequency of vibrations of a wire without ice. According to this work, the calculated value of this coefficient is of the order of one. More reliable is the experimental determination of the natural frequency. According to [9], the experimentally determined value is 0.65 – 1 Hz.

In order to perform an effective effect on the ice shell of the wire, it is recommended to use a generator with a smoothly varying frequency within 0.2 – 3 Hz and select the frequency directly on this section of the line. At the same time, you can judge the effectiveness of the impact visually. For resonance oscillations, it is necessary to close the end of the line and two wires formed a circuit consisting of two parallel conductors, of sufficient size to skip current with a frequency equal to half the natural frequencies of the system. In this case, under the influence of the ampere force that occurs when the interaction of opposite-directed currents, the wires will repel at a frequency twice the current frequency.

4. Conclusions

The icing of power lines is a dangerous and harmful phenomenon. There are several methods of dealing with it. However, a universal effective method has not yet been developed. Pre-separation of solid ice from the wire metal can contribute to the ice destruction. Melting of the ice layer significantly reduces its adhesion to the wire metal, which facilitates its subsequent separation after mechanical action. For this purpose, the wire is loaded with high-frequency current. The current value must exceed the critical value. We made estimates of the critical current value and its transmission time.

Acknowledgements

The work was carried out with the financial support of the state task of the Russian Federation no. 0720-2020-0025 "Development of methods and analysis of ways to achieve a high level of safety and competitiveness of energy systems based on digital technologies".

References

[1] Wang Z 2017 Recent progress on ultrasonic de-icing technique used for wind power generation, high-voltage transmission line and aircraft *Energy and Buildings* **140** P 42-49

[2] Tomaszewski M, Ruszczak B, Michalski P and Zator S 2019 The study of weather conditions favourable to the accretion of icing that pose a threat to transmission power lines *International Journal of Critical Infrastructure Protection* **25** P 139-151

[3] 2008 *Atmospheric Icing of Power Networks* (Farzaneh, Masoud, Springer) p 381

[4] Levchenko I I, Zasypkin A S, Alliluyev A A and Satsuk E I 2007 *Diagnostics, Reconstruction and Operation of Overhead Power Lines in Icy Areas* (Moscow: National Research University "Moscow Power Engineering Institute") p 494

[5] Wang Y 2020 Numerical study of a droplet impact on cylindrical objects: Towards the anti-icing property of power transmission lines *Applied Surface Science* **516** 146155

[6] Wen S F, Wang Y M, Zhang Z M and Liu Y L 2019 Application of anti-icing coating based on adsorption of functional substances by microporous sphere *Progress in Organic Coatings* **137** 105320

[7] Ji K, Liu B, Cheng Y, Zhan X and McClure G 2019 Evaluation and optimization of a shock load de-icing method for transmission lines with combined ice failure criteria *Cold Regions Science and Technology* **165** 102818

[8] Sukhorukov S I, Soloviev V A, Cherniy S P and Savelyev D O 2014 *Russian Internet Journal of Electrical Engineering* **1(2)** 10–12

[9] Mikheev V P 2003 *Contact Networks and Power Lines* (Moscow: Marshrut) p 416

IOP Conf. Series: Materials Science and Engineering 1035 (2021) 012043 doi:10.1088/1757-899X/1035/1/012043

Integration of solar plants into energy systems in apartment residential buildings

D M Chudinov[1,4], A I Kolosov[1], T V Shchukina[2] and A P Zverkov[3]

[1]Department of Heat and Gas Supply and Oil and Gas Engineering, Voronezh State Technical University, 14 Moskovsky Avenue, Voronezh, 394026, Russia
[2]Department of Housing and Utilities services, Voronezh State Technical University, 14 Moskovsky Avenue, Voronezh, 394026, Russia
[3]Department of Heat and Gas Supply, Heating and Ventilation, branch of Voronezh State Technical University, 191 Svobody Str., Borisoglebsk, 397160, Russia

[4]Corresponding author: dmch_@mail.ru

Abstract. Non-traditional renewable energy resources (solar radiation, wind energy, etc.), alternative thermal and electric generators have incontestable positive qualities in comparison with fossil fuel (natural gas, coal, fuel oil), classical systems: relative environmental friendliness, vast total reserves, low operational costs, reliability, ease of management. The scale of their implementation in the Central Black Earth Region can be limited by high initial capital investments, the value of the energy resource, and low awareness of the population. The article presents the results of modeling solar plants functioning during their integration into traditional energy systems of apartment residential buildings. In the article we demonstrate technical and economic indicators of projected and modernized engineering heating services in the city of Voronezh. The technically viable period of use is spring-summer-autumn. Economic indicators put solar systems into the experimental category.

1. Introduction

Today when projecting, building, exploitation and reconstruction of morally and physically obsolete apartment residential buildings apart from providing comfortable microclimate, great attention is paid to energy efficiency. Reduction of energy consumption may be achieved by using alternative sources of energy.

To meet the needs of population the builder develops apartments with larger living areas, and with several levels on the upper floors. During the heating period it leads to substantial rise of exploitation costs which are often compared to the income of the residents. The situation may be improved through the integration of solar plants into traditional energy systems for partial coverage of heating needs.

2. The methodology of determination of gross, technical and economic potential of the integrated solar plants

The expedience to use solar plants for heating needs and hot water supply in apartment residential buildings is justified by the data of gross, technical and economic potential of solar energy [1-8].

Alternative self-renewing energy sources have a number of advantages (ubiquity, large total inventory, relative environmental cleanliness), though, for the reason of wide territorial divergence, irregularity of supply they are limited in use.

There are several methods to determine gross potential of solar radiation on horizontal and vertical surfaces, though the results of the energy resource potential assessments are not always precise. Based on the processing of years-long statistics data of actinometric stations (table 1) the equations were obtained, cloudiness in Voronezh region is determined by means of correction factor (table 2) [1, 2, 3].

Table 1. Change of total solar radiation (Q) on the horizontal surface depending on geographical latitude (φ), MJ/m^2, cloudiness adjusted.

No	Symbols	Approximation dependence	Cloudiness adjusted
1	$Q_{January}$	$-614.08 \cdot \ln(\phi) + 2586.09$	0.52
2	$Q_{February}$	$-600.56 \cdot \ln(\phi) + 2638.10$	0.59
3	Q_{March}	$-0.16 \cdot \phi^2 + 5.41 \cdot \phi + 677.87$	0.54
4	Q_{April}	$-364.69 \cdot \ln(\phi) + 2108.73$	0.58
5	Q_{May}	$-150.26 \cdot \ln(\phi) + 1444.03$	0.67
6	Q_{June}	$0,00005571 \cdot \phi^5 - 0.01486526 \cdot \phi^4 + 1.57018502 \cdot \phi^3 - -82.03553868 \cdot \phi^2 + 2117.99506113 \cdot \phi - 20707.95629264$	0.7
7	Q_{July}	$0.00163130 \cdot \phi^4 + 0.34016927 \cdot \phi^3 - 26.31315104 \cdot \phi^2 + +896.02144210 \cdot \phi - 12226.70616884$	0.67
8	Q_{August}	$-322.42 \cdot \ln(\phi) + 1988.98$	0.68
9	$Q_{September}$	$-570.53 \cdot \ln(\phi) + 2779.51$	0.61
10	$Q_{October}$	$-762.80 \cdot \ln(\phi) + 3343.38$	0.52
11	$Q_{November}$	$-650.10 \cdot \ln(\phi) + 2760.50$	0.38
12	$Q_{December}$	$-623.75 \cdot \ln(\phi) + 2596.38$	0.40

Table 2. Adjusted coefficient of gross potential attenuation of solar energy (SE) for the Central-Chernozyom region of Russia.

Indicators	Month of the year											
	I	II	III	IV	V	VI	VII	VIII	IX	X	XI	XII
The Central-Chernozyom region												
The city of Voronezh (Voronezh region, 51° 43' n.l.)												
SE for cloudless sky, J/m^2	163	268	530	670	851	876	882	717	528	333	195	135
SE in conditions of cloudiness, MJ/m^2	84	142	289	385	565	620	590	473	326	176	80	50
Coefficient of cloudiness ψ	0.52	0.53	0.55	0.57	0.66	0.71	0.67	0.66	0.62	0.53	0.41	0.37
Stone steppe (Voronezh region, 51° n.l.)												
SE for cloudless sky, MJ/m^2	184	289	548	682	856	878	883	728	548	359	217	156
SE in conditions of cloudiness, MJ/m^2	98	176	314	418	599	630	609	515	344	188	86	65
Coefficient of cloudiness ψ	0.53	0.61	0.57	0.61	0.70	0.72	0.69	0.71	0.63	0.52	0.40	0.42
Kursk (Kursk region, 51° 44' n.l.)												
SE for cloudless sky, MJ/m^2	163	268	530	670	851	876	882	717	528	333	195	135

SE in conditions of cloudiness, MJ/m^2	84	172	274	372	554	605	584	475	316	165	67	52
Coefficient of cloudiness ψ	0.52	0.64	0.52	0.56	0.65	0.69	0.66	0.66	0.60	0.50	0.34	0.38
Average value of coefficient of cloudiness	0.52	0.59	0.54	0.58	0.67	0.70	0.67	0.68	0.61	0.52	0.38	0.40

Basic characteristic, determining technical potential of the integrated solar system, is the replacement rate, that indicates the proportion of heat load, provided by means of solar radiation. Replacement rate depends on dimensionless complexes, taking into account production and loss of thermal energy by solar heating system. Its detailed calculation is presented in [3].

The final choice of the system of energy supply of apartment residential building in the conditions of the Voronezh region is carried out by evaluating and comparing economic potential of alternative and traditional heat generator. Discounted costs and period of recoupment are used as economic criteria, the detailed methods of their determination are shown in [1, 2].

3. The results of modeling of solar system functioning in apartment residential building in conditions of the city of Voronezh

The facility under consideration is a new apartment residential building or a building subject to reconstruction due to moral or physical wear out, situated in the city of Voronezh. The basic characteristics of the building are represented in table 3.

Table 3. Basic parameters, characterizing the facility.

Title of the parameter	Value of the parameter
1. Type of the building	apartment residential building
2. Number of floors	5
3. Number of residents in an apartment	5
4. Source of heat energy	From the first to the fourth floor traditional heating system is used, on the fifth floor - the combined one
5. Azimuth of solar collector, degrees	0
6. Solar collector inclination angle, degrees	52 (for the system of hot water supply); 62 (for the heating system and hot water supply)
7. Time horizon	20 years

The key parameters of the plant demonstrating its competitive characteristics have been determined in consideration with the results of the previously conducted experiments in this sphere [1-8]. They are shown in figures 1-3.

Modeling of integration of solar system into the energy system of an apartment building is shown on the example of one apartment located on the fifth floor. The main distinctive structural elements of the combined heat supply system are: a 12 m^2 solar collector, a 0.6 m^3 water-storage tank, an automatic control and management system, antifreeze is used as a heat carrier in the solar circuit. The parameters of the alternative plant are in the optimal ratio and all the recommendations in the works [2, 4, 5, 8] are taken into account.

The diagrams and graphs presented in figures 1-3 confirm a sufficient level of gross and technical potential of solar energy in the Voronezh region. The value of the annual replacement rate of heat and hot water supply systems for solar plants is 14% and 44%, respectively. The economic efficiency of alternative heat generators constrains their large-scale implementation at the current cost of heat of 2262.71 rubles per Gcal in the city of Voronezh.

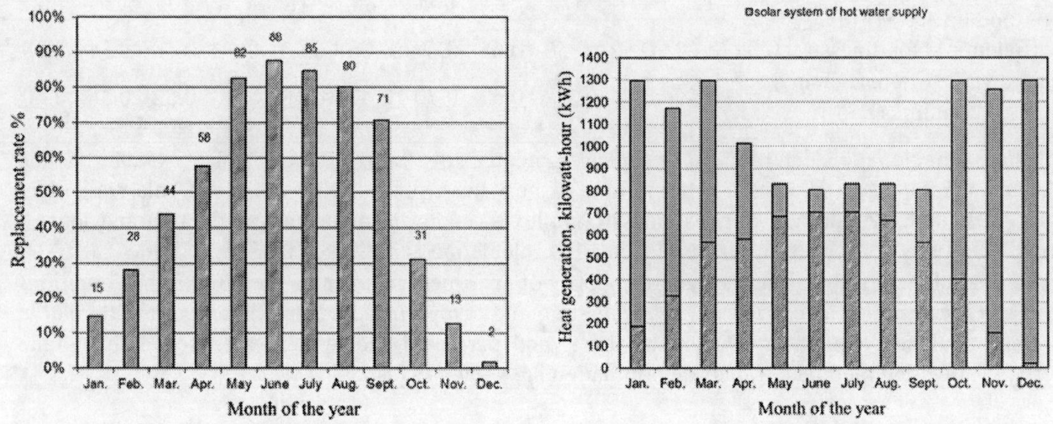

a) b)

Figure 1. The level of monthly heat generation by the solar system of hot water supply in the climatic conditions of the city of Voronezh: a) replacement rate; b) heat generation by the solar plant and terminal unit.

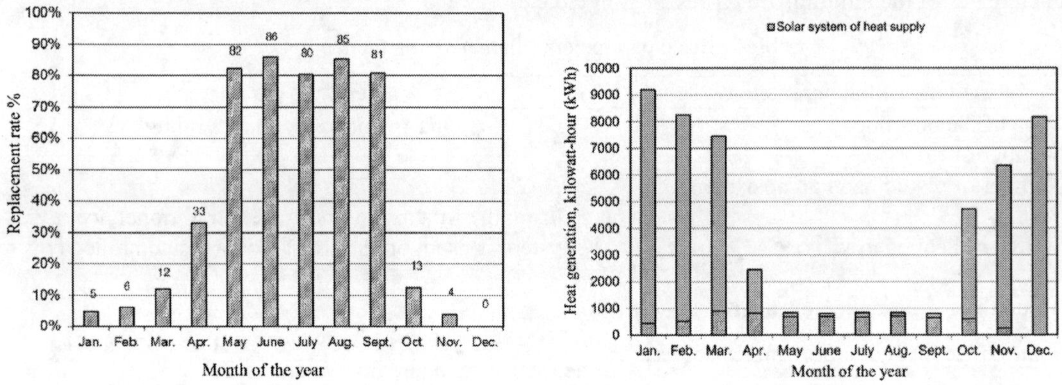

a) b)

Figure 2. The level of monthly heat generation by the solar plant of hot water supply in the climatic conditions of the city of Voronezh: a) replacement rate; b) heat generation by the solar plant and terminal unit.

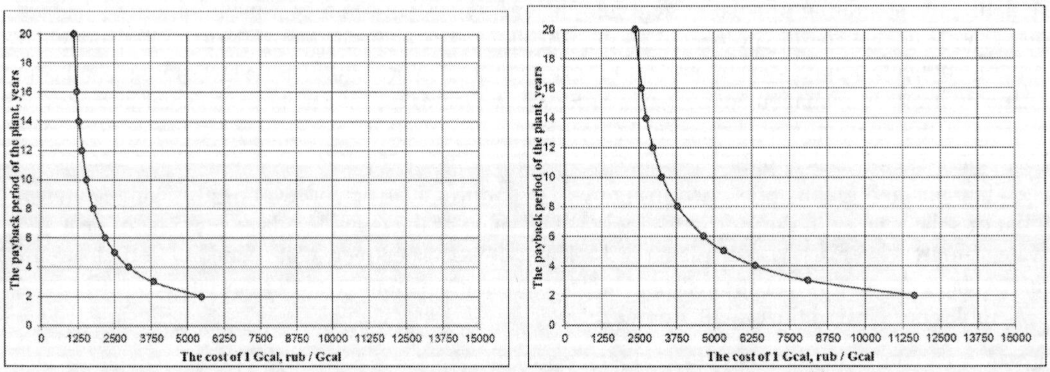

a) b)

Figure 3. Dependence of solar plant payback period a) of hot water supply, b) of heat supply on the cost of heat for the city of Voronezh.

The attractiveness and profitability of solar heat supply systems in the climatic and economic conditions of the Voronezh region can be achieved through the use of improved technical solutions (combined generation of thermal and electric energy in one design solution) [9] and lower equipment costs.

An example of the effective implementation of life cycles (design, construction, operation) of a solar system of hot water supply is a hotel located in the village of Babyakovo, Voronezh Region (figure 4). The reduction in capital investment and in payback period of the alternative heat generator was achieved through the own produced solar collectors and water-storage tank. In the summer months, the plant provides 100% coverage in the need for hot water.

Figure 4. Overall view of the solar system of hot water supply in the village of Babyakovo, Voronezh region.

4. Conclusion

In the conditions of the city of Voronezh, the use of solar systems of hot water supply is cost-effective. The replacement rate in the summer months varies from 80 to 88%, in winter - from 2 to 28%. The large-scale implementation of such energy systems is constrained by economic factors (the relatively low cost of fossil fuels, absence of soft loans, high initial capital investments), and low public awareness. An increase in the competitive characteristics of solar systems can be achieved with the global interaction of the state, developers and manufacturers. Today such facilities can be regarded as experimental. The annual increase in the cost of heat will lead to a decrease in the payback period of solar plants.

References

[1] Sotnikova O A and Chudinov D M 2007 Economic effectiveness of using solar systems of hot water supply *ABOK* **2** 88-94

[2] Kolosov A, Chudinov D and Yaremenko S 2018 Design of solar systems in high-rise buildings *E3S Web of Conferences* eds. D Safarik, Y Tabunschikov and V Murgul 02055

[3] Sotnikova O A and Chudinov D M 2005 Use of solar thermal energy in passive and active heat supply systems *Scientific Journal of Voronezh State Technical University* **1(6)** 56-63

[4] Chudinov D M, Shchukina T V and Sotnikova O A 2008 The effect of the solar plant parameters on the efficiency of the alternative heat supply of buildings *Industrial Power Engineering* **9** 44-46

[5] Sotnikova O A, Chudinov D M and Shchukina T V 2009 Ensuring necessary level of energy activity of solar systems designed for the regions of the Russian Federation *Energy Saving* **3** 74-76

[6] Sotnikova O A, Chudinov D M and Shchukina T V 2008 Economic reasons and prospects of developing solar(-generated) heat supply *Industrial Power Engineering* **6** 50-52

[7] Chudinov D M, Shchukina T V, Petrikeeva N A and Popova N M 2019 Energy saving actions in apartment houses *High Technologies in Construction Complex* **1** 32-36

[8] Sotnikova O A, Chudinov D M, Petrikeeva N A and Popova N M 2018 Reconstruction of the engineering equipment of residential buildings *Scientific and Technical Problems of Improving and Developing Gas-and-Power Supply Systems* **1** 216-223
[9] Shchukina T V, Chudinov D M, Petrikeeva N A and Popova N M 2017 Combined generation of heat and electrical energy in solar installations *Scientific Journal of Voronezh State University of Architecture and Civil Engineering* **1** 118-121

IOP Conf. Series: Materials Science and Engineering 1035 (2021) 012044 doi:10.1088/1757-899X/1035/1/012044

Effectiveness of the joint scheme of steam-power and combined-cycle power units using the example of the Nevinnomyssk State Regional Thermal Power Plant

V V Shaposhnikov[1], D N Batko and Ya O Mihalko

Department of Heat Power Engineering, Kuban State Technological University, 2 Moskovskaya Str., Krasnodar 350072, Russia

[1]Corresponding author: shaposhnikov.valentin@gmail.com

Abstract. We attempt to solve the problem of displacing high-temperature steam bleeding in a steam turbine plant at the first regenerative heater after intermediate overheating. This can be achieved when the thermal power plant contains heterogeneous equipment, in particular, a steam turbine and combined cycle power units. The authors proposed an energy-efficient scheme for the joint operation of steam turbine unit and steam-turbine part of a combined cycle plant with the displacement of high-temperature steam onto the high-pressure feed heater (HPFH 2) using the dry steam from the medium-pressure drum of the waste boiler in order to increase the upper limit of the power plant's control range. For this research, Nevinnomyssk TPP was selected as one of the main energy facilities of the Unified Energy System of Southern Russia. We studied the joint operation of the steam turbine unit K-160-130 and the combined cycle plant CCP-410 manufactured by Siemens, Germany. We developed the methodology and calculated the energy efficiency of the new power plant operation scheme. According to the calculation results, the total increase in power amounted to 6.03 MW, while there was an increase in electrical gross efficiency by 0.64%. An assessment of the economic effect of the applied solution for the conditions of the Nevinnomyssk TPP demonstrated that the annual increase in net profit will amount to 14.47 million rubles, and the payback period of investments is 4.2 months.

1. Introduction

According to the development program of the Russian Unified Energy System for 2019 - 2025 [1], by 2025 the forecasted maximum power consumption will be about 168.983 MW. This corresponds to the average annual growth rate of maximum power consumption by 1.5%. For the South Russian UES, the projected growth rate of the load is planned to be slightly higher, which is by 2.2%. According to official data in the South Russian UES [2] for December 2019, the electricity generation amounted up to 9 048 GW/h, and the energy consumption is 9 361 GW/h. This indicates a continuing imbalance, as well as the potential for using new capacities or increasing the maximum load at the operating facilities. Despite the active development of renewable energy sources [3] and research in the field of distributed generation in Russia [4, 5], thermal power plants still play a key role in this field (as of December 1, 2019, they take 67% from all the installed capacity of power plants in Russian UES) [6].

To obtain additional peak electrical power at steam turbine power plants, various solutions can be used. One of them is to turn off regenerative high-pressure heaters or bypass them by the feed water. This method is aimed at limiting the flow of steam into the regeneration system and increasing its flow into the flow part of the turbine, which leads to an increase in its power. Depending on the type of installed turbines and steam parameters, the bypassing of the HPFH of steam-powered plants by the feed water gives a power increase by up to 12% [7, 8]. When the HPFH is turned off, an increase in the turbine power occurs with an increase in the specific heat consumption. This happens due to the reduced enthalpy of feed water, while the internal turbine efficiency decreases and the vacuum in the condenser deteriorates. In addition, the systematic shutdown of the HPFH can lead to low-cycle metal fatigue in feed pipelines and heaters.

To eliminate the above-mentioned negative effects of this method, a number of researchers proposed not to use a shutdown of the HPFH but a substitution by steam bleedings from other available sources. Such steam sources can be waste heat boilers of built-in gas-turbine units [9, 10]. When installing various types of steam turbine units at the station, one can use schemes for joint operation of steam turbine units with the replacement of high-temperature steam by low-temperature one of required pressure in order to increase power, increase reliability and efficiency [11 - 14].

Studies aimed at increasing the mobility of coal-fired TPPs have become a separate direction in the global energy sector. This is carried out due to the complete shutdown of the HPFH and heating of feed water using solar water heaters installed alongside the HPFH [15 - 17].

In this work, it is proposed to use steam from a waste heat boiler of a combined cycle gas turbine as a steam source to replace the steam bleeding for the HPFH.

2. Research object description

For our research, Nevinnomyssk TPP was chosen. It is a state power plant, one of the main energy facilities of the South Russian UES. It is located in the city of Nevinnomyssk, Stavropol Territory, North Caucasus Federal District.

The block part of the Nevinnomyssk TPP is equipped with five K-155-130 power units and one K-160-130 unit, produced by Kharkiv Turbine Plant, with gas-oil boiler units TGM-94. A distinctive feature of the thermal circuit of these power units is the fact that the manufacturer recommended damping regenerative steam bleeding no. 2, which is located first along the steam stream in the turbine flow section after an intermediate superheater. The disconnection of this bleeding is connected with the burning of the inlet steam pipes of HPFH 2 by high-temperature steam. It is recommended to use steam from the first unregulated bleeding as a heating medium in HPFH 2. The parameters of the steam from regenerative bleedings for the HPFH are shown in table 1.

Table 1. Characteristics of regenerative bleedings for high-pressure heaters K-160-130 and feed water parameters for the HPFH.

Bleeding	Steam parameters in the bleedings				Feed water
purpose	P_s, MPa	t_s, °C	t_k, °C	D, kg/s	t_{fw}, °C
HPFH 3	3.14	375	237.1	2.47	228.4
HPFH 2	3.14	375	237.1	10.15	218.5
HPFH 1	1.23	451	189.1	3.83	179.9

In 2011 CCP-410, a three-circuit binary combined cycle gas turbine unit was introduced. It consists of a gas turbine SCC5-4000F, a steam turbine SST-900/700 DRH manufactured by Siemens and a recovery manufactured by CMI company. The main performance indicators of the latter are shown in table 2. The distinctive features of CCP-410 of the Nevinnomyssk TPP are the connection of high-pressure cylinder (HPC) and low-pressure cylinder (LPC) with an electric generator through a pressure reducer/ It also has an intermediate overheating of the combined steam flow from the medium pressure circuit and after the HPC.

The distance between the combined cycle plant and the unit at Nevinnomyssk TPP is about 250 m (figure 1).

Table 2. Key performance indicators for the CMI recovery boiler.

Name of the parameter	Unit of measurement	Value
The consumption of combustion products	m^3/h	2157465
The temperature of the combustion products in front of the waste boiler (WB)	°C	593.6
The temperature of the combustion products beyond the WB	°C	94.36
The gross efficiency of the waste boiler	%	89.49
The consumption of high pressure (HP) feed water at WB	kg/s	64.23
The vapor pressure in the HP drum	MPa	10.316
The HP steam pressure beyond the WB	MPa	10.03
The HP steam temperature beyond the WB	°C	549.2
The HP steam consumption beyond the WB	kg/s	62.4
The medium-pressure (MP) feedwater consumption for WB	kg/s	13.73
The steam pressure in MP drum	MPa	2.547
The MP steam pressure behind the superheater	MPa	2.41
The MP steam temperature beyond an intermediate superheater	°C	547.6
The steam consumption at the outlet to the intermediate superheater	kg/s	74.69
The MP steam consumption beyond the WB	kg/s	10.8
The steam pressure in the low pressure (LP) drum	MPa	0.446
The LP steam pressure beyond the WB	MPa	0.43
The LP steam temperature beyond the WB	°C	291.9
The LP steam consumption beyond the WB	kg/s	11.0
The consumption of the main condensate on a gas condensate heater	kg/s	88.8

Figure 1. Satellite image of the Nevinnomyssk TPP: 1 - Steam-turbine power units K-155-130 and K-160-130, 2 - Combined-cycle plant CCP-410.

3. Proposed solution

The layout and composition of the equipment of the Nevinnomyssk TPP allow applying the scheme of joint operation of the turbine units of the steam power unit K-160-130 and the steam turbine part of the combined cycle plant CCP-410 with the displacement of high-temperature steam bleeding to the regenerative high-pressure steam boiler from the recovery boiler [18, 19]. It is shown in figure 2.

In this scheme, steam is displaced from the first regenerative bleeding of the turbine going to HPFH 2, with steam from the medium-pressure drum of the waste boiler CCP-410. To maintain mass balance, condensate is returned from the K-160-130 condenser to the combined-cycle unit

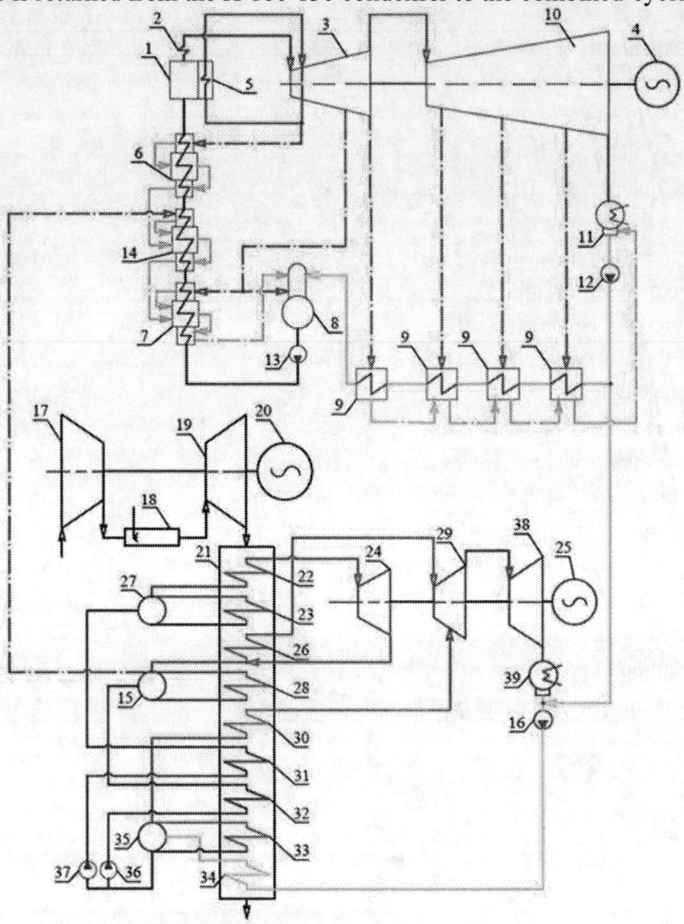

Figure 2. Schematic diagram of the joint operation of the steam turbine unit K-160-130 and combined cycle plant CC3-410 of the Nevinnomyssk state thermal power station:
1 – steam boiler; 2 – boiler superheater; 3 – steam turbine high pressure cylinder; 4 – steam turbine power generator; 5 – steam superheater; 6 – the third high pressure heater; 7 – the first high-pressure heater; 8 – deaerator; 9 – group of low pressure heaters; 10 – low pressure cylinder of steam turbine unit; 11 – condenser of the steam turbine unit; 12 – condensate electric pump of steam turbine unit; 13 – power electric pump of steam turbine unit; 14 – the second high pressure heater; 15 – medium pressure drum; 16 – condensate pump of a combined cycle gas unit; 17 – compressor; 18 – combustion chamber; 19 – gas turbine; 20 – the first electric generator of a combined cycle gas unit; 21 – waste boiler; 22 – high pressure superheater; 23 – high pressure evaporator; 24 – high-pressure cylinder of a combined-cycle gas unit; 25 – the second electric generator of the steam-gas block; 26 – medium-pressure superheater; 27 – high pressure drum; 28 – medium pressure evaporator; 29 – medium pressure cylinder of a combined cycle gas turbine unit; 30 – low pressure superheater; 31 – high pressure economizer; 32 – medium pressure economizer; 33 – low pressure evaporator; 34 – gas condensate heater; 35 – low-pressure drum with integrated deaerator; 36 – medium-pressure feed electric pump of a combined cycle gas turbine unit; 37 – high-pressure feed electric pump of a combined-cycle gas unit; 38 – low pressure cylinder of vapor-gas block; 39 – condenser of combined-cycle gas unit.

4. Methodology and results of thermodynamic efficiency calculation

The calculation is made under the following assumptions:

— the calculation is carried out at rated power, condensation mode, unnamed fuel consumption and outdoor temperature +15 °C;

— the relative internal turbine efficiency is assumed to be constant;

— the effect of changes in steam flow rate on its parameters in the flow part of the turbines is not taken into account;

— the effect of changes in steam flow on the vacuum in condensers is not taken into account;

— steam losses in the cycle and boiler blowdown are not taken into account;

— the influence of changes in the flow rates of the main and secondary condensate, feed water for steam flow rate on the other regenerative heaters (except for HPFH 1, 2 and 3) are not taken into account.

When using the scheme of joint work of the steam turbine block K-160-130 and the combined cycle gas turbine unit CCP-410, the steam that was used to heat the feed water in HPFH 2 will remain in the flow part of the turbine. Due to the fact that there are restrictions on the steam flow from the middle-pressure drum of the CCP waste boiler and the steam has a lower pressure than the displaced one (table 1), the temperature of the feed water after HPFH 2 will decrease. This will lead to an increase in the consumption of heating steam from the first bleeding for HPFH 3 for reheating feed water in front of the boiler.

The increase in the consumption of heating steam from the first bleeding for HPFH 3, kg/s:

$$\Delta D_1 = \frac{G_{FW} \cdot C_P \cdot \Delta t_{FW.UND.HPFH2}}{i_{HPFH3} - t_{C.HPFH3} \cdot C_p}, \qquad (1)$$

where G_{FW} – the feed water consumption, $G_{FW} = 135$ kg/s (487 t/h);

C_p – the water heat capacity, $\frac{kj}{kg \cdot K}$;

i_{HPFH3} – the enthalpy of the steam from the first regenerative bleeding of CTP aimed for HPFH heating 3, kJ/kg;

$t_{C.HPFH3}$ – the steam condensate temperature from the first regenerative bleeding of CCP aimed for HPFH 3 heating, ° C;

$\Delta t_{FW.UND.HPFH2}$ – the value of feed water underheating in HPFH 2 when it is heated by steam from a CCGT medium-pressure drum, °C:

$$\Delta t_{FW.UND.HPFH2} = t_{FW.HPFH2} - t^{CCP}_{RW.HPFH2}, \qquad (2)$$

where $t_{FW.HPFH2}$ – the feed water temperature after HPFH 2 when it is heated with steam from CTP regenerative bleeding, $t_{FW.HPFH2} = 218.5$ °C;

$t^{CCP}_{FW.HPFH2}$ – the feed water temperature after HPFH 2 when it is heated by steam from a CCP medium-pressure drum, °C:

$$t^{CCP}_{FW.HPFH2} = t_{FW.HPFH1} + \frac{D_{MP} \cdot (i_{MP} - C_p \cdot t_{C.MP}) + D_{HPFH3} \cdot C_p \cdot (t_{C.HPFH3} - t_{C.MP})}{G_{FW} \cdot C_p}, \qquad (3)$$

where D_{HPFH3} – the consumption of the heating steam from the regenerative bleeding of STP aimed for heating HPFH 3 before reconstruction, kg/s;

i_{MP} – the vapor enthalpy at the outlet of the CCP medium-pressure drum, kJ/kg;

$t_{C.MP}$ – the steam condensate temperature from a CCP medium-pressure drum, $t_{C.MP} = 224.9$ ° C;

D_{MP} – the steam consumption of a medium-pressure CCP drum, kg/s (the maximum allowable flow rate for Nevinnomyssk TPP is 10.8 kg/s);

$t_{FW.HPFH1}$ – the feed water temperature after HPFH 1, $t_{FW.HPFH1}$=179.9°C.

To assess the power gain in a steam turbine installation, it is necessary to determine the heat drop acquired in the superheater or to recalculate the enthalpy of steam after overheating. In the calculation, we take into account that the thermal power of the intermediate superheater does not depend on the steam consumption in it.

Steam consumption in the superheater before reconstruction, kg/s:

$$D_{S.SH} = G_{FW} - D_{HPFH3} - D_{HPFH2}, \quad (4)$$

where D_{HPFH2} – the consumption of heating steam from the regenerative bleeding of STP aimed into HPFH 2 for heating before reconstruction, kg/s.

The heat transfer acquired in the superheater before reconstruction, kJ/kg:

$$\Delta i_{S.SH} = i_{S.SH} - i_{HPFH3}, \quad (5)$$

where $i_{S.SH}$ – the steam enthalpy after an intermediate superheater before the reconstruction, kJ/kg.

The thermal power of a reheater, kW:

$$Q_{S.SH} = D_{S.SH} \cdot \Delta i_{S.SH}. \quad (6)$$

The steam consumption in the reheater after reconstruction, kg/s:

$$D_{S.SH}^{NEW} = G_{FW} - D_{HPFH3} - \Delta D_1. \quad (7)$$

The heat drop acquired by a reheater after reconstruction, kJ/kg:

$$\Delta i_{S.SH}^{AFTER.R.} = \frac{Q_{S.SH}}{D_{S.SH}^{NEW}}. \quad (8)$$

The total increase in power of the steam turbine block after reconstruction, MW:

$$\Delta N_{STP} = \Delta N_{HPFH2} - \Delta N_{HPFH1}, \quad (9)$$

where ΔN_{HPFH2} – the additional power received by the steam flow released during the replacement of the steam that went to HPFH 2 taking into account the loss of power for heating the feed water in HPFH 3, MW:

$$\Delta N_{HPFH2} = (D_{HPFH2} - \Delta D_1) \cdot \left(i_{HPFH2} + \Delta i_{S.SH}^{AFTER.R.} - i_{C.STP}\right) \cdot 10^{-3} \cdot \eta_{M.STP} \cdot \eta_{E.STP}, \quad (10)$$

where D_{HPFH2} – the steam consumption going to HPFH 2 when it is heated by steam from the regenerative STP bleeding, kg/s;

i_{HPFH2} – the steam enthalpy going to HPFH 2 when it is heated by steam from the regenerative STP bleeding, kJ/kg;

$i_{C.STP}$ – the steam enthalpy at the exit of the K-160-130 turbine, kJ/kg;

$\eta_{M.STP}$ – the mechanical efficiency of K-160-130, $\eta_{M.STP} = 0.99$;

$\eta_{E.STP}$ – the electrical efficiency of K-160-130, $\eta_{E.STP} = 0.98$.

ΔN_{HPFH1}– the STP power loss after reconstruction due to additional steam consumption from the third steam bleeding extraction at HPFH 1 for feed water reheating, MW:

$$\Delta N_{HPFH1} = \Delta D_3 \cdot (i_{HPFH1} - i_{C.STP}) \cdot 10^{-3} \cdot \eta_{M.STP} \cdot \eta_{E.STP}, \quad (11)$$

where i_{HPFH1} – the steam enthalpy going to HPFH 1 when it is heated by steam from the regenerative STP bleeding, kJ/kg;

ΔD_3 – the additional consumption of heating steam from the third bleeding on HPFH 1, kg/s:

$$\Delta D_3 = D_{HPFH1}^{NEW} - D_{HPFH1}; \quad (12)$$

where D_{HPFH1} – the steam consumption going to HPFH 1 when it is heated by steam from the regenerative STP bleeding, kg/s;

D_{HPFH1}^{NEW} – the steam consumption going to HPFH 1 when it is heated by steam from the regenerative STP bleeding after the reconstruction, kg/s;

$$D_{HPFH1}^{NEW} - \frac{G_{FW} \cdot C_p \cdot (t_{FW.HPFH1} - t_{FW.D}) - (D_{HPFH3} + \Delta D_1) \cdot C_p \cdot (t_{C.SD} - t_{C.HPFH1})}{i_{HPFH1} - t_{K.HPFH1} \cdot C_p}, \quad (13)$$

where $t_{FW.D}$ – the feed water temperature exiting the deaerator, $t_{FW.D}$=158 °C;

$t_{C.HPFH1}$ – the steam condensate temperature from the third regenerative STP bleeding going for heating into HPFH 1, °C;

i_{HPFH1} – the steam enthalpy from the third regenerative STP bleeding going for heating into HPFH 1, kJ/kg.

The power loss in a combined cycle gas turbine unit will depend on a decrease in the steam flow entering the medium-pressure cylinder.

$$\Delta N_{CCP} = D_{MP} \cdot (i_{MP} - i_{C.CCP}) \cdot 10^{-3} \cdot \eta_{M.CCP} \cdot \eta_{E.CCP}, \quad (14)$$

where $i_{C.CCP}$ – the steam enthalpy at the exit of the steam turbine of the combined cycle plant, kJ/kg;

$\eta_{M.CCP}$ – the mechanical efficiency of the CCP steam turbine, $\eta_{M.CCP} = 0{,}99$;

$\eta_{E.CCP}$ – the electric efficiency of the CCP steam turbine, $\eta_{E.CCP} = 0{,}98$.

The total increase in power from the application of a joint scheme, MW:

$$\Delta N_{SUMM} = \Delta N_{STP} - \Delta N_{CCP}. \quad (15)$$

The results of the study are presented in table 3.

Table 3. The results of the study of the joint operation scheme of the steam turbine block K-160-130 and combined cycle plant CCP-410 of the Nevinnomyssk TPP.

Parameter	Value
The reduction of steam flow from the first bleeding of K-160-130 turbine for regenerative heating of feed water, t/h	33.4
The increase in steam consumption from the third bleeding of the turbine K-160-130 for regenerative heating of feed water in HPFH 1, t/h	0.58
Reducing the steam flow entering the CCP medium-pressure superheater, t/h	38.9
The change in power while reducing the consumption of heating steam in the first STP bleeding, MW	+ 10.344
The change in power with increasing consumption of heating steam in the third STP bleeding, MW	– 0.156
The change in power of a combined cycle gas turbine unit, MW	– 4.161
The total power increase, MW	6.027

Studies of the joint operation scheme of the K-160-130 steam turbine unit and the combined cycle gas turbine unit CCP-410 of the Nevinnomyssk District Thermal Power Station were carried out under the nominal mode and constant fuel consumption and demonstrated that the use of the joint operation scheme allows obtaining additional power without reducing the efficiency of the power plant, as it happens the case with the simple shutdown of the HPFH.

The increase in the average electrical gross efficiency of the block K-160-130 and CCP-410 after reconstruction will be:

$$\eta_{AVE} = \frac{\Delta N_{SUMM}}{Q_P^H \cdot (B_{STP}^{NOM} + B_{CCP}^{NOM})} \cdot 100, \quad (16)$$

where Q_P^H – the fuel calorific power, MJ/m3, $Q_P^H = 34.1$ MJ/m^3;

B_{STP}^{NOM} – the nominal fuel consumption of the block K-160-130, $B_{STP}^{NOM} = 11.61$ nm^3/s (41800 nm^3/h);

B_{CCP}^{NOM} – the rated fuel consumption of CCP-410, $B_{CCP}^{NOM} = 16.21$ nm^3/s (58370 m^3/h).

$$\eta_{AVE} = \frac{6.027}{34.1 \cdot (11.61 + 16.21)} \cdot 100 = 0.64\%.$$

5. Methodology and results of calculating the economic effect

To assess the economic effect of the joint scheme, we accept the following assumptions and initial data:

— reconstruction involves laying steam and condensate pipelines 300 m long;

— the bleeding for HPFH 2 will be carried out only during the operation period at a technical maximum (average 6 h/day for a year). Additionally generated electricity will be fully sold;

— the increase in electric capacity will not be taken into account when calculating payments for installed capacity;

— operating costs remain constant;

— according to the data specified in order no. 1625/19 dated 12.12.2019 of the Federal Antimonopoly Service, in the first quarter of 2020, for Nevinnomyssk TPP, the tariff rate for electric energy supplied by steam turbine units is 1274.21 rubles / (MW·h); energy sold by a combined-cycle plant is 1096.13 rubles / (MW·h).

According to preliminary estimates, the cost of implementing the joint operation scheme of the K-160-130 and CCP-410 units is 5.1 million rubles, from which 2.9 million rubles are the cost of materials and 2.2 million rubles are the cost of work.

The calculation was carried out according to WD 53-34.1-09.321-2002. "Methodology for rapid assessment of the economic efficiency of energy-saving measures at thermal power plants". We used the simple criteria of economic efficiency, i.e. without taking into account the time factor. The results of an express assessment of the economic effectiveness of the proposed solution are shown in table 4.

Table 4. The results of calculating the economic effect.

Parameter	Value
The annual increase in electricity supply from K-160-130 mln kW * h/year	22.31
The annual decrease in electricity supply from CCP-410 mln kW * h/year	9.11
The growth in retained earnings from increased sales, mln rubles/year	18.44
The income tax increase, mln rubles/year	3.69
The increase in depreciation, mln rubles/year	0.28
The increase in operating costs, mln rubles/year	0
The annual increase in net profit, mln rubles/year	14.47
The payback period, years	0.35

6. Conclusion

The study showed the effectiveness of joint operation scheme of a steam power unit and a combined cycle plant during the period of technical maximum. So, in the conditions of the Nevinnomyssk TPP, the displacement of high-temperature steam bleeding unit onto the high-pressure regenerative heater (HPFH 2) of the K-160-130 steam turbine unit with dry steam from the medium-pressure drum of the CCP-410 waste boiler allows increasing the maximum electric power of the power plant by 6.03 MW at a slight increase in total electrical efficiency.

The application of the joint scheme in the conditions of the Nevinnomyssk TPP is economically justified. Its implementation will lead to an increase in annual net profit by 14.47 million rubles, with a payback period of investment - 4.2 months.

Acknowledgements

The study was carried out under a grant of the President of the Russian Federation for state support of young Russian scientists - candidates of science (Competition - MK-2020), performer MK-395.2020.8.

Reference

[1] Order of the Ministry of Energy dd 02.02.2019 N 174 "On approval of the development scheme and program for Russian Unified Energy System for 2019-2025" https://www.so-

ups.ru/index.php?id=1196&tx_ms1cdu_pi1%5Bdt%5D=31.12.2019
[2] Official site of JSC "System operator of the unified energy system" https://www.so-ups.ru/index.php?id=1196&tx_ms1cdu_pi1%5Bdt%5D=31.12.2019
[3] Devlet-Geldy G K and Golikov V D 2019 Green economy: a new vector of public-private partnership in the breakthrough development of Russia *Economics and Business: Theory and Practice* **2** 28-34
[4] Efimov N N, Papin V V and Bezuglov R V 2016 Determination of rotor surfacing time for the vertical microturbine with axial gas-dynamic bearings *Procedia Engineering* **150** 294-99
[5] Efimov N N, Papin V V and Bezuglov R V 2016 Micro energy complex based on wet-steam turbine *Procedia Engineering* **150** 324-29
[6] Information review "Russian Unified Energy System: Interim Results" https://so-ups.ru/fileadmin/files/company/reports/ups-review/2019/ups_review_1119.pdf
[7] Barinberg G D and Dlugoselsky V I 1998 115 MW cogeneration turbines as part of CCP-170 *Thermal Engineering* **1** 16-19
[8] Borovkov V M and Koshelev S M 2005 Optimization the operation of the regeneration system of cogeneration turbine units in regimes with minimal passes of steam into the *Problems of Energy* **7-8** 9-14.
[9] Galushko V F 2007 Modernization of the block part of the Krasnodar CCP gas turbine superstructure *Energysaving and Watertreatment* **5** 33-34
[10] Teplov B D and Radin Yu A 2019 Improving flexibility and economic efficiency of CCGT units' operation in the conditions of the wholesale electricity market *Thermal Engineering* **5** 323-30
[11] Biryukov B V, Sherstobitov I V and Dashevskii M F 1980 Thermal power station, patent of the USSR no 775357 appl. 01.02.1979, publ. 30.10.1980
[12] Sherstobitov I V, Biryukov B V, Kalinin V E and Bychikhin PN 1983 Thermal power station, patent of the USSR no 1000559 appl. 03.07.1981, publ. 28.02.1983
[13] Biryukov B V, Sherstobitov I V and Galushko V F 1983 Thermal power station, patent of the USSR no 1048134 appl. 22.04.1982, publ. 15.10.1983
[14] Biryukov B V and Sherstobitov I V 1993 Thermal power station, patent of the USSR no 1802177 appl. 22.01.1991, publ. 15.03.1993
[15] Zhao Y, Liu M, Wang C, Wang Z, Chong D and Yan J 2019 Exergy analysis of the regulating measures of operational flexibility in supercritical coal-fired power plants during transient processes *Applied Energy* **253** 113487
[16] Mata-Torres C, Zurita A, Cardemil J M and Escobar R 2019 Exergy cost and thermoeconomic analysis of a Rankine cycle + multi-effect distillation plant considering time-varying conditions *Energy Conversion and Management* **192** 114-32
[17] Adibhatla S and Kaushik S C 2017 Exergy and thermoeconomic analyses of 500 MWe sub critical thermal power plant with solar aided feed water heating *Applied Thermal Engineering* **123** 340-52
[18] Shaposhnikov V V, Biryukov B V, Batko D N and Trofimenko A A 2019 Thermal power plant, patent of the Russian Federation no 2691881 appl. 06.07.2018, publ. 18.06.2019
[19] Shaposhnikov V V, Biryukov B V, Batko D N and Mihalko Ya O 2020 A review of the methods for obtaining additional peak power of steam power plants by shutting down and removing of regenerative bleed *Industrial Power Engineering* **2** 21-33

IOP Conf. Series: Materials Science and Engineering 1035 (2021) 012045 doi:10.1088/1757-899X/1035/1/012045

Modeling of the cooling process of long pipelines for transportation of liquid hydrogen

O V Kalyadin[1,3] , A V Sergeev[1], A A Grebennikov[2] and A Yu Lopatin[1]

[1] Department of Solid State Physics, Voronezh State Technical University, Moscow Avenue, 14, Voronezh, 394026, Russia
[2] Basic research and educational center «Physics and technology of thermoelectric phenomena», Voronezh State Technical University, Moscow Avenue, 14, Voronezh, 394026, Russia

[3]Corresponding author: kaljadin@gmail.com

Abstract. A simplified model of cooling long insulated pipelines with liquid hydrogen flowing through them has been developed. The model allows us to determine the flow parameters at different points in time and to estimate the pipeline cooling time. When developing the model, the following assumptions were made: the linear velocities of the liquid and vapor are equal, the phases are in a state of thermodynamic equilibrium, the dependencies obtained for a single-phase flow are applicable to determine the friction coefficient of a two-phase flow. Based on the model, an automated algorithm for calculating the cooling process of a long pipeline with cryogenic components was developed. It allows us to obtain data for plotting the temperature fields of the pipeline walls and the flow of the transported cryogenic liquid at different times, as well as to determine the cooling time of the pipeline and the moment of the onset of a stationary single-phase flow. The calculation results are in good agreement with the experimental data.

1. Introduiction

Testing of liquid rocket engines (LRE) is a comprehensive test for equipment, components and systems of an engine to establish compliance of their characteristics with the specified tactical and technical requirements [1]. Tests of the LRE are a complex and multifaceted system of measures that allows us to obtain reliable information about the operation of all LRE components and to make an informed decision about its installation in a launch vehicle. Unlike other components or systems of the launch vehicle, LRE tests are characterized by some features related both to the specifics of the processes occurring in them and to operating conditions.

The most important feature that significantly limits the number of possible tests is their high cost, which is associated largely with the high cost of fuel components and the uniqueness of the equipment used [2]. In this regard, high informativity of tests, rational planning and possible minimization of both the total number of tests and fuel losses in each of them is necessary. To solve this problem, it is necessary to work out the optimal modes of difficult-to-control technological processes, which include the process of preparing a bench system for testing. However, to carry out such processes under different initial conditions in order to collect statistical data and to obtain the optimal mode is a very expensive and difficult task. Therefore, the modeling of physical processes actually occurring in the system is the only right decision [1,2].

2. Features of the cooling process of pipelines during transportation of cryogenic liquids through them

It is necessary to cool the main pipeline to achieve the required flow rate of cryogenic products. Cooling can be carried out both by liquid and gas. The choice of the phase state of the cryogenic product is determined by the purpose of the system, the requirements for the temporary and dynamic characteristics of the transition regime, the permissible levels of temperature stresses in the structural elements in liquid systems, cooling is usually performed directly by liquid products. The cooling stage of the walls of the main pipeline is characterized by intensive steam generation. This is due to the fact that the initial wall temperature is close to the ambient temperature; therefore, its value significantly exceeds both the saturation temperature (T_s) and the overheat limit temperature (T_{ol}) - the upper limit of the existence of cryogenic products in liquid form. Thus, cryogenic liquids cannot exist near the pipeline walls, while walls temperature is high, because they instantly turn into steam.

When cryogenic liquids enter a warm pipeline, the pipeline walls are stably blocked by the vapor film, i.e. film boiling is realized. With a decrease in the wall temperature below the T_{ol} temperature, a film boiling crisis occurs, and then it is replaced first by transitional and then bubble boiling. Bubble boiling degenerates into convective heat transfer as the temperature of the pipe wall approaches the saturation temperature. Boiling of cryogenic liquids at the stage of pipelines cooling is realized under various modes of movement of the vapor-liquid mixture. The film boiling is characterized by the presence of a vapor film in the near-wall region; therefore, the modes of motion of vapor-liquid mixtures are called reversed. The most typical of cooling conditions are the rod, shell and dispersed reversed modes. As wall temperature decreases, conditions for wetting the walls with liquid appear, so the reversed regimes are replaced by the usual ones, which are characterized by the presence of a liquid phase in the near-wall region. The cooling and filling of extended main pipelines with cryogenic liquids are accompanied by fluctuations in flow rate and pressure, which are caused by intensive steam generation.

It was found by experimental studies [3] that the thickness of the vapor film arising during the flow of underheated liquid is so small that it practically does not constrain the movement of the liquid jet. Moreover, the presence of a vapor film leads to a decrease in shear stresses, since the viscosity of the vapor is an order of magnitude lower than the viscosity of the liquid. The liquid warms up as it moves through the pipeline, respectively, the pressure in the pipeline increases and the filling rate decreases. If the length of the pipeline is large, then the front of the liquid warms up to the saturation temperature at some distance. In this case, the pressure in the pipeline and in the tank becomes the same, so the flow of liquid into the pipeline stops. At the same time, the core structure of the flow is destroyed due to the heating of the liquid and the increase in vapor content.

When a liquid is heated to a saturation state, almost the entire heat flux from the walls goes to evaporation. The steam generation rate and, consequently, the pressure in the pipeline continue to increase, despite the cessation of the new portions of liquid. The pressure in the pipeline can exceed the inlet pressure from 1.5 to 2 or more times. As the liquid evaporates and the vapor escapes, as well as the liquid, is partially displaced back into the vessel, the steam generation rate in the pipeline decreases, and after some time conditions for a new portion of the liquid to enter are created in it. A new flow of liquid leads to a smaller increase in pressure, since a fairly high pressure has remained in the pipeline and a part of the liquid has remained. During the entire period of cooling and filling of extended pipelines, fluctuations in the flow rate of the incoming liquid and pressure in the pipeline have a much lower amplitude compared to the first cycle.

An analysis of the experimental data shows that the steady flow of liquid occurs almost simultaneously with the cooling of the wall. The liquid appears at the outlet of the pipeline before the pipe is completely cooled, so a vapor-liquid mixture leaves the pipeline for some time. The entire cooling period τ can be divided into two stages. The first stage τ_1 is from the beginning of the liquid supply until the vapor-liquid mixture appears at the outlet of the pipeline. The second stage τ_2 is from the moment the vapor-liquid mixture appears at the outlet until the wall is completely cooled and the stationary flow rate is established. The ratio between the duration of individual stages and the total cooling time (as well as between the length of the evaporation zone and the total length of the pipeline) can be very different. Three types of pipelines are distinguished - long, medium and short, depending on the relative duration of the individual cooling stages. In long pipelines, the time of the liquid appearance at the outlet does not differ much from the time of the complete cooling of the wall. The length of such pipelines far exceeds the evaporation zone ($\tau \approx \tau_1, l_e \ll l$). The duration of the individual stages is commensurate with each other for medium pipelines. The evaporation zone in such pipelines is slightly less than their length at the beginning of cooling. It increases with decreasing wall temperature and after some time exceeds the total length ($\tau \approx \tau_1 + \tau_2, l_e \ll l$). In short pipelines, the time of the appearance of the liquid at the outlet is negligible compared to the total cooling time ($\tau \approx \tau_2, l_e > l$). The length of such pipelines is insufficient for the complete evaporation of the liquid.

The relative length of the pipelines has the main influence on the ratio between the duration of individual stages and the total cooling time. As shown by the analysis of experimental data obtained on nitrogen and oxygen, long pipelines include pipelines that are longer than $(1{,}5 \div 2) \cdot 10^3$ calibers. short ones include pipelines shorter than $0{,}5 \cdot 10^3$ calibers. In the practice of creating large cryogenic systems, the length of the main pipelines exceeds several thousand calibers; therefore, the cooling of long pipelines is of the greatest practical interest.

3. Model description

At the initial stage of cooling, the temperature of the pipeline walls significantly exceeds the saturation temperature of cryogenic liquids. Therefore, their partial or complete evaporation occurs at the outlet of the pipeline. When transporting cryogenic liquids along warm pipelines their partial or complete evaporation occurs. Therefore, the temperature distribution in the walls and the flow parameters are determined by solving a system of equations describing the change in the flow parameters in the single-phase and two-phase flow sections. The exact solution to this problem is associated with great difficulties, since the laws and forms of joint motion of a liquid and gas are much more complex and diverse than the laws and forms of motion of homogeneous flows. The discrete volumes of each phase contain a large number of molecules, so the movement of the medium inside any volume can be determined by the usual differential equations of continuity, motion, and energy. However, in two-phase flows, in addition to the external boundaries formed by the channel walls, there are internal interfaces, which are variable in time and space. Force and thermal interactions appear at the interface; therefore, the equations of hydrodynamics for each of the phases must be supplemented with equations describing the mechanical and thermal interaction of phases at the interfaces. From a practical point of view, in addition to determining liquid losses, an interesting task is to estimate the time for complete cooling of cryogenic pipelines and establish a steady flow rate of cryogenic liquid.

To calculate the cooling time of the pipeline, it is necessary to solve a system of differential equations of continuity, motion, flow energy, and thermal conductivity for walls with different boundary conditions on the inner and the outer surfaces of the pipe. In the case when the cooling of the pipeline proceeds without changing the aggregate state of the substance, the following expressions are used as the initial equations:

- flow continuity equation

$$\frac{\partial \rho}{\partial \tau} + \frac{\partial (\rho W)}{\partial z} = 0; \tag{1}$$

- equation of motion

$$-\frac{\partial R}{\partial z} - \frac{4\sigma}{d_{\text{int}}} = \rho \frac{\partial W}{\partial \tau} + \rho W \frac{\partial W}{\partial z}; \tag{2}$$

- energy equation

$$\rho \frac{\partial i}{\partial \tau} + \rho W \frac{\partial i}{\partial z} = \frac{4q}{d_{\text{int}}}; \tag{3}$$

- heat conduction equations for pipe walls

$$\frac{\partial T_w}{\partial \tau} = a\left(\frac{\partial^2 T_w}{\partial R^2} + \frac{1}{R}\frac{\partial T_w}{\partial R}\frac{\partial^2 T_w}{\partial z^2}\right); \tag{4}$$

$$\lambda_w \frac{\partial T_w(R_{\text{int}}\tau)}{\partial R} = q_{\text{int}} = a[T_w(R_{\text{int}}\tau) - T]; \tag{5}$$

$$-\lambda_w \frac{\partial T_w(R_{\text{int}}\tau)}{\partial R} = q_{ext} \tag{6}$$

q_{int}, q_{ext}, heat flux density on the inner and the outer surfaces of the pipe; R_{int}, R_{ext}, inner and outer pipe radii; T_w, T, the temperature of the wall and the one of the flow.

When describing the process of cooling thin-walled insulated pipelines by a flow of a cryogenic liquid with partial or complete evaporation, the form of writing the heat conduction equation for the pipe walls (4) - (6) is preserved, but the equations of continuity (7), motion (8) and flow energy (9) change due to the formation of a vapor-liquid mixture

$$\frac{\partial}{\partial \tau}\left((1-\varphi)\rho_l + \varphi\rho_g\right) + \frac{\partial}{\partial z}\left((1-\varphi)\rho_l W_l + \varphi\rho_g W_g\right) = 0; \tag{7}$$

$$g\left((1-\varphi)\rho_l + \varphi\rho_g\right) - \frac{\partial p}{\partial z} - \frac{4\sigma}{D} = (1-\varphi)\rho_l\frac{\partial W_l}{\partial \tau} + \varphi\rho_g\frac{\partial W_g}{\partial \tau} +$$

$$+(1-\varphi)\rho_l W_l\frac{\partial W_g}{\partial \tau} + \varphi\rho_g W_g\frac{\partial W_g}{\partial \tau} + (W_g - W_l)\left(\frac{\partial}{\partial \tau}(\varphi\rho_g) + \frac{\partial}{\partial z}(\varphi\rho_g)W_g\right) \tag{8}$$

$$\frac{\partial}{\partial \tau}\left((1-\varphi)\rho_l i_l + \varphi\rho_g i_g\right) + \frac{\partial}{\partial z}\left((1-\varphi)\rho_l i_l W_l + \varphi\rho_g i_g W_g\right) = \frac{4q}{d_{\text{int}}} \tag{9}$$

φ, volumetric steam content of the flow (in the evaporation section $0 < \varphi < 1$, in the gas flow section $\varphi = 1$); q, heat flux density on the inner surface of the pipe; σ, shear stress on the wall.

However, the system of equations remains open, therefore additional experimental dependencies are required to solve it. These dependencies should describe the relationship between the volumetric vapor content, the velocities of each of the phases, shear stresses on the wall, the temperature of the vapor phase and walls, the intensity of heat transfer from the walls to the product. Such dependencies were obtained only for some special cases. They are not universal due to the complex nature of two-phase flows. The absence of additional general equations makes solving the system of initial equations impossible. In this regard, methods for determining the characteristics of the process of cooling pipelines by cryogenic liquids are based on obtaining semi-empirical and empirical relationships from experimental data.

In this investigation, a homogeneous model was used to solve the considered problem approximately [4,5]. Its main assumptions are based on the assumption that the linear velocities of vapor and liquid are equal, the phases are in thermodynamic equilibrium, and the dependencies obtained to determine the coefficient of friction of single-phase flow are applicable to two-phase flow. In the homogeneous model, two-phase flow is considered as a single-phase flow, in which the specific volume in each section is related to the mass vapor content and the specific volume of each of the phases, i.e.

$$\overline{V}_s = \overline{V}_l(1-\varphi) + \overline{V}_g\varphi \tag{10}$$

In this case, the system of equations (1 - 6) can be used as a calculated one. We will assume that the mass flow rate is known, the thermal conductivity of the wall in the direction of motion is zero and is infinite in the direction of the normal, the wall on the outside is ideally isolated from the heat influx from the environment. These assumptions are justified from a technical point of view, they can significantly simplify the system of initial equations and its solution. In view of the above and taking into account that $di = c_p dT$ we transform the system (1 - 6) to the form

$$\frac{\partial T}{\partial \tau} + W\frac{\partial T}{\partial z} = \frac{4\alpha(T_w - T)}{\rho_g c_{pg} d_{\text{int}}} \tag{11}$$

$$\frac{\partial T_w}{\partial \tau} + \frac{\alpha(T_w - T)}{\rho_w c_{pw}\delta} = 0 \tag{12}$$

The temperature distribution in the gas and the wall at the initial moment of time and the gas temperature at the inlet are set as the boundary conditions $T(0,z) = f(z)$; $T_w(0,z) = \varphi(z)$; $T(\tau,0) = \psi(\tau)$.The heat transfer coefficient included in the system is calculated by the formula

$$\alpha = \frac{\lambda_{\delta_g}}{d_{\text{int}}} Nu \tag{13}$$

where the Nusselt number, taking into account the homogeneous flow model [5]

$$Nu = 0{,}0065\,\mathrm{Re}^{0.8}\left(\frac{c_{pg}(T_0 - T_{\text{int}})}{r}\right)^{-1/6} \tag{14}$$

Functional dependencies are obtained as a result of solving the system of equations. They make it possible to obtain the distribution of wall and flow temperatures for any moment in time in each section of the pipeline. Thus, it is possible to estimate the cooling time of a random section of the pipeline or the pipeline as a whole. In addition, the obtained dependences, plotted for different points in time, can be used to calculate the current value of the average wall temperature or plot the curves of the change in the average temperature $T_{ws}(\tau)$ and the thermal resource of the walls $Q_w(\tau)$ in any section under various initial conditions

$$T_{ws} = \frac{1}{L_{cool}} \int_0^{L_{cool}} T_w(z)dz \tag{15}$$

$$Q_w(\tau) = (T_{ws}(\tau) - T)c_{pw}(T_{ws})\rho_w\delta \tag{16}$$

4. Modeling results

An automated algorithm for calculating the cooling process of a long pipeline with cryogenic components is developed on the basis of the proposed model. A pipeline made of steel was taken as the object of modeling. The pipeline is insulated with screen-vacuum insulation. Total pipeline length is 272.5 m, bore is 96 mm, wall thickness is 2 mm. The total mass of the locking equipment located on the pipeline is 246 kg. Liquid supercooled hydrogen is fed into the pipeline under an excess pressure of 0.2 MPa, the hydrogen temperature at the inlet is 19 K. The ambient temperature is 293 K.

As a result of calculations, the distribution of hydrogen temperature and wall temperature along the length of the pipeline, as well as at different points in time, were obtained. These results are shown in figures 1, 2 for some sections and points in time.

IOP Conf. Series: Materials Science and Engineering 1035 (2021) 012045 doi:10.1088/1757-899X/1035/1/012045

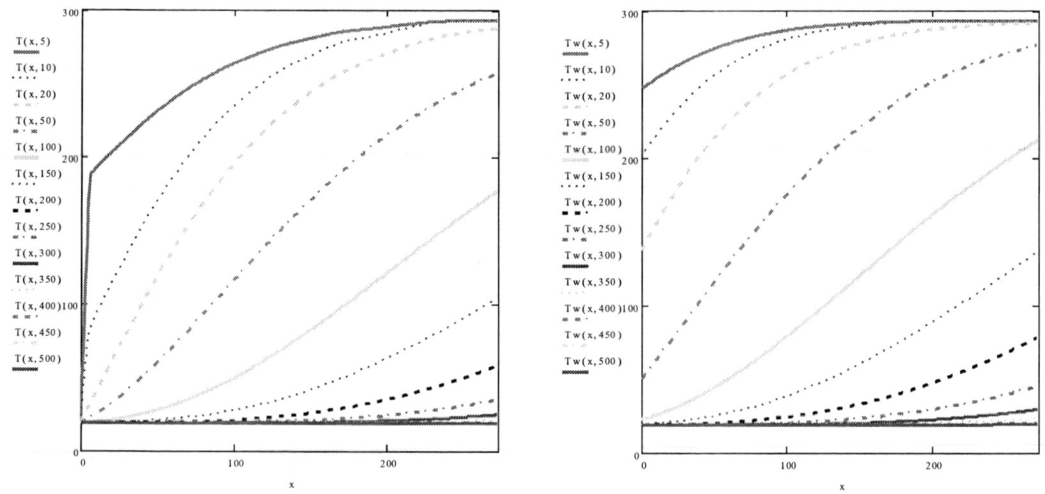

a) b)

Figure 1. Distribution of the temperature of the flow (a) and the temperature of the wall (b) along the length of the pipeline at different points in time.

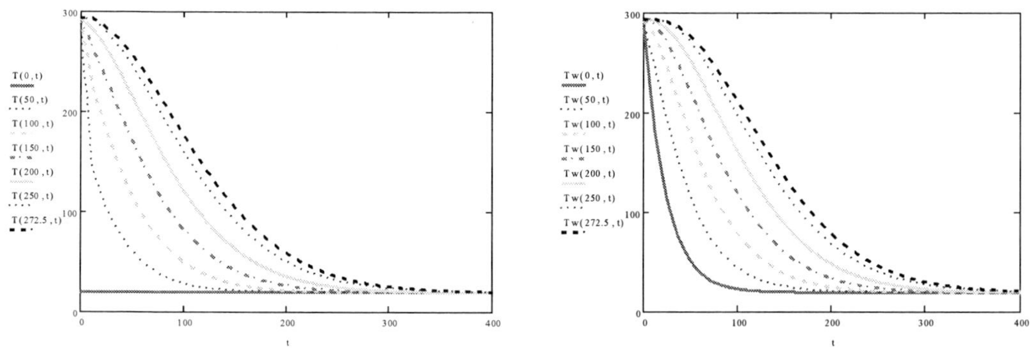

a) b)

Figure 2. Time dependence of the temperature of the flow (a) and the temperature of the wall (b) in different sections of the pipeline.

The current value of the average wall temperature according to the formula (15) can be calculated using the temperature fields plotted for different points in time along the length of the cooling pipes (figure 2) $T_w = T_w(x)$. Figure 3 shows the time dependence of the average wall temperature along the pipe length $T_{ws}(\tau)$. Based on the obtained data, the time dependences of changes in the thermal resource of the wall for the entire pipeline and its individual sections at a distance of 5, 50, 150 m from the inlet section were calculated using formula 16 (figure 4):

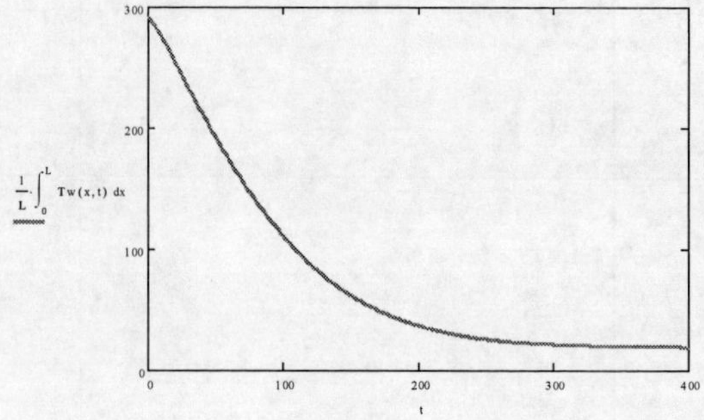

Figure 3. Time dependence of the average wall temperature.

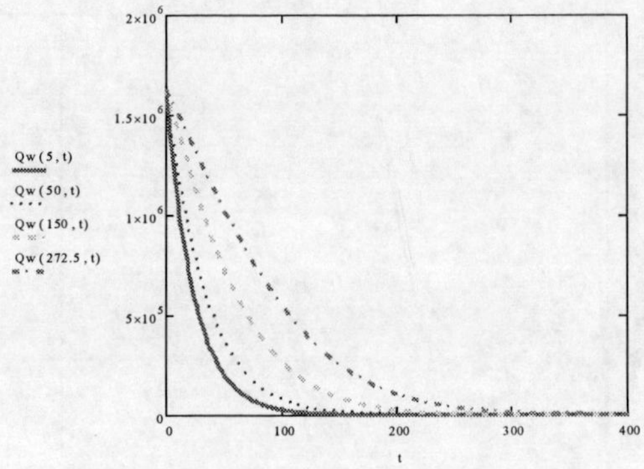

Figure 4. Time dependence of the thermal resource of the wall for different sections of the pipeline.

It can be seen from the figures that the calculated cooling time of the pipeline is ~ 400 s, which is in good agreement with the data obtained as a result of experiments on the operating stands of the KBKhA.

5. Conclusion

A mathematical model of cooling of long pipelines with cryogenic components has been developed. Based on this model, an automated calculation algorithm was created. It allows us to obtain data for plotting the temperature fields of the pipeline walls and the flow of the cryogenic liquid at different points in time, as well as to determine the time of pipeline cooling and the moment of the onset of a stationary single-phase flow. Using this model under various initial and boundary conditions, it is possible to determine the optimal mode of pipeline cooling and achieve minimum losses of cryogenic components with minimum time costs.

References

[1] Galeev A G 2010 *Basis of Structure of Test Benches for Testing Liquid Propellant Rocket Engines and Propulsion Systems* (Peresvet: Scientific Testing Center of Rocket Space Industry)

[2] Abildaeva K Zh 2018 The development of methods for cooling cryogenic component of rocket fuel - liquid hydrogen when refueling tanks of advanced launch vehicles and boosters

[3] Sudarchikov A M 2018 Critical flow of liquid in a long channel *Technical Physics. The Russian Journal of Applied Physics*

[4] Filin N V and Bulanov A B 1985 *Liquid Cryogenic Systems* (Leningrad: Mashinostroenie)

[5] Belyakov V P 1982 *Cryogenic Engineering and Technology* (Moscos: Energoizdat)

IOP Conf. Series: Materials Science and Engineering 1035 (2021) 012046 doi:10.1088/1757-899X/1035/1/012046

Intensification of heat transfer in chaotic modes

S L Podvalny and E M Vasiljev

Voronezh State Technical University, Moscow Avenue, 14, Voronezh, 394026, Russia

Abstract. This paper presents the results of numerical modeling and comparative studies of the effectiveness of diffusion, convective, and turbulent heat transfer modes in a liquid layer. Here the heat transfer physical model was used which is typical for reactors with an internal energy source. Conditions for the continuity of the fluid flow and the equations of heat transfer, analytical and simulation models of these modes are obtained based on the Navier-Stokes equation. Quantitative characteristics have been found using the simulation model, confirming the expected efficiency indicators ratios of these processes in the course of heat transfer. Convective and turbulent heat transfer processes properties are analyzed as control objects basing on the Comsol Multiphysics modeling system. Recommendations have been obtained concerning the multi-alternative algorithms synthesis for automatic control of these processes. In particular, it is advisable to use the heat flux density as a control action, whereas the liquid particle velocity dispersion should be used as an adjustable value. Corresponding control system model demonstrated the efficiency of the proposed algorithms.

1. Introduction

Widespread use of the heat transfer process in modern industrial technologies involves the continuous improvement of its efficiency indicators [1-5]. Most publications describing the activity in this direction are devoted mainly to the constructive changes in the heat transfer devices that give a certain growth of heat transfer coefficient and heat flux density [6-11]. At the same time, the significant and as yet insufficiently studied possibilities for improving these characteristics lie in the implementation of automatic heat transfer control which provides the required process quality indicators under the unsteady operating conditions.

The results of studies accumulated by now in hydromechanics indicate the possibility to increase heat transfer efficiency by means of internal turbulization occurring in the liquid layer under certain conditions. It is established that such a condition is raising the temperature gradient [12,13]. A small temperature gradients heat transfer occurs due to the motion of the liquid particles at the molecular level, i.e., the diffusion. As far as the gradient increases, the convective thermal waves arise involving significant volumes of liquid. With further temperature gradient growth, convective waves collapse causing the disordered, chaotic fluid motion which can be identified as turbulence.

Thus, the heat exchange process is characterized by the possibility of occurrence of qualitatively different, critical internal operating modes and respectively the goal can be set to control these modes using the automation.

This article describes the results of a quantitative study of the critical modes' emergence pattern basing on the corresponding mathematical and simulation models, as well as a corresponding control system for them.

2. Physical and mathematical models of the heat transfer process

A horizontal liquid layer is offered as a specific physical heat transfer scheme, with layer height $h = 5 \cdot 10^{-2}$ m, heated from below by a heat source with heat flux density q. Solid wall with thickness $\delta_w = 3 \cdot 10^{-3}$ m and heat transfer coefficient $\alpha = 500$ W/(m^2·K) is used as heat sink to the external environment with a constant temperature $T_{ex} = 20$ °C, as shown in figure 1; $\lambda_w = 100$ W/(m·K) is the wall thermal conductivity coefficient. This model is surely the best possible variant describing the heat exchangers with an internal energy source (reactors, electric heaters).

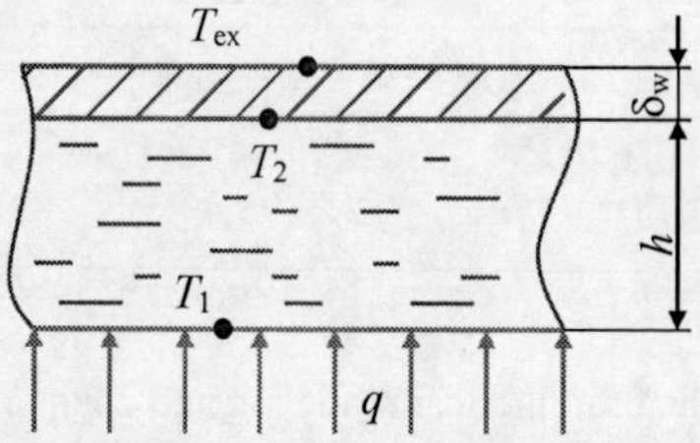

Figure 1. Heat transfer process physical model.

For mathematical description of a fluid motion and heat transfer in this physical model, the author used the Navier-Stokes equation [14-17]:

$$\frac{\partial \mathbf{v}}{\partial t} + (\mathbf{v}\nabla)\mathbf{v} = -\frac{1}{\rho}\nabla p + \nu\nabla^2\mathbf{v} + \mathbf{g};$$

the fluid flow continuity equation:

$$\frac{\partial \rho}{\partial t} + \nabla(\rho\mathbf{v}) = 0;$$

and heat transfer equation:

$$\frac{\partial T}{\partial t} + \nabla(T\mathbf{v}) = k\nabla^2 T,$$

where: $\mathbf{v}(x,y,z,t)$, $p(x,y,z,t)$, $\rho(x,y,z,t)$, $T(x,y,z,t)$ - velocity, pressure, density and temperature fields in the liquid layer varying in the space xyz and in time t; $\mathbf{g}$ - acceleration of gravity in the gravitational field, $g = 9.8$ m/sec^2; k - thermal diffusivity of liquid, $k = 5.8\cdot10^{-8}$ m^2/sec; ν - kinematic viscosity, $\nu = 2.6\cdot10^{-6}$ m^2/sec.

Linear density/temperature dependence is adopted in this model: $\rho = \rho_0 - \rho_0\gamma(T - T_0)$, where: γ - the temperature coefficient of volume expansion, $\gamma = 7.3\cdot10^{-4}$ 1/K; T_0 – the temperature for which the reference value of density ρ_0 was determined: for $T_0 = 100$ °C, $\rho_0 = 832$ kg/m^3 was chosen.

Boundary conditions correspond to a horizontal liquid layer enclosed between the side walls. The walls are either thermally insulated for modeling the diffusion mode or isothermal when heat transfer simulation in a moving fluid is made. All velocity components on the wetted surfaces of the side walls are equal to zero.

Here the two-dimensional model is adopted: horizontal coordinate x=[0; 0.3] m, vertical y=[0; 0.05] m; any changes in velocity, pressure, density, and temperature along the z coordinate are absent.

3. Critical modes study

The resulting mathematical description of the heat transfer process was used to build its simulation model in the Comsol Multiphysics package. Simulation results allowed us to establish that there are three possible modes of heat transfer in the considered system, viz., the diffusion, convective and turbulent modes.

Figure 2 shows the distribution of the temperature field in the diffusion mode.

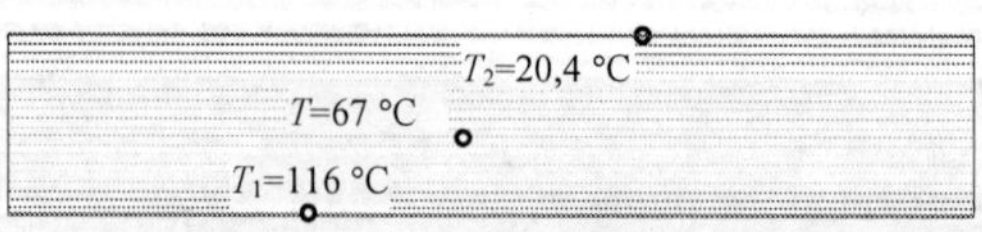

Figure 2. The temperature field in diffusion mode.

This mode occurs at very small temperature gradients and its practical value is small, however, it is considered here with the purpose to compare heat transfer quality indicators in all three critical modes.

In the convective mode the occurrence of alternating ascending and descending fluid flows is observed along with the diffusion heat transfer process. These flows generate stable convective waves (figure 3) which provide, in comparison with the diffusion mode, a significant increase in the intensity of mass and heat transfer.

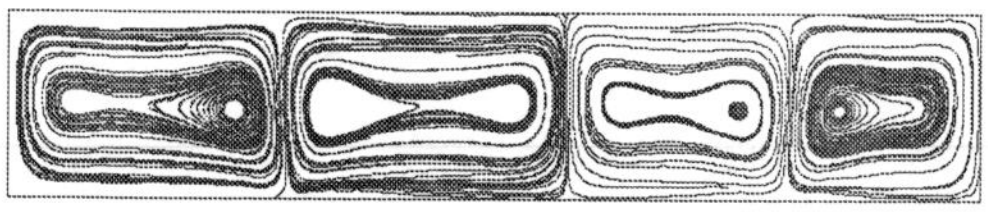

Figure 3. The fluid flow lines in the steady state convective heat transfer.

The presence in the convection mode of zones with an alternating direction of fluid circulation leads to a corresponding periodicity of the temperature field with a repeated increase and decrease of temperature relative to its average value along two coordinate axes. In figure 4 the areas with higher temperatures correspond to shaded areas with closely spaced isotherms.

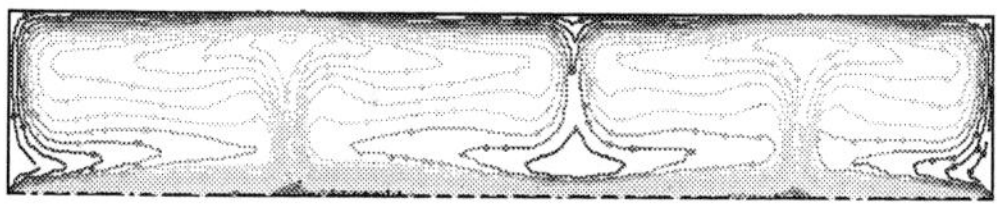

Figure 4. Temperature field in convection mode.

Figure 5 shows the velocity changes during the transient process (about 4000 sec) and after its completion at a point on the $x0y$ plane with the coordinates $x = 0.04$ m and $y = 0.005$ m. Steady state convection is characterized by the stationary value of fluid particle velocity at each point.

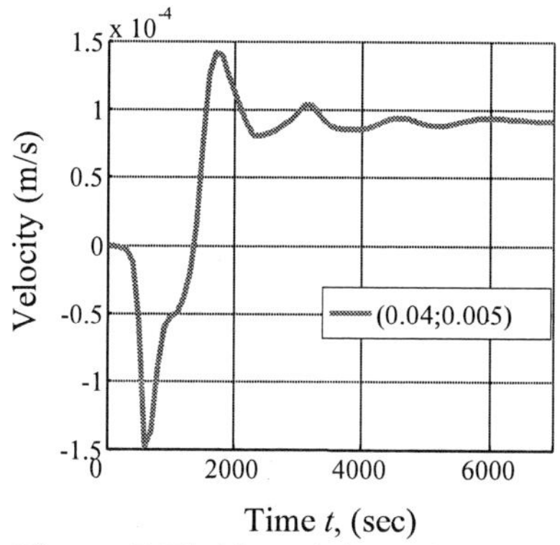

Figure 5. Fluid particles velocity change at the stage of convective heat transfer.

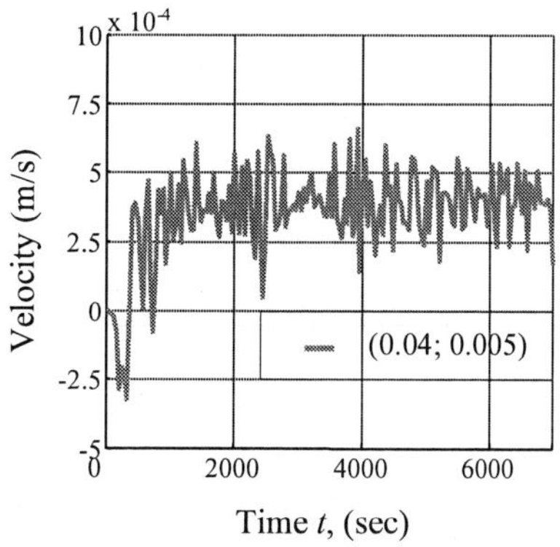

Figure 6. Change in the fluid particles velocity in the turbulent heat transfer mode.

As soon as the heat flux density q increases, the lower wall temperature increases, the unevenness of the temperature field increases, and convective waves gradually collapse. Fluid flow becomes turbulent, as shown in figure 7.

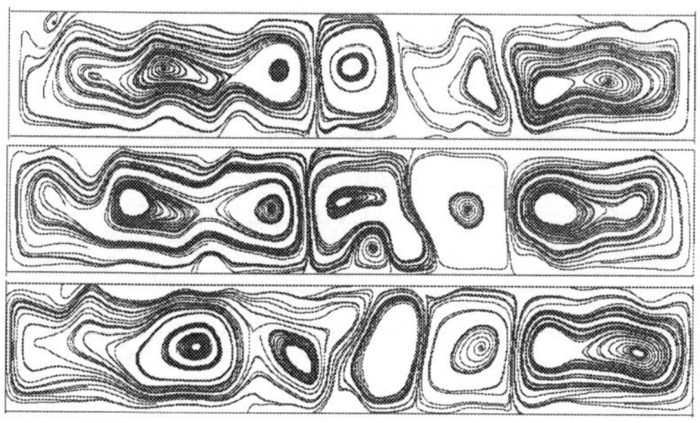

Figure 7. Fluid flow lines in unsteady mode of turbulent heat transfer at successive time point with 40 s intervals.

Temperature field unevenness becomes chaotic in the turbulent mode [18-22] which leads, in comparison with the convection mode, to the spatial alignment of time-average temperature values (figure 8), and helps to gain heat exchangers operational reliability.

Figure 8. Temperature field distribution in turbulence mode.

The turbulent flow is characterized by chaotic pulsations of the particle velocity which provide the rapid transfer of the momentum and heat, as in figure 6.

An average fluid particles velocity in the process of turbulent heat transfer is several times higher than the convective transfer velocity; whereas the amplitude of the velocity pulsations exceeds significantly the range of molecular motions (see figures 5 and 6). Thus, one should expect an increase in the efficiency of turbulent heat transfer in comparison with the convective mode. For the quantitative comparison of heat transfer quality indicators in all these modes, a heat transfer coefficient was determined for each of them.

Due to the continuity of the heat flux, we can write:

$$q = r(T_1 - T_2) \text{ – for fluid layer ;}$$

$$q = \frac{\lambda_w}{\delta_w}(T_2 - T_3) \text{ – for upper wall ;}$$

$$q = \alpha(T_3 - T_{ex}) \text{ – for the boundary with the environmen t.}$$

In these expressions, T_1, T_2 and T_3 are the x-coordinates of the temperature of the surface heated from below and the lower and upper surfaces of the wall, respectively, in accordance with figure 1; where r is the heat transfer coefficient of the liquid layer.

Using these conditions for all simulated heat transfer modes, the expression was obtained:

$$r = \frac{T_2 - T_{ex}}{T_1 - T_2} \cdot \frac{1}{\left(\frac{\delta_w}{\lambda_w} + \frac{1}{\alpha}\right)}$$

and the heat transfer coefficient values shown in table 1 were calculated.

Table 1. Comparative characteristics of critical heat transfer modes.

Heat transfer mode	T_1 (°C)	T_2 (°C)	q (W/m^2)	r (W/(m^2·K))
Diffusion	116	20,4	200	2.06
Convection	35	20,4	200	13.49
Turbulence	150	24	2000	15.73

Simulation results reflect well the physical features of the considered modes and confirm the expected gain in the heat transfer efficiency in a chaotic mode.

4. Heat exchange process control system

The analysis of critical modes of the heat transfer process makes it possible to draw the following conclusions regarding the possible structure of its control system:

the goal of control contains two practically significant alternatives, namely: maintaining the either convective or turbulent mode of heat transfer in the system, i.e. one of two critical modes;

it is advisable to choose the heat flux density q as a control action;

the distinctive feature by which one can recognize the convective or turbulent mode of heat transfer is the amplitude of pulsations of the fluid particles velocity. Therefore, variance $\sigma^2(t)$ of this velocity can be taken as a controlled quantity;

the most significant disturbing influences are the ambient temperature T_{ex} and the heat transfer coefficient α of the apparatus walls both of which inevitably change during its operation.

Thus, for the implementation of multi alternative control, a control system with a variable structure should be synthesized containing two closed loops for the deviation of the controlled variable and providing for their switching to select the heat transfer mode [23-27].

Figure 9 shows a variant of the proposed control system structure. In this version, a heat flux with some calculated nominal density q_n is supplied to the system input and the instantly calculated dispersion $\sigma^2(t)$ of the velocity $w(t)$ is supplied to the of nonlinear controllers inputs (any component $\mathbf{v}(x,y,z,t)$ or its module can be selected as $w(t)$).

Static characteristic of controller R_t (the upper one in figure 9) provides a growing heat flux density at small $\sigma^2(t)$ values, i.e., this regulator is designed to maintain turbulent heat transfer. Controller R_c decreases the nominal value of density q_n for large $\sigma^2(t)$, i.e. providing the maintenance of the convective heat transfer mode.

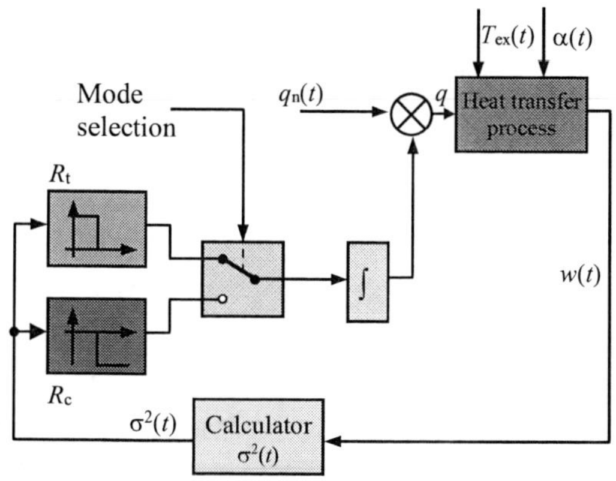

Figure 9. Heat exchange control system structural diagram.

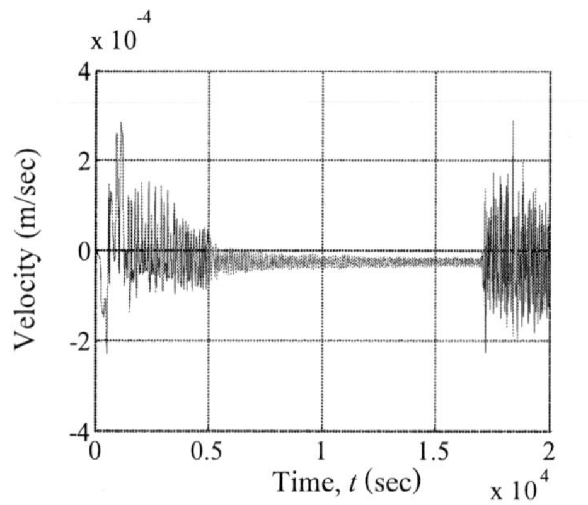

Figure 10. System performance check in the turbulent heat transfer stabilization mode.

The linear part of the regulators is common for both modes being an integrating link which provides the property of astatism in the system with respect to the disturbances $T_{ex}(t)$ and $\alpha(t)$.

The ability of the system to fend off disturbances and control heat transfer modes is illustrated in figure 10 which shows the change in the velocity of the fluid particles at the selected sensor installation point ($x = 0.06$ m, $y = 0.005$ m) in the time intervals with different modes listed in table 2.

Table 2. Situations in which the system performance was checked.

Time interval (sec)	$T_{вех}$ (°C)	α (W/m²·K)	q (W/m²)	System condition	Heat transfer mode
0-4000	20	500	1000	Open	Turbulent
4000-17000	50	250	1000	Open	Convective
17000-20000	50	250	2000	Closed	Turbulent

Attention should be paid to the expectedly noticeable inertia of the system which is about 1000 sec for this object.

5. Conclusion

Heat transfer intensification task which is traditionally solved by means of structural changes in technological equipment can be solved successfully on the basis of the liquid coolant critical states process control. Such conditions include in particular the internal turbulization which occurs in the liquid layer under certain conditions.

The results of a comparative analysis of convective and turbulent heat transfer modes allowed to establish that:

the quantitative criterion which makes it possible to distinguish between the indicated modes is, the absence or presence, respectively, of chaotic pulsations of fluid particles velocity. In this case, the amplitude of the velocity pulsations significantly exceeds the range of the molecular motions of liquid, i.e. the specified criterion has high selective properties;

it is advisable to use the density of the heat flux supplied to the heat exchanger from the heat source as a control action, providing stabilization of the selected condition of the coolant.

Controlled maintenance of the required heat transfer mode can be successfully performed on the basis of multi alternative control. For the task discussed above such control is implemented in the form of a system having a variable structure.

6. References

[1] Al-Maghalseh M and Mahkamov K 2018 Methods of heat transfer intensification in PCM thermal storage systems: review paper *Renewable and Sustainable Energy Reviews* **92** 62-94

[2] Qiu L, Sang D, Feng Y, Huang H, and Zhang X 2020 Study on heat transfer of process intensification in moving bed reactor based on the discrete element method *Chemical Engineering and Processing - Process Intensification* **151** 107915

[3] Klemeš J J, Wang Q-W, Varbanov P S, Min Zeng M and Walmsley T G 2020 Heat transfer enhancement, intensification and optimization in heat exchanger network retrofit and operation *Renewable and Sustainable Energy Reviews* **120** 109644

[4] Ji W-T, Fan J-F, Zhao C-Y and Tao W-Q 2019 A revised performance evaluation method for energy saving effectiveness of heat transfer enhancement techniques *International Journal of Heat and Mass Transfer* **138** 1142-53

[5] Baranyuk A V, Nikolaenko Yu E, Rohachov V A, Terekh A M and Krukovskiy P G 2019 Investigation of the flow structure and heat transfer intensity of surfaces with split plate finning *Thermal Science and Engineering Progress* **11** 28-39

[6] Korpyś M, Dzido G, Al-Rashed M H and Wójcik J 2020 Experimental and numerical study on heat transfer intensification in turbulent flow of CuO–water nanofluids in horizontal coil *Chemical Engineering and Processing - Process Intensification* **153** 107983

[7] Abu-Hamdeh N-H, Bantan R A R and Alimoradi A 2020 Study of heat transfer intensification through new types of tubular heat exchangers with rod bank insert *Chemical Engineering and Processing - Process Intensification* **152** 107912

[8] Alimoradi A, Mohammad Olfati M and Maghareh M 2017 Numerical investigation of heat transfer intensification in shell and helically coiled finned tube heat exchangers and design optimization *Chemical Engineering and Processing: Process Intensification* **121** 125-43

[9] Guo K, Liu B, Li X, Liu H and Liu C 2016 Flow pattern construction-based tubular heat transfer intensification using calculus of variations *Chemical Eng. Science* **1522** 568-78

[10] Anxionnaz-Minvielle Z, Cabassud M, Christophe Gourdon C and Tochon P 2013 Influence of the meandering channel geometry on the thermo-hydraulic performances of an intensified heat exchanger/reactor *Chemical Engineering and Processing: Process Intensification* **73** 67-80

[11] Hasanpour A, Farhadi M and Sedighi K 2017 Intensification of heat exchangers performance by modified and optimized twisted tapes *Chemical Engineering and Processing - Process Intensification* **120** 276-85

[12] Landau L D and Lifshitz E M 1987 *Fluid Mechanics* vol 6 (Oxford: Pergamon Press) p 554

[13] Landa P S 2001 *Regular and chaotic oscillations* (Berlin: Springer-Verlag) p 397

[14] Vabishchevich P N and Samarskii A A 2002 Monotone finite-difference schemes on triangular grids for convection-diffusion problems *Comput. Math. Math. Phys.* **42(9)** 1317–30

[15] Samarskii A A 2001 *The theory of difference schemes* (New York–Basel: Marcel Dekker) p 761

[16] Samarskii A A and Vabishchevich P N 2007 *Numerical methods for solving inverse problems of mathematical physics* (Berlin-New York: Walter de Gruyte) p 438
[17] Wichterle K and Marek M 2020 Solutions of the Navier-Stokes equation *Transport and Surface Phenomena* (Amsterdam: Elsevier) pp 109-19
[18] Podvalny S L and Vasiljev E. M 2019 Multi-altemativity information technologies in thermal processes control systems *11th IEEE Int. Conf. on Application of Information and Communication Technologies* (Moscow: IEEE) 8687074
[19] Wichterle K and Večeř M 2020 Chapter Sixteen: Extension of balances to turbulent flows *Transport and Surface Phenomena* (Amsterdam: Elsevier) pp 211-18
[20] Katopodes N D 2019 Turbulent Flow *Free-Surface Flow* (Oxford: Butterworth-Heinemann) pp 566-650
[21] Nikolaos D. Katopodes N D 2019 Turbulent Mixing and Dispersion *Free-Surface Flow* (Oxford: Butterworth-Heinemann) pp 840-907
[22] Zhou Y 2017 Rayleigh–Taylor and Richtmyer–Meshkov instability induced flow, turbulence, and mixing *I Physics Reports* **720–722** 1-136
[23] Podvalny S L, Vasiljev E M and Barabanov V F 2014 Models of multi-alternative control and decision-making in complex system *Automation and Remote Control* **75(10)** 1886-91
[24] Podvalny S L and Vasiljev E M 2015 Evolutionary principles for construction of intellectual systems of multi-alternative control *Automation and Remote Control* **76(2)** 311-17
[25] Podvalny S L and Vasiljev E M 2015 A Multi-alternative approach to control in open systems: origins, current state, and future prospects *Automation and Remote Control* **76** 8 1471-99
[26] Podvalny S L and Vasiljev E M 2020 The cybernetic concept of multialternativity in the evolution of biological systems *J. of Physics: Conf. Series, Applied Mathematics, Computational Science and Mechanics: Current Problems* **1479** (Voronezh, Russian Federation: IOP Publishing) 012109
[27] Podvalny S L and Vasiljev E M 2020 Simulation of the multialternativity attribute in the processes of adaptive evolution *Cyber-Physical Systems: Advances in Design & Modelling. Studies in Systems, Decision and Control* **259** (Cham: Springer Nature) pp 167-78

IOP Conf. Series: Materials Science and Engineering 1035 (2021) 012047 doi:10.1088/1757-899X/1035/1/012047

Digital twin for smart electricity distribution networks

S L Podvalny and E M Vasiljev

Voronezh State Technical University, 14 Moscow Avenue, Voronezh, 394026, Russia

Abstract. This paper presents the results of composing a digital model of critical states of the power distribution network with intelligent control. A specific example of a power network with thirty transformer substations connected by main and reserve cables is discussed. The analysis of critical events occurring in the power system was carried out, and a set of decision rules was formed to counter these events. A practical need was noted for regular updating and expansion of the list of critical events and corresponding parrying rules. It is shown that it is expedient to use faceted active neural networks with a structure similar to that of the controlled object as a digital twin model of intelligent control of the system under consideration. The use of such neural networks based on the rule "one event - one ensemble of neurons" provides the digital duplicate model with the ability to unlimited cumulative expansion, similar to the evolutionary property of natural biological systems.

1. Approaching the problem

Currently, the urgent task of the power industry is the development of smart power systems [1-7]. Such systems shall replace the existing automated control systems for power distribution networks which perform mainly the information service for dispatching personnel and practically do not solve the problem of power systems operational control in critical modes.

The solution to this problem involves the creation of a digital twin of the distribution network, i.e. a model that reproduces on-line its critical states and makes the decisions on the prompt elimination of these conditions without involving a human operator [8-13].

The traditional approach to the creation of such intelligent systems is to develop two models: the model of the control object itself, actually, the digital twin of the actual network, and the decision-making model that reproduces the actions of the dispatcher.

In this work, both models are combined into a single digital model that simulates the structure of a distribution power system in the form of a faceted neural network [14-18] containing the decision-making elements in case of critical situations [19-23].

This approach to the construction of a digital twin of a controlled object simplifies greatly the possibility of its constant updating in the process of power system modification and development.

2. Description of the original system

Let us consider a power distribution network consisting of an input high voltage substation and thirty low voltage transformer substations, as shown in figure 1.

Substations are connected to each other and to the input high voltage substation by main cables (shown as continuous lines in figure 1) and backup cables (dashed lines).

The reserve cables are connected to the network when the main ones are disconnected in such a way as to preserve the power supply to most consumers and to prevent at the same time the technological propagation of the failure in the form of a sequentially increasing chain of disconnections due to uneven load distribution.

For this power supply system, the task can be set to develop an intelligent digital network twin, which in case of the critical event sends a command to connect the corresponding backup line.

To analyze the properties of the original system it is advisable to compile a list of critical situations and the respective necessary controls. This will make it possible to analyze the performance of the constructed final neural network quantitatively with the simultaneous occurrence of either several critical events or an unforeseen situation.

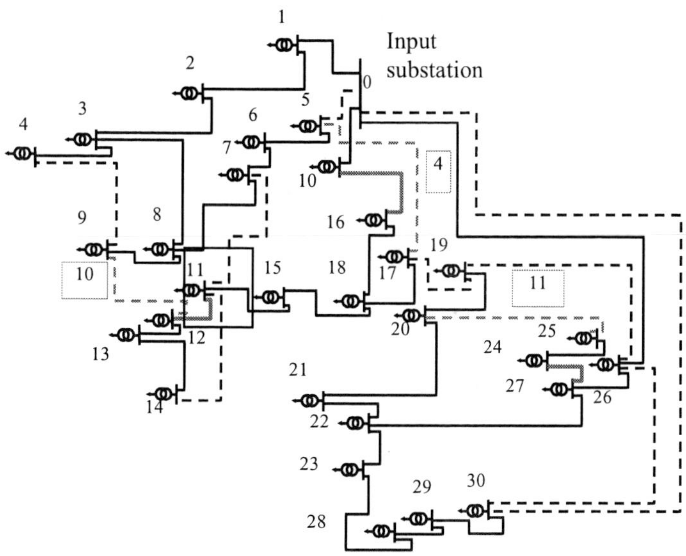

Figure 1. Power network diagram.

The numbers of these events with the corresponding substations numbers connected with the failed line are shown in table 1.

Table 1. Critical events list (fragment).

	The numbers of critical events												
	1	2	…	10	11	12	13	14	15	16	17	18	…	30
Substations numbers	0-1	1-2	…	0-26	26-27	27-22	23-28	28-29	29-30	23-22	22-21	21-20	…	24-25

Disconnection of one or another line connecting two substations directly can lead to the subsequent de-energizing of the substation chain.

In table 2 a fragment of the matrix of substation numbers is shown that is de-energized when critical events occur. The number of rows in table 2 equals the number of possible critical events and the number of columns equals the number of substations.

Symbol “One” in the *ij* cell means that when a critical event *i* occurs the substation *j* will be de-energized.

Table 2. Critical events matrix.

Event number	De-energized substation numbers 1	2	3	4	5	6	7	8	9	10	11	12	13	14	15	16	17	18	19	20	21	22	23	24	25	26	27	...	30
1	1	1	1	1	1	1	1	1	1	0	0	0	0	0	0	0	0	0	0	0	0	0	0	0	0	0	0	...	0
2	0	1	1	1	1	1	1	1	1	0	0	0	0	0	0	0	0	0	0	0	0	0	0	0	0	0	0	...	0
3	0	0	1	1	1	1	1	1	1	0	0	0	0	0	0	0	0	0	0	0	0	0	0	0	0	0	0	...	0
4	0	0	0	1	0	0	0	0	0	0	0	0	0	0	0	0	0	0	0	0	0	0	0	0	0	0	0	...	0
5	0	0	0	0	1	1	1	1	1	0	0	0	0	0	0	0	0	0	0	0	0	0	0	0	0	0	0	...	0
...	...	...	...	...	...	...	...	...	...	...	...	...	...	...	...	...	...	...	...	...	...	...	...	...	...	...	...	...	...
29	0	0	0	0	0	0	0	0	0	0	0	0	0	0	0	0	0	0	0	0	0	0	0	1	1	0	0	...	0
30	0	0	0	0	0	0	0	0	0	0	0	0	0	0	0	0	0	0	0	0	0	0	0	0	1	0	0	...	0

Every known critical situation can be associated with the respective decision rule. The decision rule indicates the number of at least one backup line which connection will reduce the number of de-energized substations. It is important to note that the set of critical situations and associated decision rules is fundamentally inexhaustible. This circumstance determines high requirements for the generalizing properties of the decision-making systems.

A fragment of the decision rules matrix is given in table 3.

Table 3. Decision rules matrix.

Event number	Substation numbers with pluggable backup lines (backup line numbers) 4-9 (1)	7-11 (2)	0-5 (3)	5-17 (4)	11-14 (5)	17-19 (6)	19-26 (7)	0-30 (8)	30-26 (9)	9-12 (10)	20-25 (11)
1	0	0	1	0	0	0	0	0	0	0	0
2	0	0	1	0	0	0	0	0	0	0	0
3	0	0	1	0	0	0	0	0	0	0	0
4	1	0	0	0	0	0	0	0	0	0	0
5	0	0	1	0	0	0	0	0	0	0	0
...	...	...	...	...	...	...	...	...	...	...	...
29	0	0	0	0	0	0	0	0	0	0	1
30	0	0	0	0	0	0	0	0	0	0	1

The above power distribution network characteristics make it possible to move on to the task of constructing its digital twin possessing the decision-making intellectual properties when any critical situation arises in the network.

3. Designing and research of an object-oriented facet neural network

A faceted neural network is characterized by the fact that for each event an ensemble (set) $\{f,s\}$ of facet features f is formed, whose set of values s defines the specific object $a(f,s)$:

$$a \in A,\ |A| = z,\ \ f = \{f_1 f_2 ... f_n\},$$

$$s = \{s^{f_1}, s^{f_2}, ..., s^{f_n}\},\ \ s^{f_i} = \{s_1^{f_i}, s_2^{f_i}, ..., s_t^{f_i}\},\ i = \overline{1,n}.$$

For the problem under consideration, the list of facets $f = \{f_1 f_2 ... f_n\}$, n=30 forms the columns of table 2; the number of the event objects z = 30 are the rows of this table; the number of values of each feature t = 2: $s^{f_i} = \{0;1\}$.

Graphic illustration of the network fragment faceted organization is shown in figure 2. In this figure, the ensembles (sets) of neurons F perform the logical operation "and". The numbering of neurons corresponds to the event number in tables 1 and 2.

A fundamental feature of the resulting model is its structural similarity to the object model, i.e. the model structure represents the object environment. For example, if the main power line 11-12 fails the substations 12, 13 and 14 will be de-energized, which corresponds to event 26. In this case, the decision will be made to turn on the backup line 10 (see figure 2). At the same time, from figure 1 it follows that the specific event related to the same situation is the de-energization of substations 13 and 14. Event 27 corresponds to this case and the model makes the decision to turn on and activate the backup line 5. Thus, using the model will result in the connection of two backup lines numbered 10 and 5. An analysis of the circuit in figure 1 shows that in case of a failure of the mainline 11-12 both solutions shown above are equivalent.

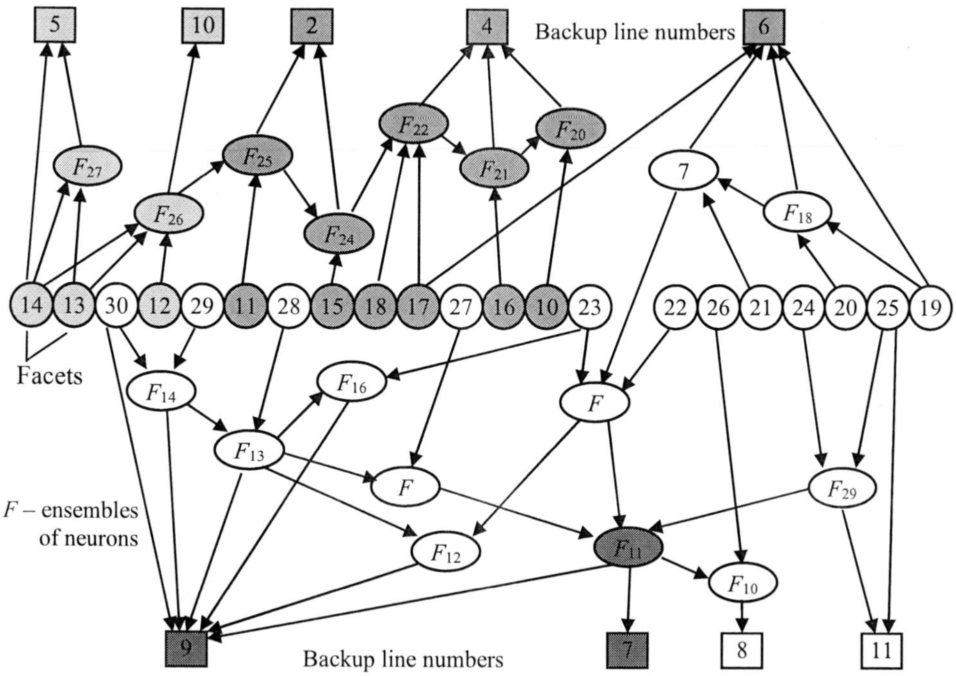

Figure 2. Distribution network faceted model.

It follows from the analysis that the constructed model has generalizing properties to the extent allowed by the structure of the object. For example, if situation 20 is taken into account when training the network (see F_{20} in figure 2), then it will be sufficient for parrying the correct critical situations 21,22,24,25,26,27,28 that were not considered during training since all of them are the particular cases of the general situation 20.

It should be noted that the appearance of generalizing properties in a facet neural network is due to the presence of hierarchical relationships in it, i.e. it is multileveled.

The training procedure in such a neural network includes:
adding a new object in the form of an ensemble (set) of values of $a(f,s)_{z+1}$ features (z+1 row in table 2);
inclusion of each feature – the facet of an object into the network structure of this feature.

At the same time, the relationships formed previously in the network are not destroyed, which excludes the possibility of the retraining phenomenon occurrence in it. The block character of this training is reduced to a simple build-up of information about the new situations and it does not require any multi-parameter optimization. Therefore, the network retains its high selective properties, i.e., for each emergency situation, only one solution is generated with the imposition of these solutions when several failures occur simultaneously. For example, if events 10 and 11 occur simultaneously, the model will generate solutions 7, 8 and 9, which will completely restore power supply (see figure 2).

The characteristics of the model's structural change when new events appear in the training set are shown in figure 3.

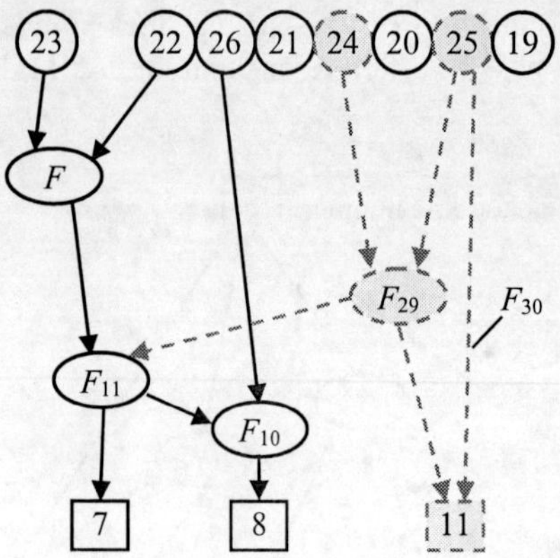

Figure 3. Rebuilding the network when the new events 29 and 30 are trained (disconnection of substations 24, 25).

When new events 29 and 30 appear (switching off the substations 24, 25) the neural network structure is implemented without changing the previously formed structural connections, i.e. the training process is exclusively cumulative.

4. Conclusion

The task of constructing digital twins of objects with intelligent control faces the problem of constant updating and expansion of all components of the model. This includes both possible reconfigurations of the control object itself and the list of critical events and decision rules that fend off these events.

In other words, the digital twin must be capable of continuous development in the form of cumulative evolution.

It is proposed to solve this problem on the basis of facet-type neural networks with the following evolutionary properties:

a hierarchical decision-making structure that ensures the implementation of high generalizing properties in the network, that are necessary when new critical situations appear;

memory organization according to the facet principle "one event - one ensemble" facilitates the addition of any number of events to the system without any overtraining effect;

the modular structure of the system which allows the integration of new ensembles of neurons into the network additively, i.e., similar to the "uncomplicated simplicity" of natural biological systems.

References

[1] Dkhili N, Eynard J, Thil S and Grieu S 2020 A survey of modelling and smart management tools for power grids with prolific distributed generation *Sustainable Energy, Grids and Networks* **21** 100284

[2] O'Dwyer E, Pan I, Acha S and Shah N 2019 Smart energy systems for sustainable smart cities: Current developments, trends and future directions *Applied Energy* **237** 581-97

[3] Kakran S, Saurabh and Chanana S 2018 Smart operations of smart grids integrated with distributed generation: A review *Renewable and Sustainable Energy Reviews* **81** 524-35

[4] Ponce-Jara M A, Ruiz E, Gil R, Sancristóbal E and Castro M 2017 Smart grid: assessment of the past and present in developed and developing countries *Energy Strategy Reviews* **18** 38-52

[5] Abdi H, Beigvand S D and La Scala M 2017 A review of optimal power flow studies applied to smart grids and microgrids *Renewable and Sustainable Energy Reviews* **71** 742-66
[6] Nashivochnikov N V, Bolshakov A A, Lukashin A A and Popov M. 2020 The system for operational monitoring and analytics of industry cyber-physical systems security in fuel and energy domains based on anomaly detection and prediction methods *Cyber-Physical Systems: Industry 4.0 Challenges. Studies in Systems, Decision and Control* vol **260** (Cham: Springer Nature Switzerland AG) pp 261-73.
[7] Allegrini J, Orehounig K, Mavromatidis G, Ruesch F and Evins R 2015 A review of modelling approaches and tools for the simulation of district-scale energy systems *Renewable and Sustainable Energy Reviews* **52** 1391-1404
[8] Boje C, Guerriero A, Kubicki S and Rezgui Y 2020 Towards a semantic construction digitalt: directions for future research *Automation in Construction* **114** 103179
[9] Tao F, Qi Q, Wang L and Nee A Y C 2019 Digital twins and cyber–physical systems toward smart manufacturing and industry 4.0: Correlation and Comparison *Engineering* **5(4)** 653-61
[10] Ayani M, Ganebäck M and Ng A H C 2018 Digital twin: applying emulation for machine reconditioning *Procedia CIRP* **72** 243-48
[11] Frazzon E M, Agostino Í R S, Broda E and Freitag M 2020 Manufacturing networks in the era of digital production and operations: a socio-cyber-physical perspective *Annual Reviews in Control* **49** 288-94
[12] Pethuru Raj P and Surianarayanan C 2020 Digital twin: the industry use cases *Advances in Computers* **117(1)** 285-320
[13] Tao F, Zhang M and Nee A Y C 2019 Background and concept of digital twin *Digital Twin Driven Smart Manufacturing* (London: Elsevier) pp 3-28
[14] Frazão X and Alexandre L A 2014 Weighted convolutional neural network ensemble *Progress in Pattern Recognition, Image Analysis, Computer Vision, and Applications. Lecture Notes in Computer Science* vol 8827 (Cham: Springer Nature Switzerland AG) pp 674-81
[15] Kandel E R, Dudai Y and Mayford M R 2014 The molecular and systems biology of memory Cell **157(127)** 163-186
[16] Podvalny S L and Vasiljev E M 2017 The principle of multi-alternativity in intelligent systems. Active neural network models *Proc. Computer Science* **103** 410-15
[17] Cichy R M and Kaiser D 2019 Deep neural networks as scientific models *Trends in Cognitive Sciences* **23(4)** 305-17
[18] Meshalkin V, Puchkov A, Dli M and Lobaneva Y 2020 Deep neural networks application in models with complex technological objects *Cyber-Physical Systems: Advances in Design & Modelling. Studies in Systems, Decision and Control* vol 259 (Cham: Springer Nature Switzerland AG) pp 291-300
[19] Podvalny S L, Vasiljev E M and Barabanov V F 2014 Models of multi-alternative control and decision-making in complex system *Automation and Remote Control* **75(10)** 1886-91
[20] Podvalny S L and Vasiljev E M 2015 Evolutionary principles for construction of intellectual systems of multi-alternative control *Automation and Remote Control* **76(2)** 311-17
[21] Podvalny S L and Vasiljev E M 2015 A Multi-alternative approach to control in open systems: origins, current state, and future prospects *Automation and Remote Control* **76(8)** 1471-99
[22] Podvalny S L and Vasiljev E M 2020 The cybernetic concept of multialternativity in the evolution of biological systems *J. of Physics: Conf. Series, Applied Mathematics, Computational Science and Mechanics: Current Problems* 1479 012109 (Voronezh, Russian Federation: IOP Publishing)
[23] Podvalny S L and Vasiljev E M 2020 Simulation of the multialternativity attribute in the processes of adaptive evolution *Cyber-Physical Systems: Advances in Design & Modelling. Studies in Systems, Decision and Control* **259** (Cham: Springer Nature Switzerland AG) pp 167-78

IOP Conf. Series: Materials Science and Engineering 1035 (2021) 012048 doi:10.1088/1757-899X/1035/1/012048

Digital control of structurally unstable power facilities using a polynomial approach

V I Zakhvatov, S L Podvalny and A V Mikhailusov

Voronezh State Technical University, 14 Moscow Avenue, Voronezh, 394026, Russia

Abstract. The problem of improving the quality of control of structurally unstable power facilities is viewed based on a polynomial controller with a dynamic change in settings using the inverse pendulum on a carriage as an example. The mathematical model of the control object was built;a polynomial controller was synthesized using the method of symbolic computation of systems of differential equations with an additional input for specifying the geometric mean root from the outside.Reverse pendulum model on a carriage and a model of regulators were created using the MATLAB Simulink visual modeling environment to analyze the quality and parameters of the created control system. As a result of modeling, direct dependence of the position setting time of pendulum on the value of the geometric mean root was obtained which confirmed the possibility of external control of the speed of the control system. The possibility to control the control actions level to prevent saturation of executive devices was investigated and confirmed. In fact, a system of external parametric control of the processes dynamics was created and the described regulator does not require adaptation and training.

1. Introduction

In connection with the constant complexity growthof the designed systems, the controlmeans for the dynamic processes are getting more and more complex. And a particular difficulty is the management of structurally unstable power objects thatare often of a non-minimal-phase nature. This type of objects is also called complex, since its control necessitates the creation of regulators capable of operating in different modes to ensure the maximum area of attraction. Examples of such objects are balancing mechanisms, various types of motors, the movement of cranes with a suspended load, etc.

Improving the quality of complex objects controloften requires the control of the dynamics of processes, and the control of control actions themselves to prevent saturation of executive devices [1, 2].

A classic example of a complex object is a reverse pendulum on a carriage. Different positions of the pendulum require different control modes, which makes it impossible to use a static regulator in case of need in the maximum area of attraction.

2. Problem formulation

In this work, it was required to carry out the process of synthesizing a polynomial controller with its settings changing dynamically thus making possible an adjustment to the control object parameters changes without the re-synthesis of the control system. External parametric control was supposed to allow the parametric control of system parameters.

To test the control quality it was required to build a model of a reverse pendulum on a carriage using the MATLAB Simulink, standard solutions in the form of PID controllers were used in addition, to compare the results of these solutions with the resulting control system.

3. Initial conditions

To build a model of the system it was required to enter the initial parameters. First of all, a carriage with a reverse pendulum, shown in figure 1 is controlled by the force F. The model shows the acting forces and, it should be noted that the direction of the pendulum angle, relative to the vertical axis, is counterclockwise. And the growth of the horizontal coordinate of the carriage movement is directed from left to right (a similar system is discussed in [3]).

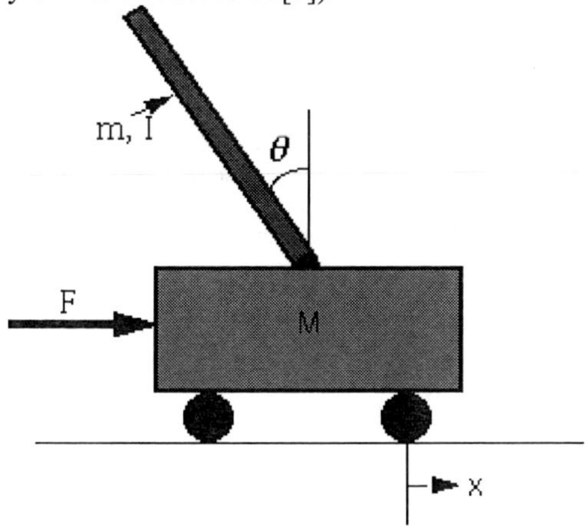

Figure 1. Direction of forces.

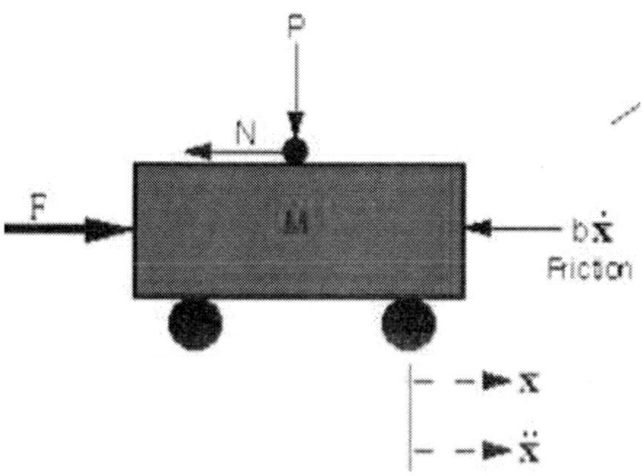

Figure 2. Carriage forces.

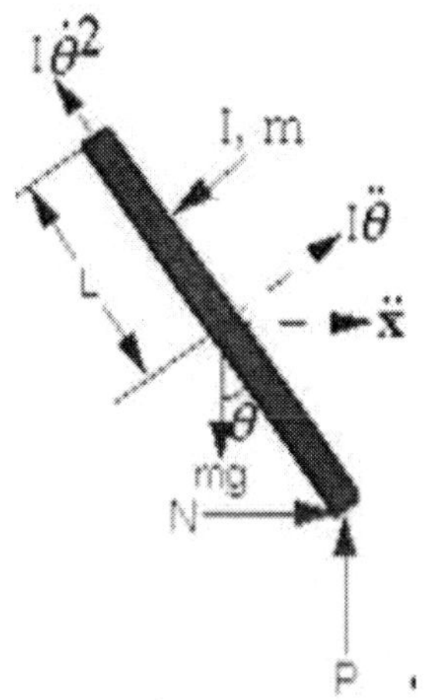

Figure 3. Pendulum forces.

4. Building the mathematical model of pendulum

Figures 2 and 3 show the forces acting on the pendulum and on the carriage.

The described physical system has two degrees of freedom both of them are represented by the coordinate of the carriage position and the pendulum deflection angle. Here, guided by Newton's laws, the equations (1, 2) were written for both degrees of freedom.

$$\frac{d^2x}{dt^2} = \frac{1}{M}\sum_{cart} F_x = \frac{1}{M}\left(F - Nb\frac{dx}{dt}\right) \tag{1}$$

$$\frac{d^2\theta}{dt^2} = \frac{1}{I}\sum_{pend} \tau = \frac{1}{I}(NL\,cos(\theta) + PL\,sin(\theta)) \tag{2}$$

5. System linearization

Since the stabilization occurs in the vicinity topendulum's zero deflection angle, severalsubstitutions could be made:

After substitutions, new equations of motion (3) and (4) were obtained.

$$(I + ml^2)\ddot{\phi} - mgl\phi = ml\ddot{x} \tag{3}$$

$$(M + m)\ddot{x} + b\dot{x} - ml\ddot{\phi} = u \tag{4}$$

For further linearization, we apply the Laplace transforms for the equations (5) and (6).

$$(I + ml^2)\Phi(s)s^2 - mgl\Phi(s) = mlX(s)s^2 \tag{5}$$

$$(M + m)X(s)s^2 + bX(s)s - ml\Phi(s)s^2 = U(s) \tag{6}$$

6. Getting the transfer function

As soon as linearized equations were obtained, it became possible to compose the transfer function of the system of the reciprocal pendulum on the carriage which is necessary for the future for the synthesis of the controller.

Since the input of the system is U(s) (applied force), X(s) is the carriage position, and the system output is Φ(s) (pendulum deflection angle). The coordinate equation (7) was compiled.

$$X(s) = \left[\frac{(I+ml^2)}{ml} - \frac{g}{s^2}\right]\Phi(s) \tag{7}$$

For more convenient recording, a replacement (8)was made

$$q = [(M + m)(I + ml^2) - (ml)^2] \tag{8}$$

Transfer function after substitution (9).

$$\frac{\Phi(s)}{U(s)} == \frac{\frac{ml}{q}s^2}{s^4 + \frac{b(I+ml^2)}{q}s^3 - \frac{(M+m)mgl}{q}s^2 - \frac{bmgl}{q}s} \tag{9}$$

After abbreviation the complete form of the pendulum transfer function was obtained (10).

$$\frac{\Phi(s)}{U(s)} == \frac{\frac{ml}{q}s}{s^3 + \frac{b(I+ml^2)}{q}s^2 - \frac{(M+m)mgl}{q}s - \frac{bmgl}{q}} \tag{10}$$

And the equation (11) shows the carriage transfer function:

$$\frac{X(s)}{U(s)} == \frac{\frac{(I+ml^2)}{q}s^2 - \frac{mlg}{q}}{s^4 + \frac{b(I+ml^2)}{q}s^3 - \frac{(M+m)mgl}{q}s^2 - \frac{bmgl}{q}s} \tag{11}$$

A simple analysis of equations (10) and (11) shows that we are dealing with objects of non-minimal-phase type whereas an extensive number of publications are devoted to its control with the detailed simulation using the Matlab [4], wherein, as a rule, the systems of adaptive control with an observer or a reference model are usually built [5, 6]. The identification of such models is particularly difficult in these systems, which complicates significantly the final version of the closed-loop control system. Due to these circumstances, the search for simpler control algorithms is permanently in progress. It is the search for simplified control options that our further discussion is devoted to.

7. Controller synthesis

As an integral part of this task a polynomial controller synthesis was required that would satisfy all necessary conditions and have a minimum order [7, 8].

To reduce the system order the authors neglected the carriage friction which helped to obtain the transfer function of the second-order pendulum (12).

$$\frac{\Phi(s)}{U(s)} == \frac{\frac{ml}{q}}{s^2 - \frac{(M+m)mgl}{q}} \tag{12}$$

Transfer function equation after substitution (13).

$$w_O = \frac{50}{11s^2 - 343} \tag{13}$$

Since the final transfer function is of the second order, we could synthesize a first-order polynomial controller using the symbolic methods for solving systems of differential equations according to the method presented in [9].

Using the polynomial controller synthesis method presented in [8], the transfer function of the controller Wp was obtained.

In the process of synthesis, an additional parameter was introduced, which would be an additional input parameter of the final controller. This parameter is the geometric mean root of the characteristic equation (Ω), which in turn will set the system bandwidth.

The final transfer function of the controller Wp with an additional parameter Ω is shown in (14).

$$W = \frac{(0.66\Omega^2 + 6.86)s + (0.22\Omega^3 + 20.58\Omega)}{s + 3\Omega} \tag{14}$$

8. System performance analysis

Setting time was considered as the time during which the pendulum is set to a position with a deflection angle of <0.01 radian.

In fig. 4 an oscillogramis shownby means of which the pendulum installation time was estimated. In this case, the starting angle is 0.5 radian and the geometric mean root is 36.

Analyzing Figure 4, one can see that the setup time is 2 seconds. This is due to the large value of the initial deviation and the small value of the geometric mean.

The dependence of the installation time on the geometric mean root was considered. To do so, we will conduct a series of experiments with an initial deflection angle of 0.3 radians. The resulting dependence graph is shown in Figure 6.

Experimental results confirmed the dependence of the installation time on the value of the geometric mean root.

Within the framework of the developed system,the control action level can be controlled in turn by changing the geometric mean root. In fig. 6 the results of experiments are shown with a change in Ω and at a constant starting angle in the form of a dependence of the maximum level of on the value of Ω.

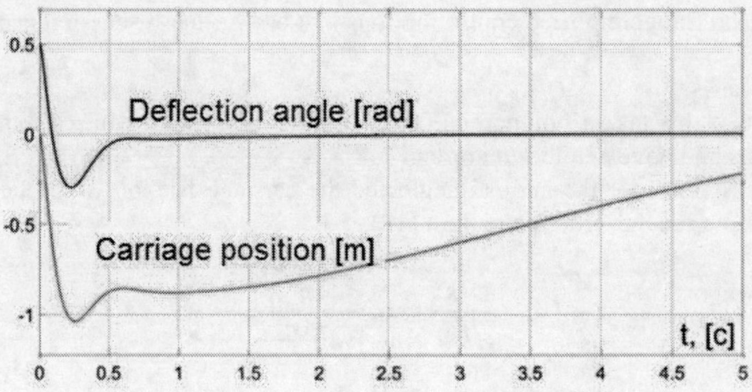

Figure 4. Pendulum output oscillogram.

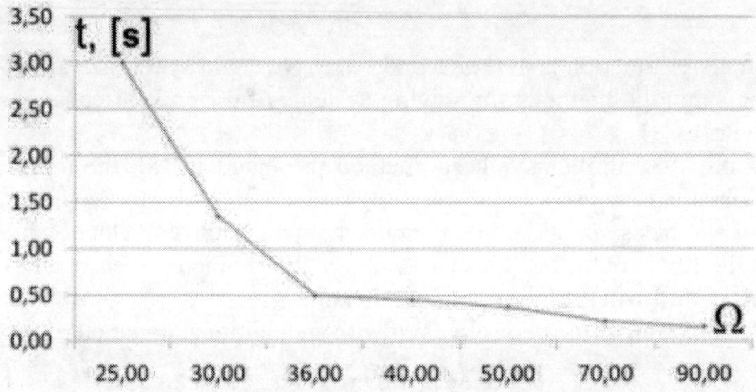

Figure 5. The dependence of the installation time by omega.

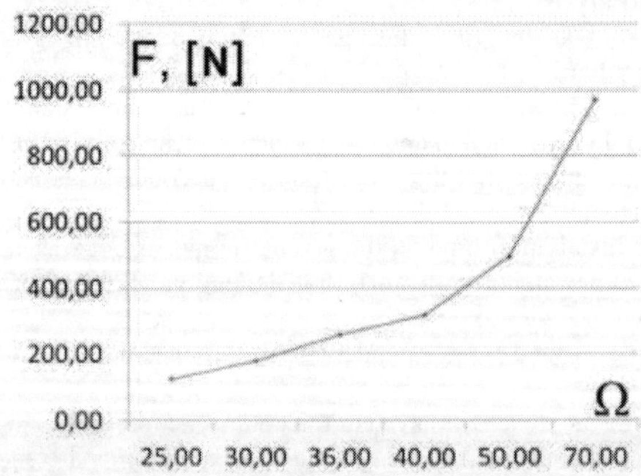

Figure 6. Dependence of exposure level on omega.

A series of experiments confirmed the possibility of controlling the maximum impact level by means of changes in the geometric mean root in the characteristic equation.

A number of experiments were carried out to outline the attraction area which showed that using a carriage regulator with its significant effect on a carriage position, the attraction area is limited by 1 radian deflection angle.

IOP Conf. Series: Materials Science and Engineering 1035 (2021) 012048 doi:10.1088/1757-899X/1035/1/012048

9. Conclusions

In this work, a control system for structurally unstable power objects (a reverse pendulum on a carriage) was created basing on a polynomial controller with the ability to change the settingsdynamically.

During the simulation of the synthesized control system and subsequent experiments, direct dependence of the installation speed on the geometric mean root was established.

The possibility to control the control actions level was also confirmed and made it possible to control the saturation of executive devices.

The comparison of the polynomial controller with the PID controller showed that at small deflection angles the controller works with comparable quality but with an increase in the initial pendulum angle, the polynomial controller works better.

With the analysis of the parameters of the considered regulator, several promising areas for further research were identified:

- improving the quality of prevention of the executive devicessaturation;
- bandwidth control based on spectral characteristics of measurement noise;
- system speed is controlled without the readjustment of the controller;
- dynamic control of the area of attraction of systems;
- development of multi-alternative systems.

In fact, the possibility of external parametric control of the processes dynamics was confirmed and the described regulator does not require adaptation and training (the settings are changing instantly).

All the tasks have been completed and the final parameters of the control system meet the specified conditions.

References

[1] Boubaker O 2013The Inverted Pendulum Benchmark in Nonlinear Control Theory: A Survey*International Journal of Advanced Robotic Systems***10(5)**

[2] Krishna N 2016 Modeling and controller design of cart inverted pendulum system using MRAC scheme SI: 1*Frontiers of Current Trends in Engineering and Tchnology*pp 21-24

[3] Goodwin G K and Grebe M E2004 *Control Systems Design*(M.: BINOM. Laboratory of Knowledge) p 911

[4] Andrievsky B R and FradkovA L 1999 *Selected Chapters of Theory of Automatic Control With Examples in Matlab Language*(St. Petersburg: The science) p 467

[5] IsermanR1984 *Digital Control Systems* (M.: Mir) p 541

[6] Kruglov S P*2016* Adaptive control of a non-minimal-phase scalar object of the second order with the provision of the given characteristics of the transient process*Scientific Bulletin of Nizhny Novgorod State Technical University* **65(4)**pp 33-53

[7] Tararykin S V2005Methodology for designing digital polynomial regulators of electromechanical systems*Vestnik IGEU* **3**pp 23-29

[8] Zakhvatov V I 2019 *Computer Algebra Methods for Polynomials in Automatic Control Problems* (VSPU. XIII All-Russian meeting on governance) pp 141-147

[9] Gaiduk A R 2012*Theory and Methods of Analytical Synthesis of Automatic Control Systems* (M.: Fizmatlit) p 360

IOP Conf. Series: Materials Science and Engineering 1035 (2021) 012049 doi:10.1088/1757-899X/1035/1/012049

Intelligent identification of telecommunication channels in the distributed systems for digital control of power facilities

S L Podvalny and M A Likhotin[1]

Voronezh State Technical University, 14 Moscow Avenue, Voronezh, 394026, Russia

[1]Corresponding author: maximus-lihotin@mail.ru

Abstract: This paper presents the results and problems of telecommunication power amplifier modeling. The research tool is the mathematical apparatus of artificial neural networks. Simulation of the device was made according to the black box principle, operating solely with input and output data of the power amplifier under study. The criterion for the quality of model estimation was the signal-to-noise ratio adopted in radio engineering. Brief results of the previous research are shown. The problems associated with the simulation of a power amplifier operation basing on neural networks were investigated. An example is given of nonlinear distortions appearing in the device under consideration. The description is given of the joint work of a neural networks group based on analog signal clustering which works according to the principle of Grossberg's adaptive resonance theory. The application results of this type of systems are shown. Discussing the problems concerning the use of this approach to simulation of the investigated processes.

1. Introduction

The domains for research work in the sphere of intelligent identification of power facilities management are quite extensive. In particular, one of the popular control methods is the use of artificial neural networks. They are widespread in medicine, economics, technology, etc., and by large and well-known corporations and companies such as Google, Microsoft, Yandex and many others. The relevance of using this mathematical apparatus will be required for quite a long time because the range of neural networks application is very diverse. Hence, the mathematical model of this type was chosen in this work as a research tool.

The problem set forth in this was formulated as follows: a power amplifier is connected to the telecommunication channel output and two data arrays are available, one of which was fed to the input to the object under study, and the other was read at its output. These two datasets format was presented in an integrated manner. It is necessary to build a neural network that, when a reference sequence of complex numbers is fed to its input, will result in the set of corresponding data at the amplifier's output which differs from the reference data by the mean square deviation value –40 dB. Whereas it is customary in radio engineering, to calculate the ratio between target and received data in decibels but not their absolute difference, the formula below is applied to calculate the simulation errors.

$$\mathrm{EVM} = 10 \cdot \log_{10}\left(\frac{\sum_{i=1}^{N}|e_i|^2}{\sum_{i=1}^{N}|x_i|^2}\right)$$

where **x** is the reference sequence output, **e** is the difference between the received data and **x**, **N** is the of the considered sample size, a number of signals.

This study is a continuation of [1], it contains the analysis of various approaches to a power amplifier simulation using the black box method and the neural networks resources. In the previous publication, the use of an ordinary two-layer network was described, its training was carried out using the well-known method of reverse error propagation. Nesterov's method was also used as a modification of this algorithm and showed a sufficient gain in the model convergence rate.

2. Simulation problems

Certain limitations are imposed on the power amplifiers development process related to the device size and power consumption. In this regard, the nonlinear distortions level increases due to the deviations of the transfer characteristic from linear.

The actual power amplifier has an output power limit determined by the type of switching circuit, the type of amplifier element, bias voltage and supply voltage. Therefore, starting from a certain value of the input power level the dependence ceases to be linear and the system gain decreases with increasing input power. Figure 1 is a diagram of the output/input power dependence in the case of a linear power amplifier (solid line) and nonlinear one (dashed line) [2].

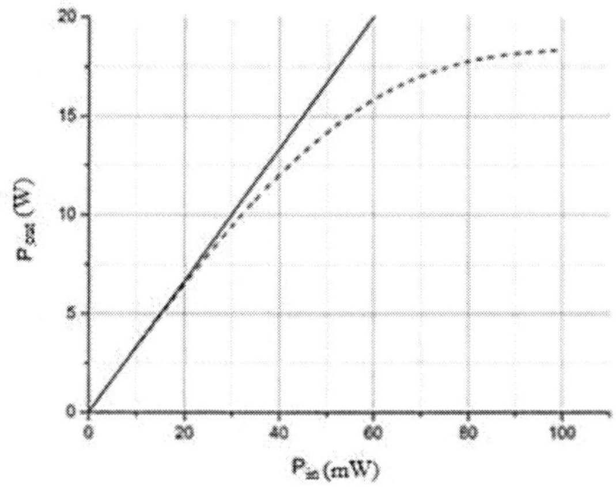

Figure 1. Output power versus input power.

Examining the input and output signals, one can see that the noisiest component is at the level of low amplitude values. Dividing the sample into its constituents does not give a significant advantage, since the neural network cannot by its nature repeat the noised values of the supplied images. Obviously, alternative methods should be used to process this data, which is the subject of this study.

3. Using the adaptive resonance theory

One can cluster the training set and select a particular neural network for each class for training. Grossberg adaptive resonance theory was chosen as a tool for this, its essence is dividing an analog signal into parts which satisfy the coefficient of "similarity". In figure 2 the examples are shown of one class of input signal and the corresponding output one.

Whereas the training in such a neural network is effected without a teacher, then in case of a high "similarity" coefficient in the result of identical classes clustering, they can be in plenty. Therefore, one can limit their set to 10. Each data batch with such limitation will be trained by a separate model of the sgdm type, since it demonstrated the best generalizing ability of such data in comparison, e.g., with RMSProp or adam [3]. Training will continue until each neural network reaches the learning limit of -30 dB relative EVM error. The architecture of such a network will be small: only 24 values of the input layer, 48 of the hidden one and 12 of the output. And accordingly, the data set will be quite large, since training models will be about 10. But, as can be seen in figure 3, this structure works very poorly.

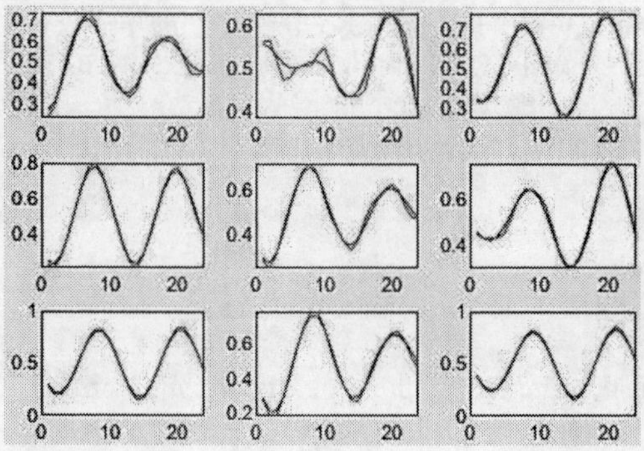

Figure 2. Example of one class of clustered input.

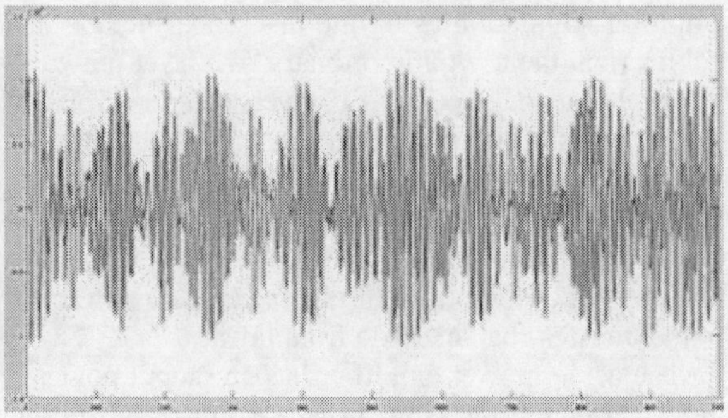

Figure 3. An example of the operation of neural networks trained for each class separately.

If we analyze the network operation solely by classes, the results will be unsatisfactory. Data in the table shown in figure 4 in the fourth column displays the relative errors for each model that is trained exclusively for its class, respectively.

{} 10x4 cell

	1	2	3	4
1	*155x24 dou...*	*155x12 dou...*	*311x12 dou...*	3.0678
2	*57x24 double*	*57x12 double*	*114x12 dou...*	2.4781
3	*95x24 double*	*95x12 double*	*190x12 dou...*	3.5681
4	*43x24 double*	*43x12 double*	*86x12 double*	3.6924
5	*132x24 dou...*	*132x12 dou...*	*264x12 dou...*	3.0956
6	*34x24 double*	*34x12 double*	*68x12 double*	3.8633
7	*13x24 double*	*13x12 double*	*26x12 double*	1.5968
8	*38x24 double*	*38x12 double*	*76x12 double*	0.8850
9	*129x24 dou...*	*129x12 dou...*	*258x12 dou...*	3.1114
10	*34x24 double*	*34x12 double*	*34x12 double*	1.5323

Figure 4. Relative errors of each model.

As a result, having analyzed the functioning of each neural network, it had turned out that each model of the class was retrained. This is due to the fact that models are trained exclusively to their patterns and, in connection with this, the generalizing ability deteriorated much more than it was intended.

If we consider functioning of the network obeying the adaptive resonance theory, then significant problems arise, e.g., if to memorizing hundreds and thousands of realizations of the same dynamic process is required, then with small values of the similarity parameter in the recognition mode, the memory of the neural network shall never get a sufficient amount of necessary information to make reasonable solutions. And at large values of this parameter similar data differing only in a small number of secondary details, are stored in memory as prototypes of different image classes [4]. Thus, this class of neural networks is quite sensitive to noise, and it leads to an increased number of errors in the simulated process.

4. Analysis of training by signal spectrum

Alternatively, one can use Fast Fourier Transform as an additional data preprocessing and then try network training by signal spectrum.

In figures 5, 6 an example of a signal spectrum is shown for 48 values of the real and imaginary parts, respectively.

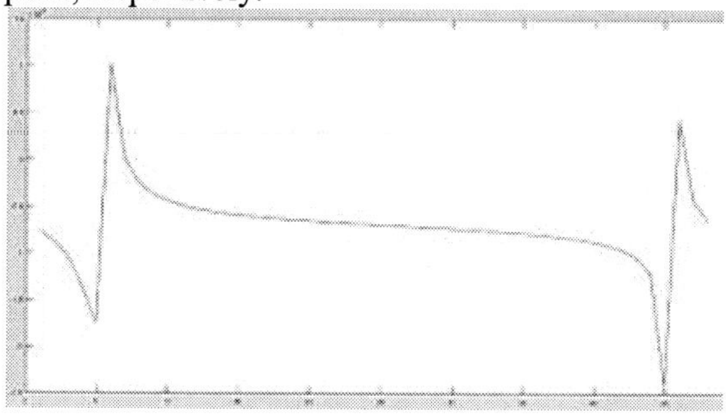

Figure 5. An example of a signal spectrum of the real part.

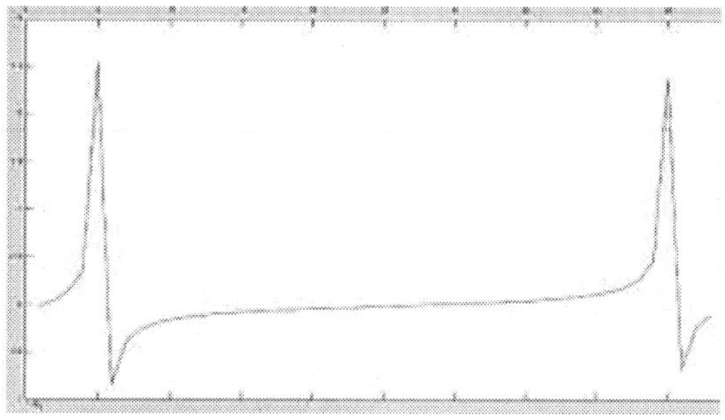

Figure 6. An example of the spectrum of the imaginary part.

Such data will be fed to the network inputs. Accordingly, the entire sample is normalized before training, and the inverse Fourier transform of readings at the network outputs will be performed to check the quality of the model. It was found out that with the previous architecture the model demonstrates a rather unsatisfactory result. Relative error reached the value of approximately -2 dB for test data. In figure 7 both the target data and received data of the test sample are shown.

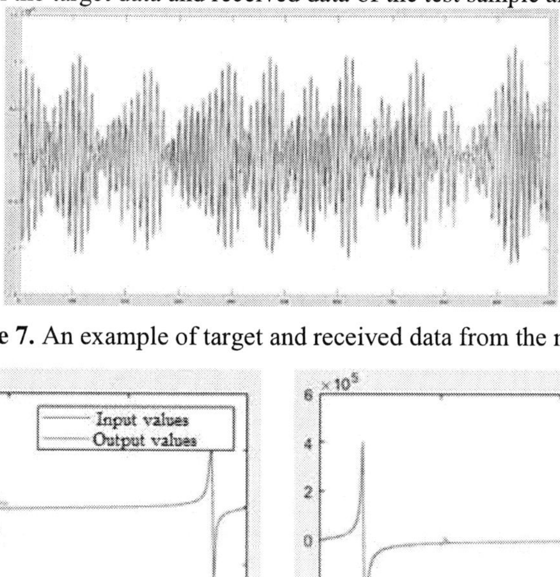

Figure 7. An example of target and received data from the model.

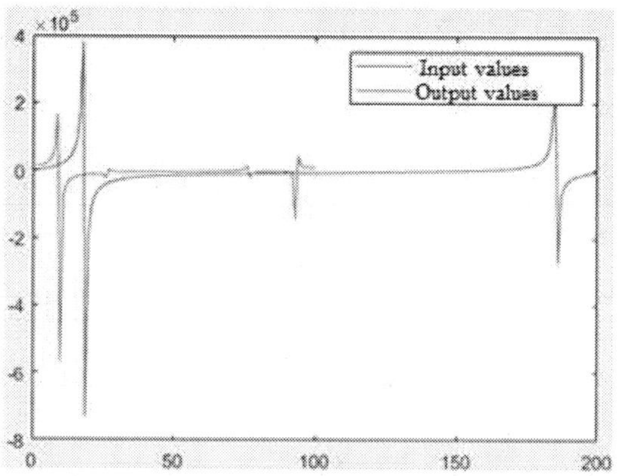

Figure 8. An example of input and output data bursts in the model.

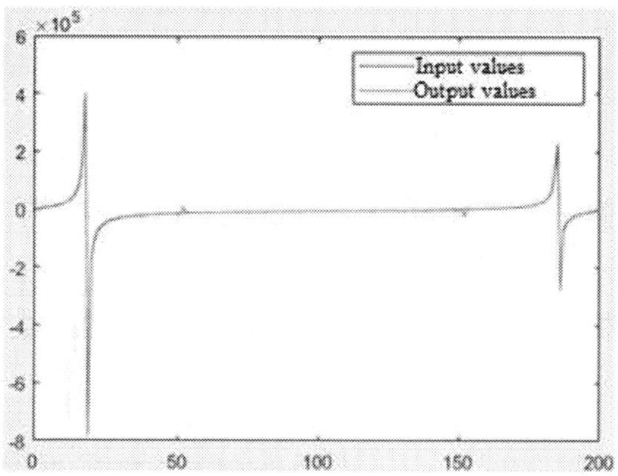

Figure 9. An example of input and output data bursts in the model.

Analyzing the training sample, it was found that the signal spectrum at the input and output of the network is modified noticeably, as shown in figure 8.

It happens due to the different dimensions of the Fourier transform vector. In this regard, the neural network inputs and outputs will be the same. An example after such a transformation is shown in figure 9.

And accordingly, after training, the model demonstrated relative error in the -14.3 dB test dataset. An example of target and acquired data is shown in figure 10.

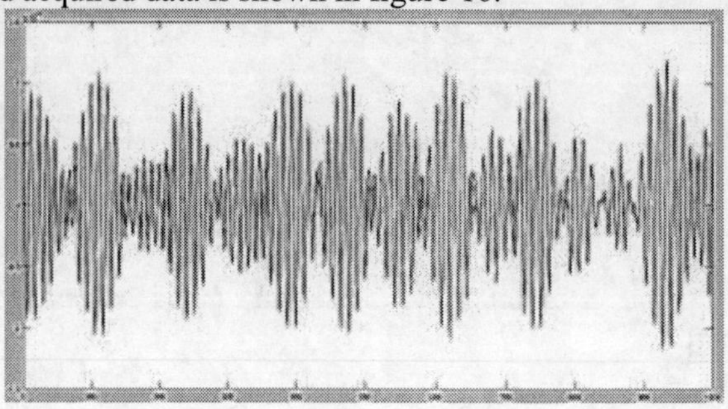

Figure 10. An example of target and received data from the model.

To train the model by the signal spectrum the model architecture, e.g., the number of neurons in the layers was expanded. The input and output layers will then consist of 200 neurons, whereas the neurons number in the hidden layer is 800. We will also increase twice the number of epochs, i.e. to 1200. The table shows the summary indicators of the model with such an architecture.

Table 1. Relative error Indicators for training by signal spectrum.

Theining algorithm	Data mixing	Relative error, dB
	Never	-21.3379
sgdm	Each epoch	-20.8667
	Before training	-20.9133
	Never	-19.2077
adam	Before training	-19.9591
	Each epoch	-16.7946
	Never	-13.097
rmsprop	Every epoch	1.02794
	Before training	-13.0204

Figures 11, 12 show examples of a target and obtained data in the test sample using the sgdm and rmsprop algorithms, respectively.

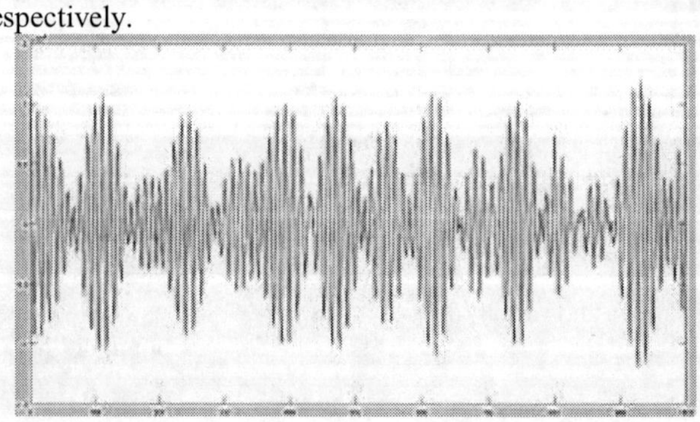

Figure 11. An example of target and received data from a model trained with the sgdm algorithm.

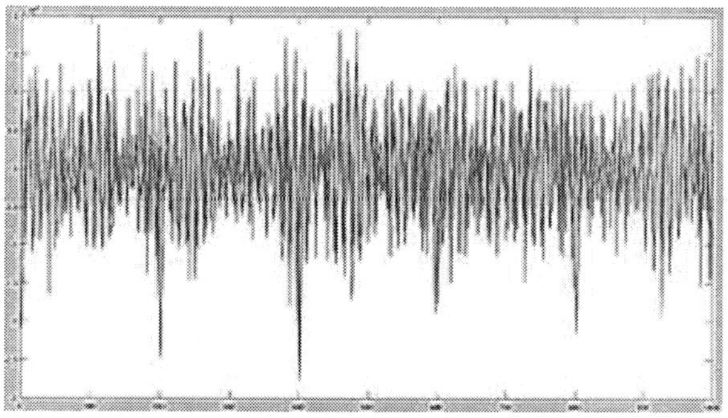

Figure 12. An example of target and received data from a model trained with the rmsprop algorithm.

The table shows that these algorithms demonstrate comparable results while the rmsprop algorithm in some examples showed a relative error order of magnitude worse than training on a quasi-periodic signal dataset.

5. Conclusion

Therefore, we can conclude that operating with input and output data only is not enough to achieve high accuracy of the power amplifier model since in the signal under investigation the noisiest is the output in comparison with the input.

For example, in order to analyze the operation of a neural network, an output signal from a power amplifier (without noise) was fed to its input and the input signal was trained. The results of this model reached an accuracy of about -37 dB for the test dataset despite the significant nonlinearity of the amplifier.

It was found out that neural networks approximate the input signal very well, recognize the dynamics of data changes, but they are unable to learn the noisy signal component which limits its capabilities when such a process is simulated. In the future, the research of the considered algorithms of neural network simulation may be continued using the output signal filtering procedures.

References

[1] Podvalny S and Likhotin M 2018 Investigating the capabilities of artificial neural networks for the simulation of a telecommunication amplifier operation *J. Information technologies for modeling and control* pp 417-423

[2] Shutov V 2015 *Microwave power amplifiers linearization by digital predistortion* (physical-math. sciences degree) p 146

[3] URL: Rapidly converging modern learning algorithms for neural networks do not guarantee the achievement of the best generalizing ability [Electronic resource]: Access mode: World Wide Web. http://neuropro.ru/memo346.shtml

[4] Zakovorotny A 2016 New architectures and learning algorithms for neural networks of adaptive resonance theory *J. Scientific Result. Series "Information Technology"* pp 4-11

Complex engineering object digital twins – power engineering smart transformation basic concept

M G Zhabitskii[1], V E Melnikov and O V Boyko

High Engineering School, National Research Nuclear University MEPhI (Moscow Engineering Physics Institute), Kashirskoe sh., 31, Moscow, 115409, Russia

[1]Corresponding author: jabitsky@mail.ru

Abstract. The formation of information models and digital twins is a general element of the modern description of complex engineering objects. Their presence is already becoming a requirement of national and international technical regulation and contractual relations. The stage of forming separate digital twins within different stages of the life cycle has been passed and the approach has proven to be effective and in demand. The most striking example is BIM technologies. Now it is important to switch to complex end-to-end information models and digital twins for the full life cycle of complex engineering objects. This innovation becomes a universal tool for the interaction of all stakeholders in high-tech industries. Competence in technologies for working with complex information models and digital twins on the full life cycle of complex engineering objects is already a key competitive factor for high-tech engineering and operating companies in power production. This is a proactive trend in the formation of intellectual energy data space. It will lead to another cycle of improving the efficiency, safety, and economy of the industry and businesses.

1. Introduction

The fourth industrial revolution technologies [1] caused the transformation of engineering and business approaches to managing systems that make up the modern technological environment. The most important component of any modern technological sphere is the energy infrastructure and energy system. Power production is a system of systems (SOS) and includes some other full scale systems as elements. Different main subsystems can be identified in some details:
-energy generation subsystem which includes energy production facilities (thermal powerplants, nuclear powerplants, hydro powerplants, etc.);
-linear network power distribution infrastructure, main and local power lines (overhead lines with supports, underground cable lines, and others);
-electrical voltage transformer and power distribution system (which include electrical substation by different types, such as Transmission substation, Distribution substation, Collector substation, Converter substations, Switching station and some other specific types);
-end-users networks and equipment;
-other energy infrastructure territory facilities (for example, energy storage facilities, fuel and component warehouses, repairing facilities and other objects).

Most of these subsystems and their component objects are complex engineering objects (CEO). The most complicated and intricate CEO are nuclear power plants. And they are typical parts of the global energy generation sector. They are common in many countries, such as the USA, France, China, Japan, Russia, South Korea, Canada, Ukraine and some others. Nuclear power plants work in more than 30 countries just now. There were 451 working nuclear power reactors all over the world in the middle of 2018. Total capacity is about 394 GW [2]. Nearly 60 nuclear power reactors are under construction at the moment [3].

2. Methods

The engineering architecture, set of equipment, interaction algorithms and safety requirements for nuclear power plants are the most large-scale, complex and high-tech among all production facilities in the history of mankind. Therefore, it is advisable to implement the full life cycle of nuclear power facilities on the basis of the most modern industrial and information technologies.

The modern approach to life cycle management is based on an adequate realistic correspondence of the structure and regularities of functioning of an actual object and its digital information model. The modern concept of a digital information model varies widely among different authors and for different purposes. In our case, the most appropriate definition seems to be: "Information model (IM) is a set of structured and unstructured information containers that represent a complex reliable information source about a project (object evolution) at all or some stages of its life cycle" [4]. Partial digital information models have been used in industry and economics for several decades. They were gradually integrated into more complex ones in some cases. A remarkable example of technology development is BIM (Building Information Models). This approach is rapidly taking over areas of application in the design, construction and decommissioning of complex engineering objects [5]. Information models are elaborating on the operation and product manufacturing stage. It seems simple and natural to combine all these digital information models into a complete model that is unified for all stages of the life cycle of the CEO. In fact, this process is very difficult for many reasons. The main of them are:

-different purposes, objectives and efficiency criteria for different stages of the CEO life cycle [6,7,8,9];

-different creators and developers of digital information models (businesses and individuals) have different competencies and interests specific to each stage of the CEO life cycle;

-the engineering digital tools of various vendors are not fully compatible, especially at transmitting complex attribute information between digital information models;

-different data samples from the complete set are relevant at different stages of the CEO life cycle. Some data sets may not be used for a very long time, and then it becomes required again.

We need to associate the stages of the CEO life cycle with various information model structures and data classes in order to understand the future of CEO information modeling. The first complex stage is the creation stage. It combines design, construction and commissioning processes. The digital information model for these stages demonstrates the step-by-step model structure development and data accumulation. The base is a BIM model (Building Information Model), which primarily determines the spatial location of structures, equipment elements and engineering communications of objects. The BIM component does not represent a complete information model of investigated objects, processes and projects. Usually, we use project management systems, including classical management (time, resources, personnel, business processes). We need digital tools for system analysis and engineering (quality management, configuration management, collision management, etc.). Specialized digital tools for logistics modeling and assembling works visualization are used too. The structure of the information model thus includes a multidimensional spatial deeply attributed information about the object. Multi-level decomposition is enabled. Typically, there are four levels: object - the building - engineering system - the piece of equipment. Various engineering communications are important model elements too. All levels of the information model have a detailed system of attributes, including engineering and physical, logistics, and financial and economic parameters. The information model a priori provides the ability to track time points. The time variable is included both in the state of objects and in models of engineering and economic processes.

The CEO information model for the productive operating stage has a completely different structure. The spatial arrangement of elements is almost invariable, although it is included in the background model. However, data about the technological modes of operation are incomparably more important at this stage. The digital model must include data about operating modes, failures, and malfunctions of equipment and systems. The specialized model block describes maintenance and repair procedures. If hardware units or parts of them are replaced, the individualizing attributes at all levels of the model should be changed. In addition, the model provides for fixing the flows of material and energy resources, both irrevocably spent and circulating. A comprehensive digital model captures both engineering and physical parameters and the economic characteristics of the production process.

At the stage of reconstruction and modernization of an industrial facility, the information model is similar to the model at the construction stage. However, the starting state is not an empty production site, but a model with accumulated changes during operation. It will be necessary to perform a complex procedure of object inspection at the previous stages of technological development in order to fix this starting state.

For many objects of industry and engineering infrastructure of the human environment, the life cycle is very long. In most cases, a special decommissioning procedure is required. This is necessary even for industrial objects of standard purposes. In the case of high-risk facilities, such as those contaminated with chemicals or nuclear power facilities, the task of decommissioning is significantly complicated [10,11,12]. At the same time, the decommissioning stage will most require the full volume of data - project documentation, working documentation with the accumulation of changes and data on the course of operation. The information model inherited from previous stages of the life cycle can be effectively converted into a digital decommissioning model. This approach allows us to achieve the environmental, economic, and engineering efficiency of the final stage of the life cycle of a complex engineering object. However, the data required for decommissioning is quite different from the data obtained in the course of productive work. They can be obtained from design and operational data for special approval procedures, supplemented by the results of specialized integrated engineering surveys. Updated BIM models will be in demand again, taking into account wear, contamination, and destruction. Of particular interest is the analysis of the residual value of equipment and materials to determine economically, environmentally, and engineering efficient recycling technologies. Naturally, project management tools, information models of work production, and logistics organizations are again in demand at the final stage of the life cycle.

Russia is one of the largest exporters of turnkey nuclear power plants on the world market. It is the largest export item in the high-tech energy sector and a significant share of the world market. In recent years, national legislation in various countries has required the provision of digital information models [13]. This requirement is becoming a mandatory element of contracts for the construction of nuclear power plants and other engineering facilities on the world market. However, the legal requirements of different States are variable. Standardization of understanding information models is also not developed in engineering and business practice. This makes it difficult for the customer and the contractor to understand and interact on these points of the contract. As discussed above, the information models used at different stages of the life cycle are very different. These differences relate to both the detailed elaboration of information models and the data structure itself. The most obvious approach for the contractor is to build, use, and transmit information models of the construction stage to the customer. However, for the customer, this model is only of limited interest after the commissioning stage finish. Its need is to convert BIM into a complex digital information model of the operation stage. practice, this task has been set, but not fully resolved. BIM models are passed as information digital models. BIM models are in demand and effective for designers and builders. This format of information models is not sufficient for owners and operating organizations. In the future, they have to create a separate structure of information models, largely duplicating the earlier developed ones.

3. Results and discussion

For the transformation to intelligent energy power systems, the stages separated IM are no longer sufficient. The creation of end-to-end dynamic digital twins (DT) is required at the present stage. The digital twin includes time series of data corresponding to the instantaneous state of the modeled object. The data structure of the digital twin is based on the data structure of the digital model. At the same time, dynamic data series allow us to return and analyze the object's state at previous times. For the effective use of the digital twin, it is important current values automatically regular in addition to the data series. Due to the large amounts of data describing complex engineering objects, the modern approach is to automatically extract data about the parameters of the engineering system and write them to the database of the digital twin. Even 15-20 years ago, the volume of data generated by a complex engineering object was too large for their full digital fixation throughout the entire lifecycle. Also, the cost of sensor and data transfer systems with stable data channels was unacceptably high. Currently, the technical and economic restrictions on continuous full-scale data collection and storage of digital twins are strongly minimized during the fourth industrial revolution. The expected engineering and economic

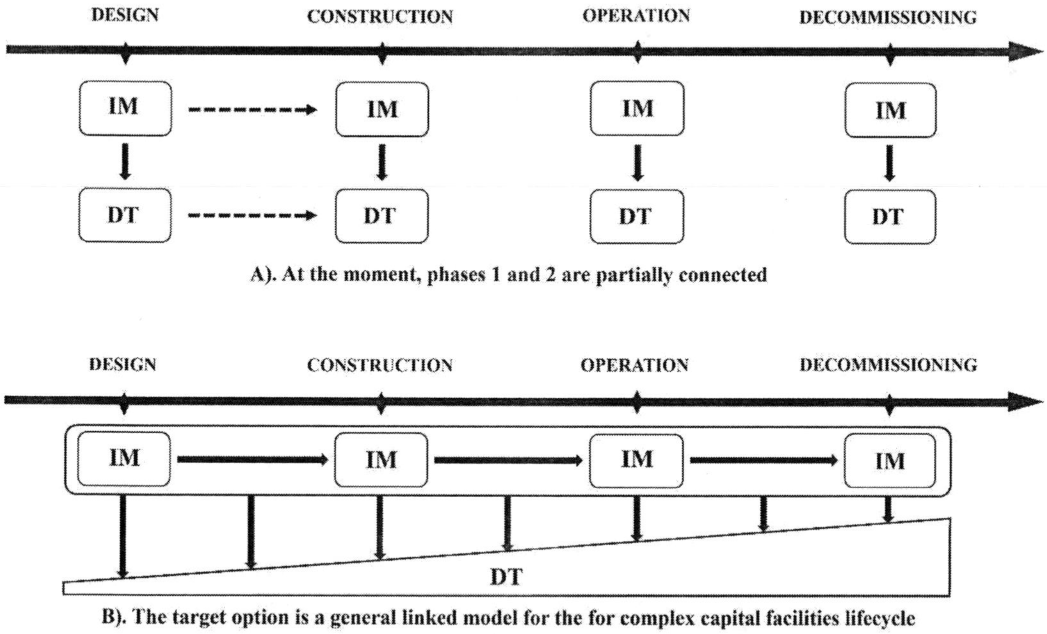

effects of their use are large enough for widespread implementation [14]. The main obstacle to large-scale implementation of digital twins throughout the life cycle of complex engineering objects is the methodological problems of integrating content-based heterogeneous data that arise during successive stages of the life cycle in a single digital twin.

Figure 1. Transformation separated information models & digital twins for different lifecycle stages to complex end-to-end ones.

The previous approach to creating information models was to create a data structure specific to each stage of the lifecycle. Then the initial data was taken selectively from the final model of the previous stage. Unclaimed data were archived and partially lost. In practice, until recently, both electronic and paper archives were used. In some cases, digital archives were referred to as unrecognized digital scans of paper documents. If it was necessary to use data from the early stages of the lifecycle, fragmentary re-entry of information was performed. The structure of the information model is re-constructed in a modified and truncated format. Selecting and re-entering some specific data of the previous digital twin is done selectively, in many cases manually. This procedure simultaneously leads to the loss of model reliability and increases the operating costs of working with information/ The described scheme for working with a digital twin is shown in figure 1, variant A. Currently, technological and methodological readiness allows us to work with a single end-to-end digital twin of a complex engineering object within the framework of the dynamic complexity of the data structure of information models throughout the full life cycle. This perspective approach is shown in figure 1, option B. This approach makes it possible to create an end-to-end digital twin of a complex engineering object for analysis procedures, predictive

modeling, and management of technological, engineering, and business processes throughout the lifecycle. Currently, work with large energy facilities is in the transition stage to a new methodology for building digital twins. This process will simultaneously improve safety and technical manageability, as well as reduce operating costs for the full life cycle of the object. Within the framework of nuclear power, this approach requires end-to-end interaction and full information openness between the subjects of design, construction, operation, reconstruction, and decommissioning. This task is currently realized and implemented by Russian GK Rosatom's divisions in the external and internal markets of nuclear power facilities. Among energy generating facilities, nuclear power stations are the points of advanced implementation of end-to-end digital twin technology on the full life cycle. This is primarily due to several specific requirements:
-security and reliability of NPP operation [15,16,17,18];
-mandatory approval of all changes in design and technology by the developer, international and national Supervisory authorities;
-a complex, highly regulated and controlled procedure for decommissioning radiation-hazardous facilities.

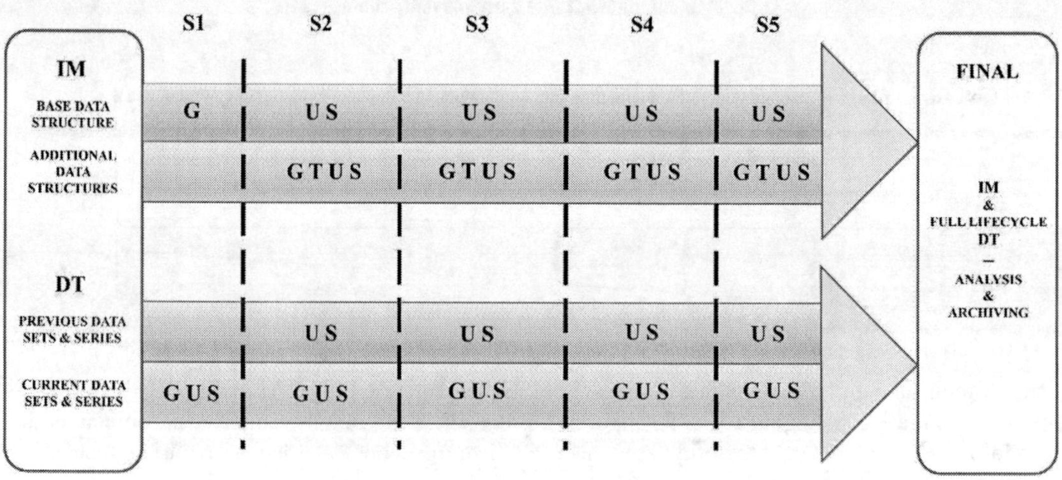

Figure 2. Information model (IM) and digital twin (DT) life cycle.
G – data generation, T – transformation, U – usage, S – storing;
S1 - design, construction and commissioning stage, S2 - productive operation stage,
S3 - modernization and reconstruction stage, S4 – after- reconstruction productive operation stage,
S5 – Decommissioning stage.

The consequence of this advanced development is a deep study and large-scale testing of the technology of digital information models and end-to-end digital twins of extremely complex objects on the example as NPP. It is also important that the market of engineering and operational services in the energy sector should include entities that have different approaches and technologies. An example of this experience will allow you to thoroughly analyze the advantages of working at a new technological level. Besides, there are real enterprises-carriers of a new generation of technology.

It is important to understand that the end-to-end approach to the formation, filling and use of digital twin data implies the accumulation of data and their dynamic structuring without loss at time intervals of about a hundred years. At various stages of the lifecycle, a full set of procedures will be implemented for working with the structure of the information model, as well as with dynamic data sets of the digital twin. These procedures include long-term storage of structured data, their usage without modification and addition, modification and addition of the data structure, as well as building up time series of data at all stages of the object's lifecycle. The general scheme of working with data and its structures within the full lifecycle is shown in figure 2.

4. Conclusion

Thus, one can conclude the world-wide transition that is taking place before our eyes from big disparate data sets to a new complex entity when describing engineering objects on their full lifecycle. Integrated information models and end-to-end digital twins are becoming universal tools for the interaction of all stakeholders in high-tech industries just now. In the energy industry as a whole and the power generation facilities sector, the nuclear power segment is currently one of the main leaders in this area. The inevitable almost complete coverage of engineering and technical infrastructure and the main business processes of the energy sector with integrated end-to-end digital twins will lead to another cycle of increasing the efficiency, safety, and economy of the industry. Competence in technologies for working with complex information models and digital twins in the full life cycle of complex engineering facilities is already a key competitive factor for high-tech engineering and operating companies in the power industry. In the foreseeable future, this trend will only increase.

References

[1] Schwab K 2017 *The Fourth Industrial Revolution* (New York: Crown Publishing Group) p 192
[2] International Atomic Energy Agency 2019 Nuclear Power Reactors in the World *Reference Data* **2** (Vienna:IAEA) p 80
[3] The Power Reactor Information System (PRIS) https://pris.iaea.org/pris
[4] ISO 19650-1-2018 Organization and digitization of information about buildings and civil engineering works, including building information modelling (BIM) — Information management using building information modeling — Part 1: Concepts and principles
[5] Volk R, Stengel J and Schultmann F 2014 Corrigendum to Building Information Modeling (BIM) for existing buildings — Literature review and future needs *Automation in Construction* **38** pp 109-127
[6] Garcia H, Aumeier S, Al-Rashdan A and Rolston B 2020 Secure embedded intelligence in nuclear systems: Framework and methods *Annals of Nuclear Energy* **140** 107261
[7] Volodin V and Tolokonskii A 2019 Concept of instrumentation of digital twins of nuclear power plants units as observers for digital NPP I&C system *Journal of Physics: Conference Series* **1391** 012083
[8] Petrenko A, Mochalov Yu, Maksimov A and Zhdanova O 2020 Integrating project and the PDEC information model *Journal of Physics: Conference Series* **1475** 012008
[9] Denisova L and Meshcheryakov V 2019 Control systems design: The technology of simulation and optimization *Journal of Physics: Conference Series* **1260** 032011
[10] Lim H.-K, Jeong K.-S and Lee K.-W 2011 Risk assessment of radiological and non-radiological hazard for accident prevention in decommissioning activities *International Conference on Manufacturing Engineering, Quality and Production Systems, MEQAPS - Proceedings* pp 237-240
[11] Kulikov K, Nizamutdinov R, Abramov A and Tsubanikov A 2009 Decommissioning and dismantling solution development for volodarsky civil nuclear fleet support ship *Proceedings of the International Conference on Radioactive Waste Management and Environmental Remediation ICEM* **2** pp 731-735
[12] Dragusin M, Pavelescu A and Iorga I 2011 Good practices in decommissioning planning and pre-decommissioning activities for the magurele VVR-s nuclear research reactor *Nuclear Technology and Radiation Protection* **26** pp 84-91
[13] Skvortsov A 2016 Overview of the international regulatory framework in the field of BIM revolution *CAD and GIS of highways* **2** pp 4 – 48

[14] Osetskaya M, Allenykh M and Varvus S 2019 The digital economy: Modern challenges in Russian nuclear power industry *Advances in Intelligent Systems and Computing* **726** pp 244-253
[15] Patterson E, Taylor R and Bankhead M 2016 A framework for an integrated nuclear digital environment *Progress in Nuclear Energy* **87** pp 97-103
[16] Lee S, Jung W and Yang J-E 2016 PSA model with consideration of the effect of fault-tolerant techniques in digital I&C systems *Annals of Nuclear Energy* **87** pp 375-384
[17] Park J, Suh Y and Park C 2016 Implementation of cyber security for safety systems of nuclear facilities *Progress in Nuclear Energy* **88** pp 88-94
[18] Liang Q, Guo Y and Peng C 2020 A review on the research status of reliability analysis of the digital instrument and control system in NPPs *IOP Conference Series: Earth and Environmental Science* **427**

AUTHOR INDEX